SUCCESSFUL BUSINESS USES
for abandoned service stations

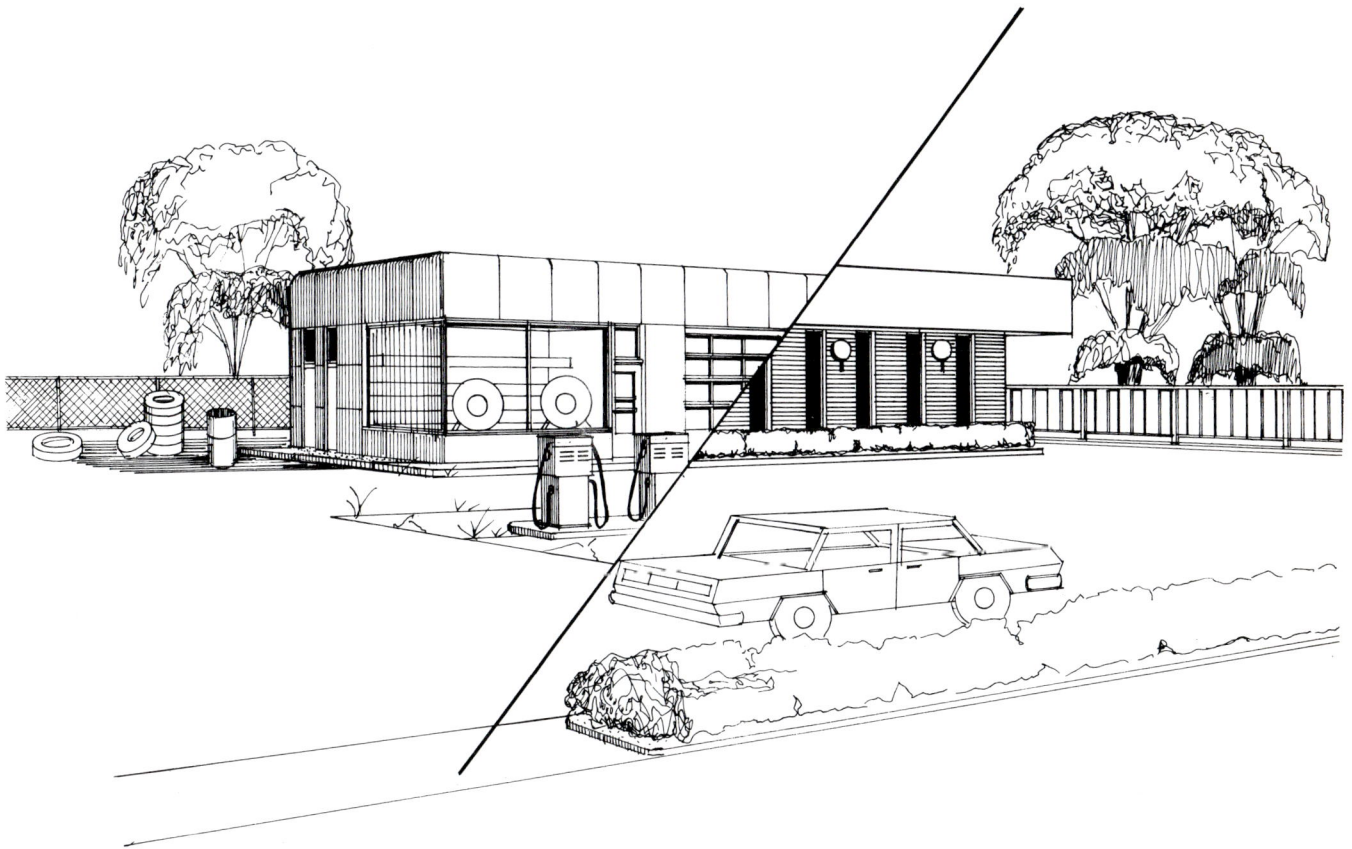

A. L. Kerth, a.i.a.

Library of Congress Catalog Card Number 82-61265
ISBN 0-9601188-2-9

Printed in the United States of America

TABLE OF CONTENTS

INTRODUCTION

INTRODUCTION

This second book has been completed at the request of many readers of the author's first book on this subject, petroleum marketers, and realtors who have viewed the author's slide presentation at various seminars. The program illustrates new business uses that have been established in abandoned service stations. There has been great interest in the photographs and background stories of these conversions. Several investors and marketers have started successful businesses based on the information provided at these seminars.

New business uses for abandoned service stations can make money. Thousands of entrepreneurs have found bargains among these commercial properties frequently situated on major intersections, in areas where vacant land was no longer available. The location is the key to many successful ventures.

Have you ever investigated the rentals for a 1200 or 1800 square foot building in a shopping center? If you can find one to lease, the rental will probably be 50% to 75% higher than an abandoned service station with comparable space. The cost to buy a 100' x 125' lot and erect a 1200 square foot building, based on 1981 costs, ranges from $150,000 to $300,000.

This book will assist the entrepreneur who is looking for a new retail business, and the owner of an abandoned service station who is seeking assistance on how to recycle a capital asset that is not yielding an income.

During periods of a sluggish economy, recycling of an abandoned service station can be the answer to the high cost of new construction. Considerable construction time can be saved and a new business use can be established at a fraction of the cost of a new building.

To develop an attractive conversion concept it is recommended that a local architect be retained to provide his expertise. The architect is familiar with energy conservation, zoning laws, building department requirements and community needs. His services can contribute to the success of a new business venture.

The preparation of this book was greatly aided by the generous cooperation of all the owners and entrepreneurs who so kindly took the time and trouble to furnish the author with necessary information. My thanks to Robert A. Connell for his assistance in editing this publication. Most of all, the author thanks his wife for her constant encouragement and able assistance.

FORECAST FOR THE FUTURE

In the first volume of "A New Life for the Abandoned Service Station", the writer predicted that 50,000 to 60,000 service stations would close over a five-year period. In 1973, the service station population totalled 226,000 and, by the end of 1980 the number had declined to 160,000. In the author's opinion, the number of service station abandonments and/or closings has not leveled off as some experts have predicted. This writer believes that the decline will continue until another 45,000 to 50,000 stations close, leaving a core of approximately 110,000 service stations to serve the needs of the consumer.

This forecast is based on the following factors:

1. Continued closing of marginal service stations.

2. The change in petroleum marketing trends; i.e., the swing to self-service facilities and loss of bay work by thousands of marginal service stations. It is difficult for a low volume full service unit to survive only on profits from gasoline.

3. Occasional petroleum shortages, higher oil prices, followed by inflation and further declines in gasoline consumption.

4. The growth of the automotive aftermarket. This has affected the bay work of the service station. The author has spoken to hundreds of new car owners and asked them where they take their car for service. Approximately 95% said "back to the dealer where the car was purchased for at least the first two years."

5. Costly environmental requirements which the average service station dealer cannot afford.

6. The change in engine design in new cars has extended the mileage between oil changes and computerization has reduced the need for many tuneups.

The author, therefore, envisions a mix of service stations for the future as follows:

1. Large service centers that provide quality and reliable full service.

2. The self-service station for the price buyer.

3. The "gas only" station with attendants for those who desire limited service.

4. The convenience store with gas and the car wash with gas for the "one stop" motorist.

Based on this forecast, and estimating an average value of $100,000 for the service stations that will be sold, approximately $4.5 billion dollars of desirable real estate will soon become available for other business purposes.

WHO'S BUYING ABANDONED SERVICE STATIONS?

Franchises, fast food restaurants, banks, insurance and realtors, convenience food store chains, retail stores such as bicycle, shoes, clothes, plant shops, animal hospitals, just to mention a few. In the author's opinion, which is based on interviews with businessmen-owners, the savings when compared with grass roots development range from 15% to 50%. Abandoned service stations in densely populated areas of cities, where vacant lots are at a premium, are the primary source for business uses in commercial areas. In 1981 the price range for a three bay building in a good business area, situated on a 125' x 100' lot, varied from $75,000 to $125,000. The 150' x 100' to 200' x 150' parcel in suburban or rural areas varied from $100,000 to $250,000, depending on surroundings. If a service station has been abandoned for more than a year, the price may be discounted by as much as 25% to 50%.

Inflation, unemployment, recessions, and the gasoline shortages, resulting from intermittent mideast unsettlement, create many problems for millions of small businessmen. The author's interviews revealed that many of the businesses were recession-proof and even improved during periods of slow economy. A list of these stable businesses and a brief description of the product and major site selection data are as follows:

1. PET FOODS: Sales by the case or by the individual can, together with gourmet pet foods and accessories.
   Site selections: Located in a suburban or urban community with 100,000 population. Large property 150' x 100' on main street, adjacent to retail business area with drive-in and walk-in trade. Parking is essential since most customers will require a car trunk and transportation. The building should contain at least 1800 square feet of floor area for a start-up store. Sufficient room for storage expansion is essential as business improves.

2. BICYCLE STORE: Sales and service of a franchised domestic brand or international bicycle. Supplemental income may be derived from the sales of winter sports equipment, roller skates and accessories.

   Site selections: In suburban area, excellent near a college or high school, railroad parking lot or in a business area where there is pedestrian traffic. The ideal property size ranges from 150' x 100' to 200' x 200' to allow for parking of customers and test riding of bicycles. Smaller sites in excellent areas should not be overlooked. If you have insufficient land to expand horizontally, go vertically; install a second floor for storage and provide a dumbwaiter. Storage of bicycle inventory is the most important factor.

3. BANKS: Feasibility analyses conducted by many major banks have revealed that thousands of abandoned service stations throughout the country are ideally situated for their expansion programs in cities, suburban and rural areas. A bank will spend from $75,000 to $150,000 on a large parcel ranging from 150' x 100' to 200' x 200' on a primary highway for a new branch office. A bank prefers one-half to one acre of land for drive-up banking facilities, parking for customers, stack-up area and landscaping. A bank will spend from $75,000 to $200,000 remodeling a building into a branch office.

   Site selection: Banks generally do not compromise with quality. They require large parcels of land with ample space for drive-up customers and parking. Sites should range from a minimum of 150' x 100' to 200' x 200'. Good visibility is an important feature.

4. FAST FOOD FRANCHISES: Abandoned service stations are carefully screened by the real estate departments of franchises. High cost of land acquisition, together with rapidly rising costs, are giving many fast food franchises second thoughts about demolishing service stations. Conversions or major alterations are now given serious consideration.

   To aid the businessman in his selection of a suitable franchise, the author recommends the Franchise Opportunities Handbook, published by the Superintendent of Documents, U.S. Government Printing Office, Washington, D.C. 20402.

   Site Selection: Urban, suburban and heavily travelled highways in rural areas. Minimum lot size 150' x 150' to 200' x 200'. Parking with room for landscaping is of major importance.

5. CONVENIENCE STORE: Because of the high cost of grass roots development, real estate agents for the food stores are screening abandoned service stations. They prefer to retain the gasoline volume; however, they will not pass up a superior location without gasoline allocation or gas permit.

   Site selection: Population density within three mile radius in a suburban area. Large property is recommended. Minimum acceptable size is 125' x 100' on a primary artery with room for parking at least eight to ten cars.

6. LIQUOR STORE: Sales of wines and liquors. Cold beer and wine should also be available for the consumer.

   Site selection: In a business area with walk-in traffic potential. Parking for eight to ten cars is required. Minimum site 100' x 50'. Sufficient land should be available for the addition of a walk-in cooler and enlargement of the storage area.

7. HEALTH FOOD STORE: Sales of vitamins, herbs, diet foods, nuts, cereals and yogurts. A floor area from 800 to 1800 square feet is recommended for the specialty food store.

   Site selection: A minimum 10,000 square foot property in a business section with parking for drive-in customers and walk-in traffic. A business district in an upper middle income area offers the best opportunity for success.

8. ICE CREAM STORE: Sales of old-fashioned ice cream in small ice cream parlors, decorated in an old-fashioned, turn of the century motif. Most stores are franchised. A 1200 to 2000 square foot building is recommended. A self-service store is unique and attracts the do-it-yourself consumers.

   Site selection: Densely populated suburban or urban areas, near apartments, townhouses or condominiums. The minimum size property recommended is 100' x 125'.

9. PLANT STORES OR RETAIL FLOWER SHOPS: Sales of potted plants and/or fresh

flowers. The rental of plants for office building interior landscaping coupled with a weekly maintenance contract provides a supplemental income for the plant store in or near a business district.

Site selection: A 12,000 square foot property with parking spaces for eight to ten customer cars situated in a commercial district. Pedestrian walk-in traffic is important to assure a successful business venture.

10. HARDWARE AND/OR ENERGY CONSERVATION CENTER: The inflationary economy will result in a thriving business for stores catering to the "do-it-yourself" homeowner. Energy conservation heads the list along with home improvements. A super-hardware store with a large area allocated to inventory, can serve the homeowner searching for lumber, flooring, paint, equipment, insulation, weatherstripping and supplemental heating equipment.

Site selecton: A 15,000 square foot property and a building containing between 2000 and 3000 square feet is recommended. Parking should be provided for at least twelve to fifteen cars. The property should be situated in a populated suburban area on a heavily travelled major artery.

FRANCHISING

A large percentage of abandoned service stations are being converted to franchised business uses. A few of these uses are the fast food, donut, dairy, plant, photocopy and automotive aftermarket stores The entrepreneur who is considering buying an abandoned service station and converting it to a franchise use should research thoroughly the franchisor. There are advantages and disadvantages that must be weighed by the businessman. To assist the reader who may be considering this concept, a brief outline of advantages and disdvantages are listed.

ADVANTAGES:

1. Because of the franchisee's limited experience in a new field, he receives the advantage of the francisor's experience.

2. Under certain conditions, a franchisor may offer financial assistance.

3. The franchisor can offer a well developed consumer image and goodwill, together with proven products and services; (the built-in community acceptance of the product). The Small Business Administration states that customers are more receptive to trying a product when they buy it from a local citizen.

4. The franchisor can offer competently designed facilities, layout of displays and fixtures based upon experience.

5. The franchisor can offer chain buying power.

6. The franchisor can offer training assistance and proven methods of doing business.

7. The franchisor can offer national or regional promotion and publicity. This is one of the most important advantages that a private owner normally cannot afford.

8. The franchisor can offer a location analysis and staff to research prospective new sites for franchisees.

DISADVANTAGES:

1. The franchisee will have to comply with standardized imposed operations. He is not his own boss.

2. The franchisee may have to pay a fee related to a percentage of income, thereby sharing his profits with the franchisor.

3. The franchisee may not have the freedom to meet local competition.

4. Contract agreements may be slanted to the advantage of the franchisor.

5. Considerable time may be spent preparing reports.

6. It may become necessary to share the burden of a franchisor's faults and mistakes.

Before investing in a franchise, investigate the franchisor, the proximity of similar existing franchises, the product or service performed, the franchise contract, the franchisor's financial backing and financial statements, the number of franchises in existance, the training program, and the length of time the franchisor has been in business.

There are a number of publications available that

provide franchise background. The Department of Commerce annually published "The Franchise Opportunity Handbook" which includes statistics on franchises, (*$6.00). The United States Government Printing Office, Washington, D.C. 20402. Two other books published annually are "The Franchise Annual", listing most franchises, (*$12.95), Info Press, 736 Center Street, Lewiston, N.Y., 10942, and "A Directory of Franchising Organizations" with limited information, (*$3.50), Pilot Books, 347 Fifth Avenue, N.Y., N.Y., 10016. Another excellent book that furnishes franchising statistics is "Franchising in the Economy", ($4.00), The United States Department of Commerce.

Many new franchises are established every week. Several newsletters are published monthly that list every new franchise and try to review their services. A number of new franchises show great promise, such as: computer stores, health food stores, insulation contractors, and stores selling energy-related products. Several recession-proof businesses, such as bicycle stores, the automotive aftermarket, the hardware and home improvement center, and locks and security products offer excellent opportunities to the enterprising entrepreneur. The franchises who sell products for the do-it-yourself consumer may be the most successful. The success of the individual franchise will depend on the location and the capability of the franchise.

SITE SELECTION GUIDELINES:

Several guidelines should be considered by the realtor, investor, small businessman, planner, appraiser or franchisee and/or franchisor in the selection of abandoned service stations for other business uses. The author has compiled general site selection data to assist the reader in his search for a suitable abandoned service station.

THE LOCATION, INGRESS AND EGRESS:

1. Size of property. Is there sufficient parking area for the type of business?

2. Where is the nearest competition with regard to the site being considered?

3. Is visibility of building a major factor?

4. Is visibility of freestanding signs adequate?

5. Is ingress and egress good?

* 1980 Price

6. Is the prospective location situated on an intersection with a traffic light?

7. If an intersection location, is there a "right turn only" lane? Would this create a problem for traffic exiting from the parking area of the premises? Many motorists, particularly women do not like to cross a right turn lane at a traffic light to gain access to a center lane of a major artery. (See drawing Fig. 1)

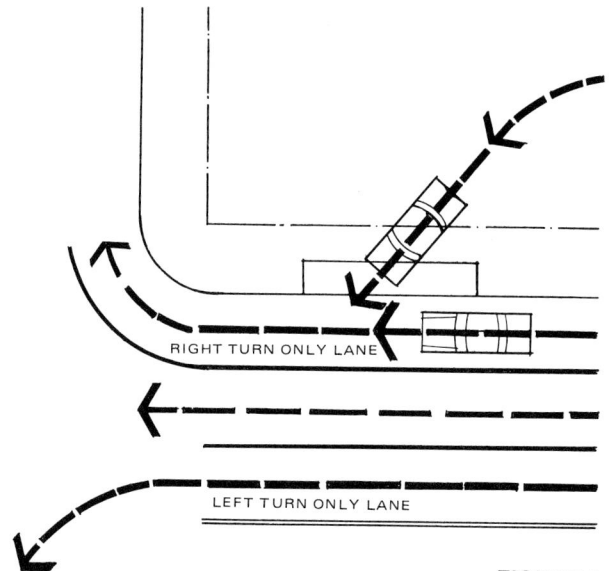

RIGHT TURN ONLY LANE

LEFT TURN ONLY LANE

FIGURE 1

8. Is the corner considered to be a "swing corner"? This is a corner where a great deal of traffic turns right. Generally a "swing corner" is considered a plus for any business site. (See Fig. 2)

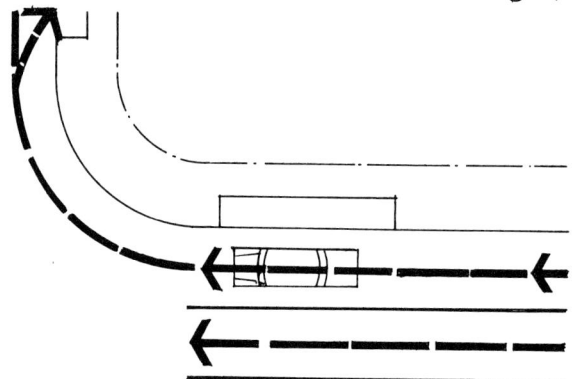

SWING CORNER   FIGURE 2

9. If this is a near corner property, is the traffic too heavy? Would it back up and block the ingress and egress ramps? (See Fig. 3)

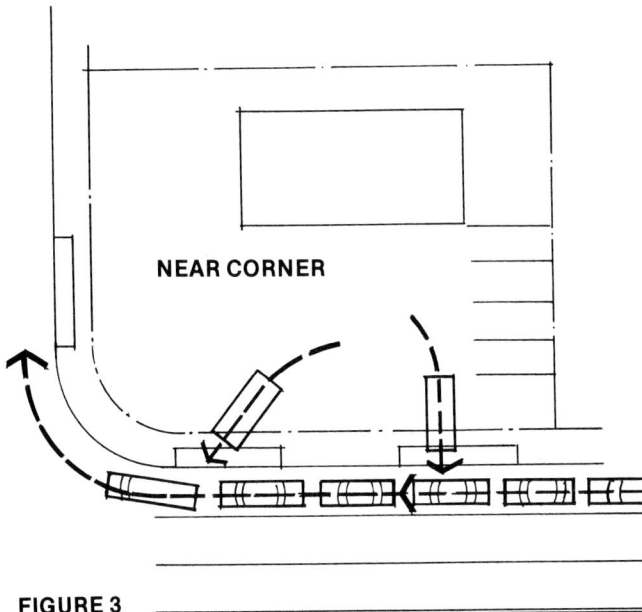

NEAR CORNER

FIGURE 3

10. Should your business be situated on a far corner on an intersection controlled by a traffic light? This type corner has a plus factor. The motorist and potential customer's vehicle comes to a stop allowing time for the driver or spouse to identify the new business situated on a far corner. Egress from far corner properties is protected by the traffic light, a very important factor in growth or densely populated areas. (See Fig. 4)

11. Have you considered interior properties which have many advantages? They are not encumbered with any of the problems described in items 7, 8, 9, or 10.

12. Is there a parking lane on the street frontage? This is important where limited on-site parking is available.

13. What is the condition of the black top, sidewalks and landscaping? Their condition will have a bearing on your remodeling costs.

14. Have you overlooked irregularly shaped properties? Many small parcels are strategically located.

15. Have you checked for possible future county or state road widenings that would have an adverse effect on your development of the property? Whenever there is a change in highways, there can be a great effect on the traffic patterns. That, in turn, can affect the efficiency of the business use.

16. Does the state or county have any plans for the building of a new highway that would bypass the site? Would this loss of transient traffic have an adverse effect on your business? Would local traffic sustain the business?

LOCAL ZONING:

Most local zoning ordinances state that a service station is considered a nonconforming business use and may be permitted only by special exception of the zoning board. The codes generally state that a non-conforming use can be changed to a new use providing that the new use is an upgrading to a higher and better use group. The following items of importance should be resolved:

1. Can you change the service station use to another business?

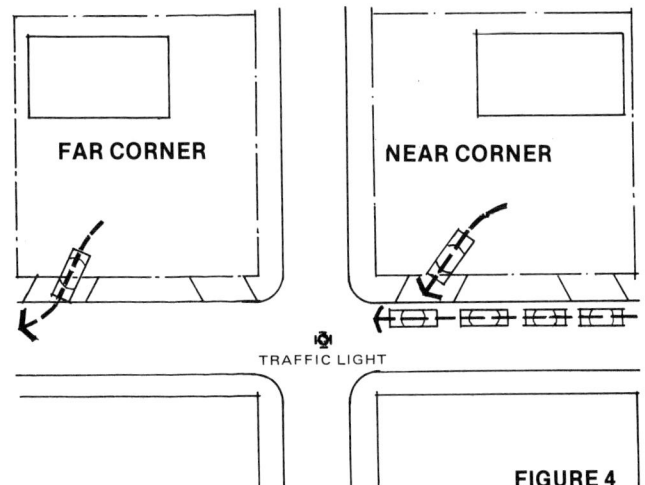

FAR CORNER　　　　**NEAR CORNER**

TRAFFIC LIGHT

FIGURE 4

2. Have the underground storage tanks been removed? If not, is the removal of tanks required by the local municipality when converting to another use?

3. Will you require a parking permit so that customers can park on the premises?

4. Will you be permitted to erect the type of identification sign required for your new business?

In the event that new zoning approvals are required, the following should help you prepare for an appearance, with a local architect and/or attorney, before the appropriate board:

1. If the building remodeling is extensive, retain

the services of a local architect to prepare preliminary plans showing the "before" and "after" conditions. An architectural rendering in color can be invaluable to illustrate the proposed extent of the remodeling.

2. Outline to the Board that upon completion of the remodeling, the new business use will result in the creation of a number of new jobs and that the new business will be an asset to the community.

3. It should also be stressed that the conversion of the building will remove the blight of any abandoned service station on the highway and in the community.

THE BUILDING

1. Is the building structurally sound?

2. How old is the building?

3. Is the present architectural style economically adaptable to the proposed new use?

4. Are the plans of the building available?

5. Is there a current certificate of occupancy available?

6. Do you have well water and a septic system? Will these utility systems be adequate for your new business use?

PROFESSIONAL ADVICE:

1. Have you secured the services of a local architect to assist you in remodeling the abandoned service station into a new business use?

2. Obtain quotations from local contractors to perform the work.

3. If you are knowledgeable about construction, you may be able to sublet work and save money.

4. Obtain a title search before purchasing to assure there are no liens or encumberances that would affect your business.

5. In the event that the local authorities have or-

dinances which limit your rehabilitation of the site and require zoning hearings, secure the services of a good local attorney and retain an architect to prepare an architectural rendering, illustrating in color the proposed use to be installed on the premises.

IMPORTANT PLANS AND DOCUMENTS FOR THE BUYER

Two or three weeks before you close title on your abandoned service station property, provide the seller's attorney with a list of plans and documents which you, as the purchaser, require. Certain items marked with an asterisk are essential. The others will save you money if you can obtain a copy of them.

*1. The original or photostatic copy of the Certificate of Occupancy. Don't close title without it.

2. The original or a photostatic copy of the Electric Underwriters Certificate.

*3. Copies of any Zoning Resolutions stipulating conditions at the time the Board granted the service station variance.

4. Copies of any sign permits.

5. Copy of any surveys, topographic or a completion survey. These drawings might indicate any easements or the location of corner stakes or permanent boundary markings. Utilize the services of the original surveyor to prepare the survey required by the title company and bank prior to the closing. Quite often an old completion survey of the structure as originally built will provide utility information with regard to the location of underground water, electric and sewage lines, cesspools, and/or septic systems and tanks.

A CHECKLIST FOR THE DO-IT-YOURSELF ENTREPRENEUR

For the do-it-yourself entrepreneur, who intends to convert an abandoned service station without retaining the services of a professional, the author has prepared a check list of items of work that should be considered to complete a building alteration:

1. 90% of service station buildings have never

been insulated. To limit the heat loss, the ceiling should be removed and 6″ to 8″ of insulation installed. The interior of all masonry walls should be insulated with the thickest insulation possible. A vapor barrier should be installed on the interior side of the insulation. Where space permits, a separate furred wall should be installed.

2. GLASS AND DOORS: The greatest heat loss in a service station occurs through the overhead door area. The overhead doors may be closed up or replaced with smaller thermopane windows to conserve heating fuel. Thermopane or insulated glass will reduce heat loss. All exterior doors must be weatherstripped and store fronts caulked. Where feasible, the introduction of coated room darkening window shades and/or venetian blinds will reduce solar glare in the summer and act as a heat-loss barrier during winter months.

3. OPEN SPACE: The removal of non-load bearing partitions creates a large open space for a store and will make it much easier to heat, cool or illuminate. An open space can be controlled from a business and security standpoint.

4. LIGHTING: Careful selection of lighting levels and electrical fixtures will result in energy conservation and lower electrical bills. The foot candle design levels must be based on the business use. For example: 75 foot candles is recommended for a business office and 30 foot candles is the level recommended for the sales room of a store.

5. HEATING AND AIR CONDITIONING: In the event the existing warm air heating unit is to be reused, a complete and thorough check of the system by a knowledgeable heating contractor is required. A series of tests must be conducted to assure that there is sufficient chimney draft; the burner flame must be checked; the smoke level tested; all ducts being reused must be vacuumed; a forced draft inducer installed on the flue stack; and outside air must be provided for the oil burner. These are just a few of the required checks that must be made to assure proper operation of the existing heating unit. In the event that a new heating and air conditioning system is to be installed on the roof, the duct work must be thoroughly insulated. Roof

mounted units permit maximum utilization of the entire building for displays, exhibits and furniture for the new business use. Since the roof units could be a source of future leaks, it is important that pitch pockets be sealed and that the roof be thoroughly inspected upon completion of work. Old underground fuel oil tanks being reused should be cleaned of all water, rust and scale by a qualified contractor.

6. ROOF: The roof of a former service station is an area where winter heat loss can be substantial. Insulation should be installed between the beams of a flat roof as outlined in Item 1. Rigid roof insulation may also be installed over the old roof and then covered with 3-or 4-ply built-up roofing. Asphalt shingle roofs on colonial and ranch buildings must be inspected to determine the need for replacement. If the roof is more than 15 years old, it is recommended that new 235# asphalt or fiberglass shingles be installed. If the old shingles are curled and dry, they should be removed completely before reshingling. Flat tar and felt builtup roofs are generally dry and cracked. Where the condition is not too severe, the roof may be revitalized with an application of a latex sealer. The roof sealer penetrates the felt and minimizes cracks in the surface. The roof should be revitalized with a sealer every three years.

7. FLOOR DRAINS: Grates should be removed and all floor drains sealed and filled with concrete to eliminate all odors.

8. LIFTS: All lift frames, rails and cylinders are to be removed. Holes are to be backfilled and concrete floors patched.

9. AREA LIGHTS FOR PARKING AREA: The former service station area lights will provide excellent illumination for a customer parking area. The fixtures are to be inspected and repaired where necessary. Lenses are to be cleaned and the light fixtures relamped where lighting levels are low. Replace defective ballasts.

10. LANDSCAPING: Overgrown plant areas should be cleaned out to determine the extent of shrubbery replacement. Evergreens, yews, juniper and pines should be salvaged, trimmed back and fertilized. Diseased and dead shrubs

should be replaced. To reduce the need for weeding, plastic cover should be spread over landscape areas, leaving cut out sections for trees. Gravel or wood bark should be placed over the plastic cover.

11. PAVED PARKING AREAS: Before any asphalt work is started, all yard drains should be inspected. The lines leading to drywells or to the storm sewer should be tested with a hose. If drywells are in use, they should be located, inspected and cleaned out. This procedure will eliminate flooding and damage to the building and yard surfaces. A flooded parking lot presents an uninviting appearance to potential customers.

The existing pavement should be preserved. The high cost of asphalt makes this mandatory. It is recommended that the asphalt be sealed with an industrial sealer. Buy the best material available if you are installing it yourself. The best and probably the most expensive sealer for a parking lot is easier to apply because it flows smoothly and covers more square feet per gallon. It costs as much to install cheap material as it does to install the most expensive. Based on 1981 costs, one of the best industrial sealers will cost approximately 8¢ to 10¢ per square foot for a large area in excess of 10,000 square feet. Rent the professional 36″ wide squeegees, which are flexible and broken in. The product can be purchased by the 55-gallon drum. When applied over a smooth asphalt, the coverage will average 100 square feet per gallon. Four men can apply two coats of a sealer on a 12,000 square foot driveway in approximately eight to ten hours. Two applications of a good sealer not only improves the appearance of the driveway, it also provides protection for the next ten years.

DRIVE-UP WINDOWS ARE IMPORTANT TO THE SUCCESS OF MANY BUSINESSES:

The American consumer is always in a hurry. Some people do not like to spend time shopping in stores and waiting on lines for service. They feel it is a waste of time. A drive-up window added to the business use not only accommodates this segment of the consumer populace, but can mean the difference between success or failure. In a recent magazine article, the owner of a chain of convenience stores stated that he had installed a drive-up window at two loca-

tions where business had declined. After several months, store sales had increased to the point that he was considering this concept for his entire chain.

This window must be located adjacent to a cashier's counter with ready access to the products being sold. Canopies, or overhangs, can provide the cover necessary to protect both customer and employee from the elements. Listed are the retail business uses where drive-up windows can contribute to profits.

1. Dry cleaner
2. Ice cream parlor
3. Bank
4. Convenience food store
5. Donut store
6. Liquor store
7. One-stop neighborhood store
8. Beverage store
9. Dairy store
10. Fast food restaurant
11. Photo film service
12. Pet food store

RECESSION-PROOF BUSINESSES:

A fluctuating economy causes increased unemployment. During these periods, franchises and small business flourish. Many of the unemployed are in search of a fresh start. Some are looking for their own business. A list of a number of businesses that are recession-proof is based on the many interviews the author has completed. All of them were found in **abandoned service stations** that had been converted by the owners or lessees. A number of the businesses increase profits during periods of recession due to the nature of the product or services provided.

1. Pet Food Store
2. Bicycle Sales and Service
3. Furniture Restoration
4. Self-Service Wand Car Wash
5. Home Improvement Center
6. Home Decorator Center
7. Hardware Store
8. Appliance Repair Center
9. Liquor Store
10. Factory Clothing Outlet
11. Deli-Bakery, Caterer
12. Convenience Food Store
13. Equipment Rental Center
14. Automotive Parts Store
15. Thrift Bakery

16. Ice Cream Parlor
17. Laundromat and Dry Cleaner
18. T.V. and Stereo Sales and Service
19. Mini Warehouse
20. Energy Conservation Store
21. Dry Cleaner and Tailor
22. Shoe Repair
23. Self-Repair Shop
24. Watch Repairs
25. Locksmith
26. Hobby Craft Shop
27. Sewing Machines, Sales and Service
28. Upholstery and Slip Covers
29. Plumbing, Electrical, Cabinetmaking Shops
30. Secretarial Services and Telephone Answering Services
31. Woodworking Shop
32. Tailor and/or Seamstress

NEW BUSINESS USES FOR ABANDONED SERVICE STATIONS

In the publication "A New Life for the Abandoned Service Station", the author included a list of 97 business uses for closed and/or abandoned service stations. There is no end to the long list of small business uses for these adaptable buildings. The structures are perfect for a small businessman with a limited budget. Most businesses can be developed in stages and considerable savings may be realized by the owner, who has the knowledge of construction skills, in the conversion of the buildings.

A supplementary list of uses is given to illustrate the versatility of these structures. This book shows how the conversion of an abandoned service station is not only faster than reconstruction, it is 25% to 75% cheaper than a grass roots development. The business uses marked with an asterisk are described and highlighted in detail in the text:

*1. Airport Auto Rental Agency
2. Ambulance, Emergency Service
*3. Appliance Store, Sales and Services
4. Automobile Painting
5. Automobile, Self Repair Shop
*6. Automotive Tune-up Shop
*7. Bakery, Thrift Store
*8. Bakery and Snack Bar
*9. Bus Terminal
*10. Camera Store
11. Candle and Burlwood Store
*12. Candy Store
*13. Carpet Store
14. Computers, Sales and Services
15. Cookie Store and Bakery
16. Custom Slip Cover Shop
*17. Cutlery and Pots, Cookware Store
18. Day Nursery
*19. Caterer - Restaurant
*20. Drapery Center
*21. Drive-Up Bank
*22. Drive-Up Dairy Mart
*23. Entertainment Center
*24. Furniture Restoration Store
25. Health Food Store
*26. Health Foods and Restaurant
27. Health Spa
28. Hobby Center
*29. Homeowner's Decorating Center
*30. Homeowner's Energy Conservation Center
*31. Ice Cream Store
*32. Jogging Shoes and Clothes
33. Manufacturer's Representative
*34. Mattress Store
*35. Mini Storage
36. Muffler Shop
*37. Multi-Business Uses
*38. One Stop Shop
*39. Pet Food Store
*40. Pet Grooming Center
*41. Picture Framing Center, Do-It-Yourself
*42. Plant Store
*43. Plastercraft Store
*44. Quick Oil Change
45. Roller Skate Sales and Repairs
46. Security Center, Keys, Locks and Burglar Alarms
*47. Senior Citizens' Center
*48. Shoe Store, Athletic, Sales and Repairs
49. Spa Sales Center
*50. Swimming Pools, Sales and Supplies
51. Telephones, Sales and Rental Store.
52. Television, Sales and Service
53. Tool and Equipment Center
*54. Towel Store
55. Toy Store
*56. Trophy Shop
*57. T-Shirt and Jeans Sales
58. Van Customizing Shop
59. Vinyl Repair Center
*60. Wedding Center
61. Truck and Trailer Rental Center

AIRPORT AUTO RENTAL AGENCY

Any abandoned service station situated on a highway adjacent to or near an aiport has the potential of being converted to an Auto Rental Agency. A two or three bay service station situated on a 150' x 175' plot can be converted into a business that will accommodate an agency. The property and building illustrated on these pages is ideally suited for this new use. The land to the rear of the building provides access to a drive thru roll-over car wash in the end bay. The center bay is to be converted into the service area for any minor or emergency repairs. The building is to be altered to include a registration counter, waiting area, space for two clerks and a store room. The store front requires minor modernization. The original architectural style should be retained.

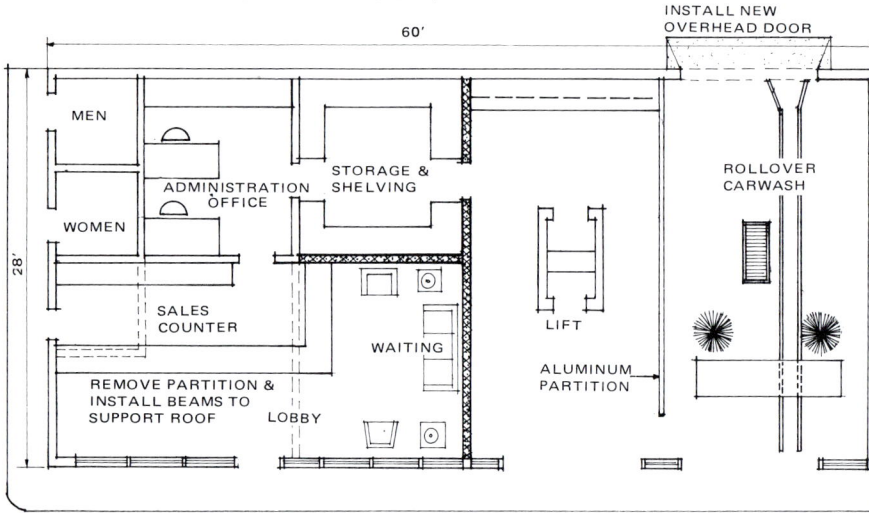

FLOOR PLAN SCALE 0' 5' 10' 15'

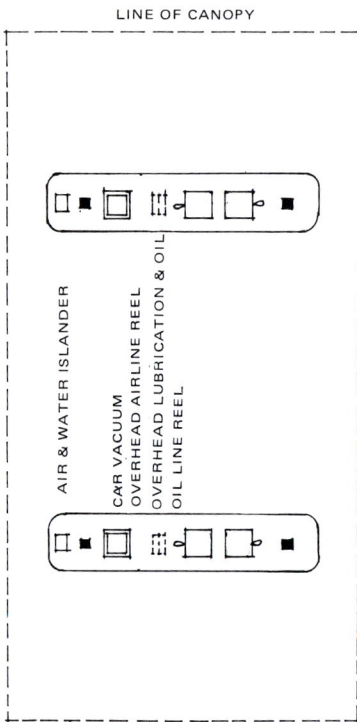

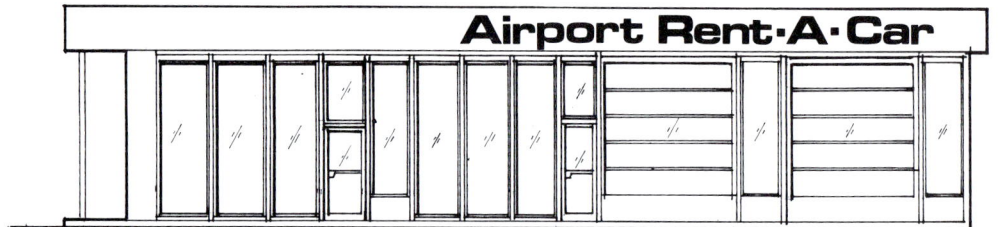

FRONT ELEVATION

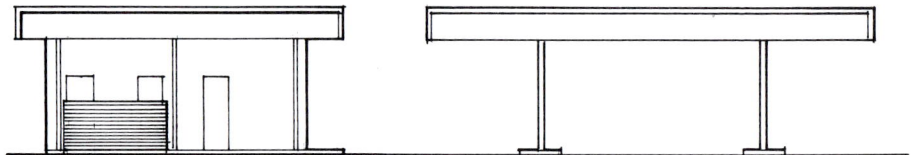

LEFT SIDE ELEVATION

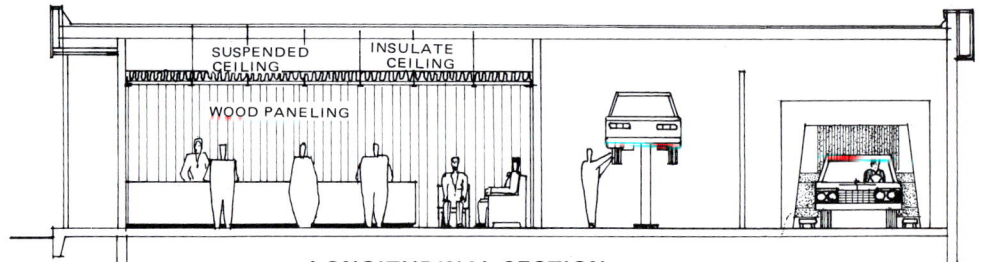

LONGITUDINAL SECTION

AIRPORT AUTO RENTAL AGENCY

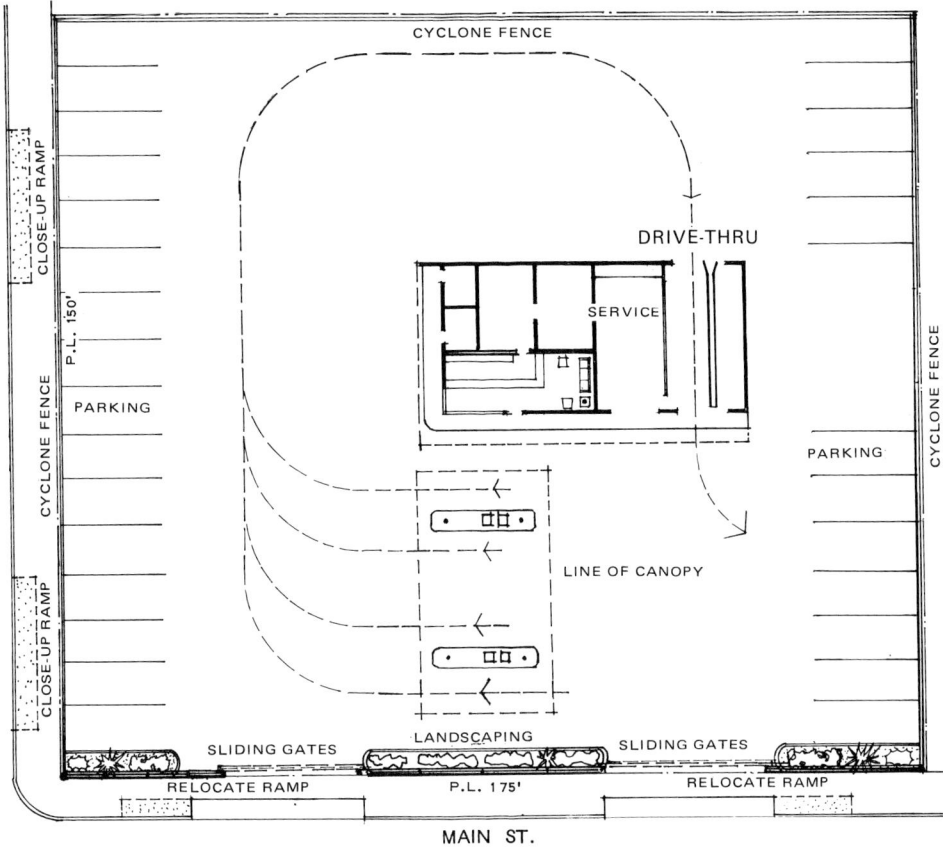

PLOT PLAN   SCALE 0' 10' 20' 30'

It is essential to maintain vehicles in A-1 condition for customer acceptance. This can be accomplished with the in-bay car wash operated by an employee. The pump islands are to be converted to a unit to provide gasoline, interior vacuuming, oil, water and air. A large parking area is mandatory. This plan provides for a twenty-nine car storage.

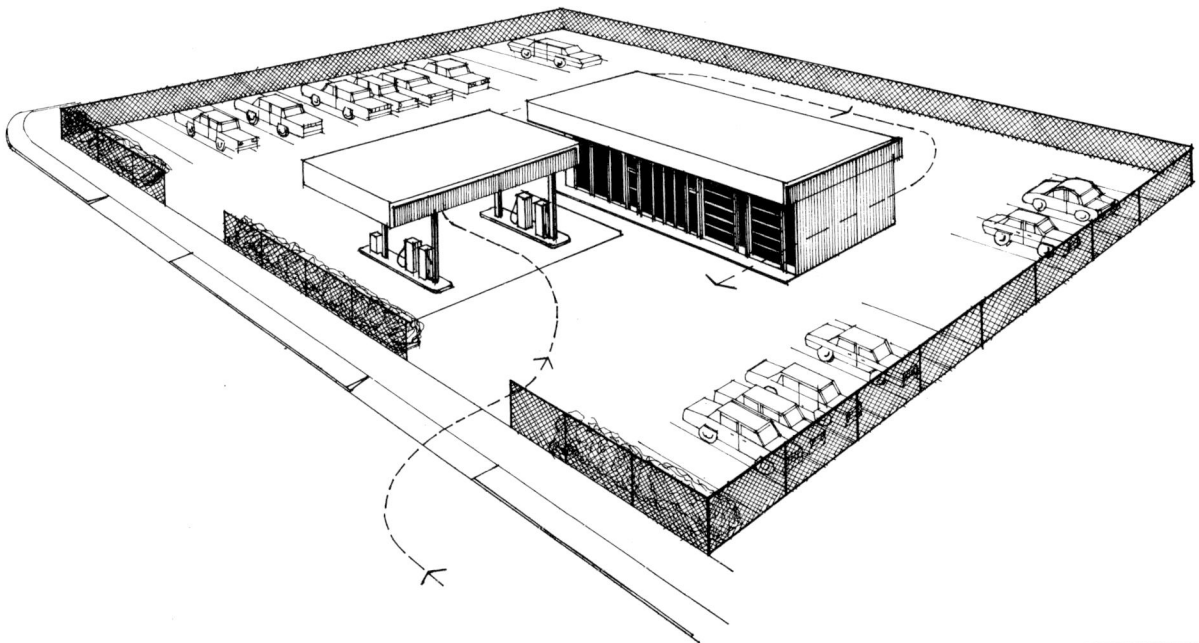

ANIMAL CLINIC

A LARGE PROPERTY WITH AMPLE SPACE FOR PARKING

THE REMODELED FORMER TWO BAY SERVICE STATION

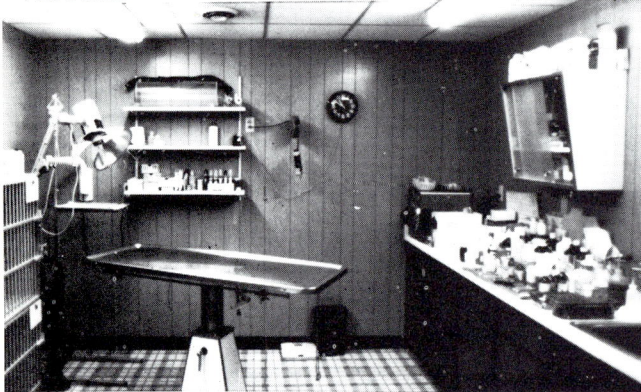

THE OPERATING ROOM

THE WAITING ROOM

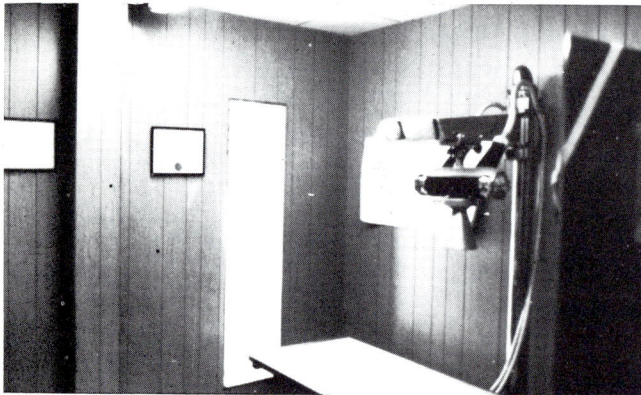

THE X-RAY ROOM

THE CAGE AREA AND WASHING SINK

Many veterinarians have discovered that an abandoned service station contains the exact space required for a thriving animal clinic. Most service station properties have a parking area and ample space for expansion--a very important factor that must be considered by anyone entering a business venture.

FLOOR PLAN

This animal clinic occupies a former two bay service station situated on a heavily travelled major artery in Colorado Springs. The veterinarian acquired this 160' x 100' corner property for $52,000. He remodeled the lubritory area into an examining room, laboratory and x-ray area, an operating room and an area devoted to cages for recuperating animals. An attractive application of vertical siding and batten strips covers the exterior of the former service station. The building alterations were accomplished at a cost of $20,000 and an additional $2500 was expended for landscaping. According to the veterinarian, a savings of $75,000 was realized by purchasing and remodeling the abandoned service station instead of erecting a new building on a vacant parcel.

APPLIANCES-SALES & REPAIRS

A service oriented business can be a very successful venture for the mechanically knowledgeable entrepreneur. An appliance store, which provides repairs and has a parts department, should be an instant success. During periods of sluggish economy, consumers tend to have appliances repaired rather than replace them. The three bay side entry building has sufficient floor space which can be subdivided into a sales display room and parts and service department. Sufficient land should be available for future expansion.

THE ABANDONED SERVICE STATION

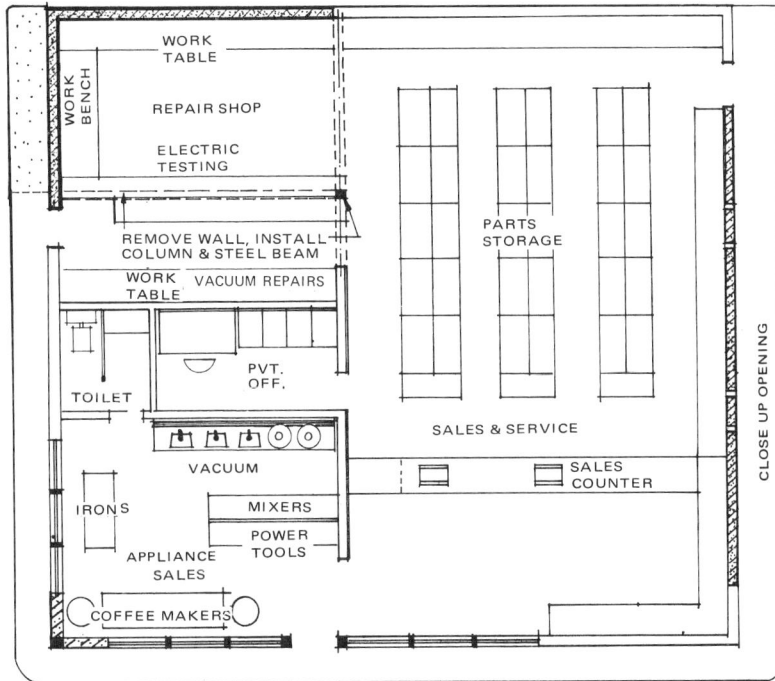

FLOOR PLAN   SCALE   0'   5   10'   15'

- WORK TABLE
- REPAIR SHOP
- ELECTRIC TESTING
- WORK BENCH
- REMOVE WALL, INSTALL COLUMN & STEEL BEAM
- WORK TABLE
- VACUUM REPAIRS
- PVT. OFF.
- TOILET
- VACUUM
- IRONS
- MIXERS
- POWER TOOLS
- APPLIANCE SALES
- COFFEE MAKERS
- PARTS STORAGE
- SALES & SERVICE
- SALES COUNTER
- CLOSE UP OPENING

LEFT SIDE ELEVATION

DIAGONAL WOOD SIDING

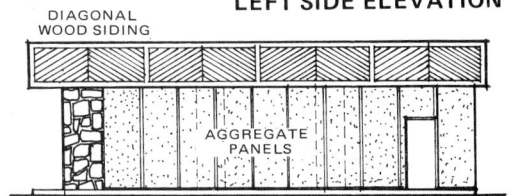

AGGREGATE PANELS

RIGHT SIDE ELEVATION

INSULATION

PARTS STORAGE

LONGITUDINAL SECTION

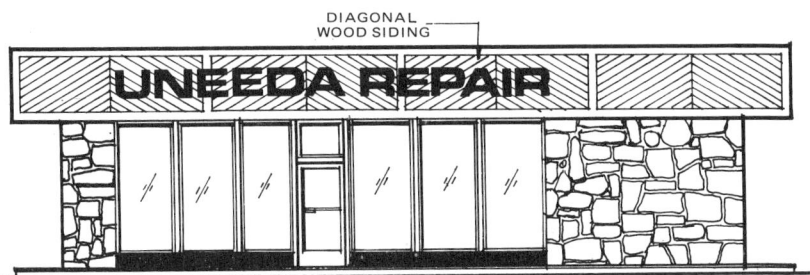

FRONT ELEVATION

DIAGONAL WOOD SIDING

UNEEDA REPAIR

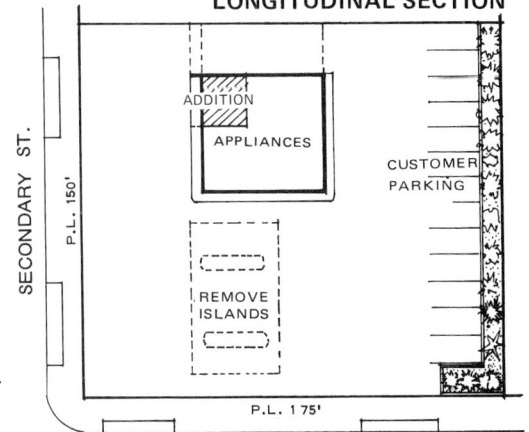

PLOT PLAN

SECONDARY ST.

P.L. 150'

ADDITION

APPLIANCES

CUSTOMER PARKING

REMOVE ISLANDS

P.L. 175'

MAIN ST.

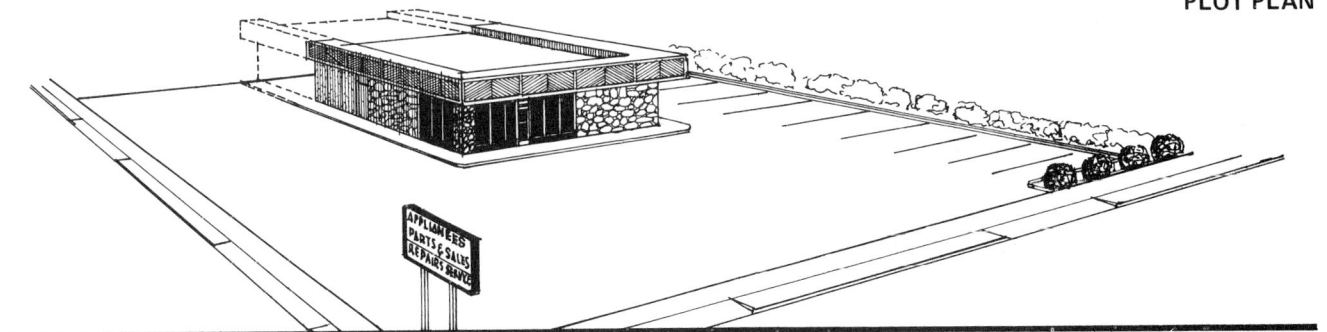

APPLIANCES PARTS & SALES REPAIRS SERVICE

AUTO ACCESSORY STORE AND TIRE CENTER

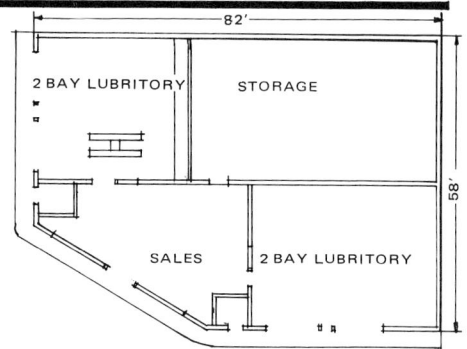

FLOOR PLAN
ABANDONED BUILDING

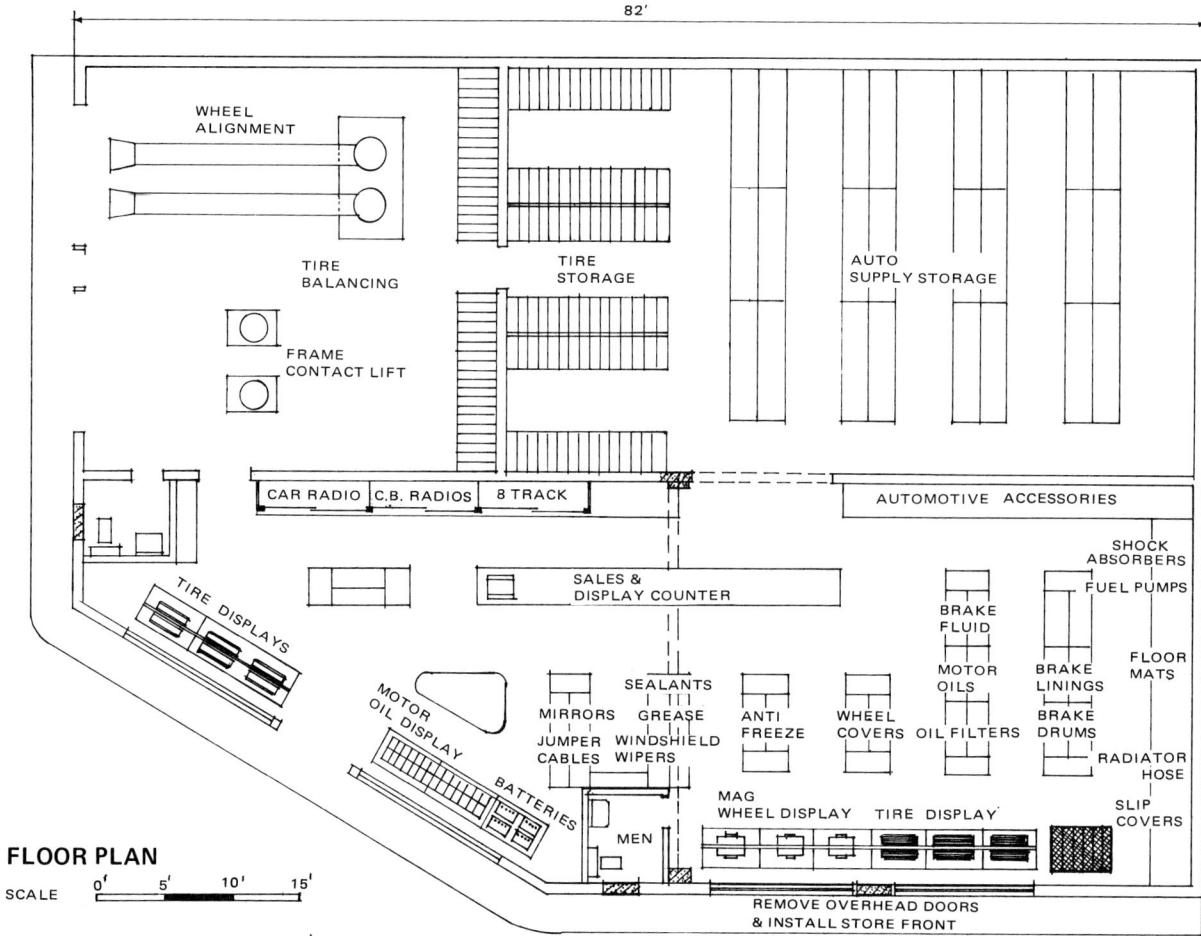

82'

58'

2 BAY LUBRITORY

STORAGE

SALES

2 BAY LUBRITORY

82'

WHEEL ALIGNMENT

TIRE BALANCING

FRAME CONTACT LIFT

TIRE STORAGE

AUTO SUPPLY STORAGE

CAR RADIO | C.B. RADIOS | 8 TRACK

AUTOMOTIVE ACCESSORIES

SHOCK ABSORBERS

FUEL PUMPS

TIRE DISPLAYS

SALES & DISPLAY COUNTER

BRAKE FLUID

FLOOR MATS

MOTOR OIL DISPLAY

SEALANTS

MOTOR OILS

BRAKE LININGS

MIRRORS | GREASE

ANTI FREEZE

WHEEL COVERS

OIL FILTERS

BRAKE DRUMS

JUMPER CABLES | WINDSHIELD WIPERS

RADIATOR HOSE

BATTERIES

MAG WHEEL DISPLAY

TIRE DISPLAY

SLIP COVERS

MEN

FLOOR PLAN

SCALE 0' 5' 10' 15'

REMOVE OVERHEAD DOORS
& INSTALL STORE FRONT

AUTOMART

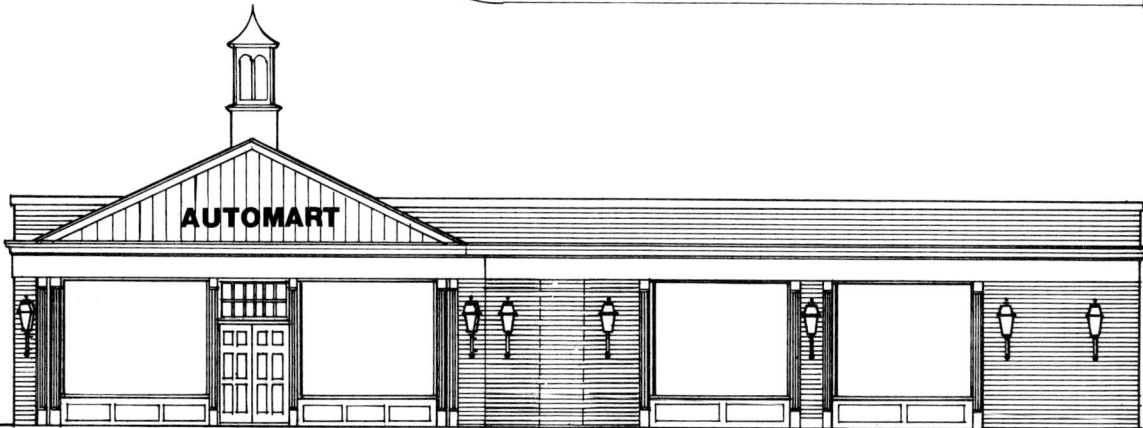

FRONT ELEVATION

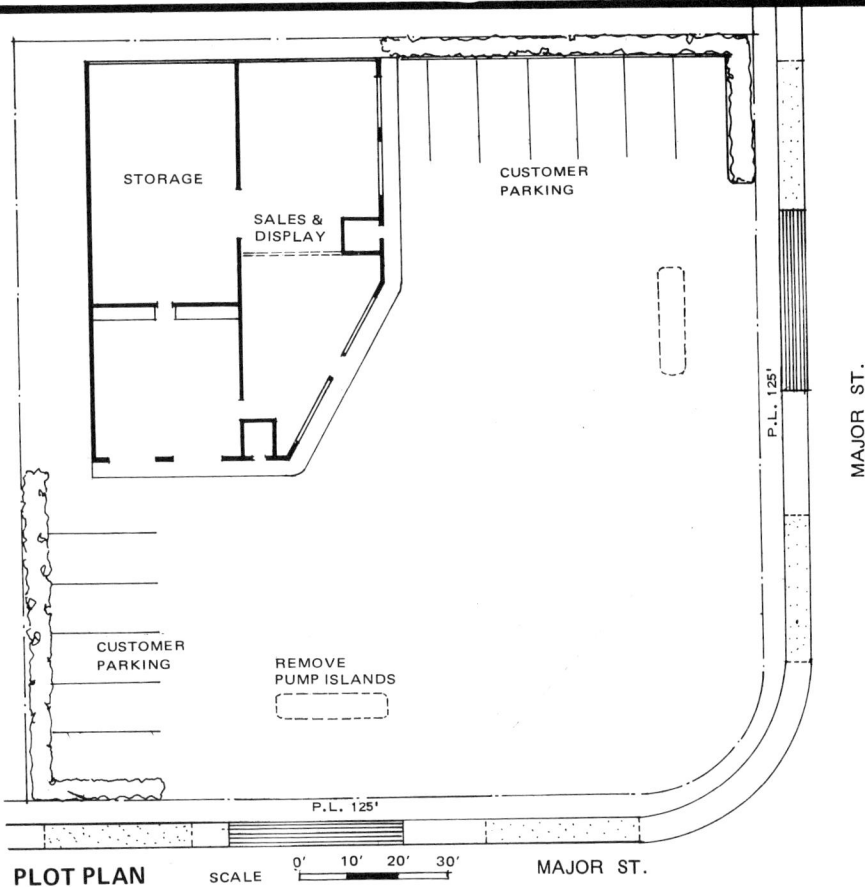

PLOT PLAN   SCALE   0'   10'   20'   30'   MAJOR ST.

Auto products income will rise from an estimated $7.2 billion in 1980 to an estimated $7.8 billion in 1981 according to the U.S. Department of Commerce's "Franchising in the Economy". This large abandoned service station will make an excellent conversion to a tire and auto accessory store. Two lubritory bays must be retained for installation of new tires, tire balancing and wheel alignment. A large sales and display area is available for the fast moving auto product lines. The large storeroom permits the owner to make quantity purchases of products that will improve profit margins. A display showroom has been positioned equidistant between the two main streets to promote tire sales. The higher costs of automobiles and auto repairs are encouraging more car owners to undertake their own repairs. Therefore, this facility has specialized in the sale and installation of tires and auto accessories.

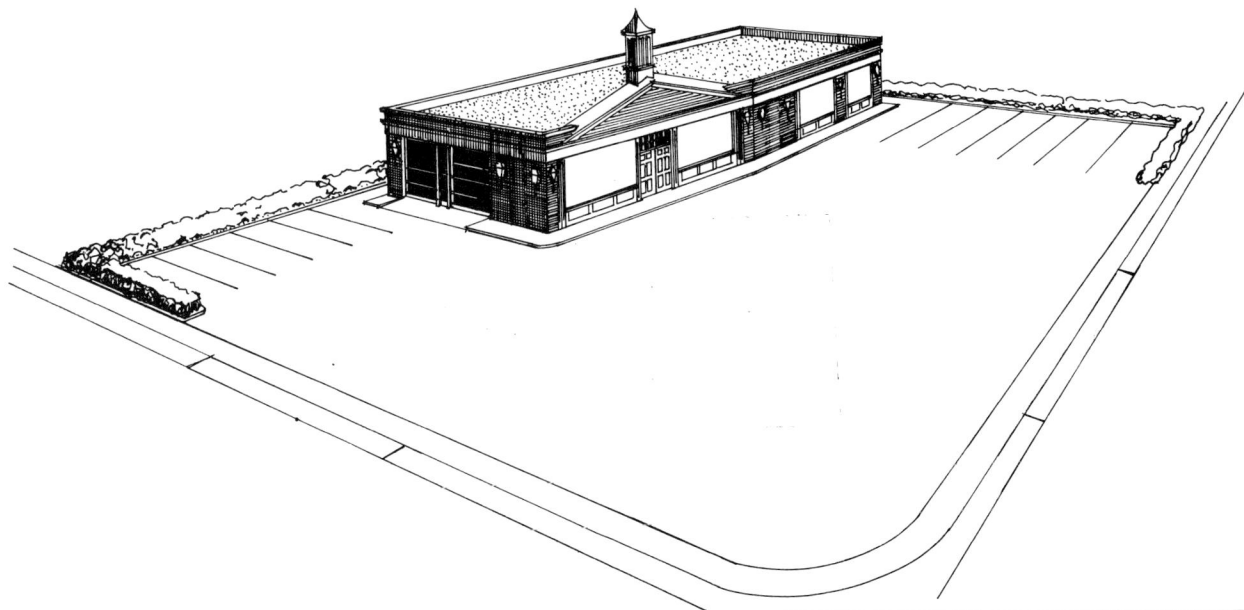

AUTOMOTIVE SERVICE, QUICK OIL CHANGE CENTER

THE DRIVE-THRU QUICK SERVICE BAYS

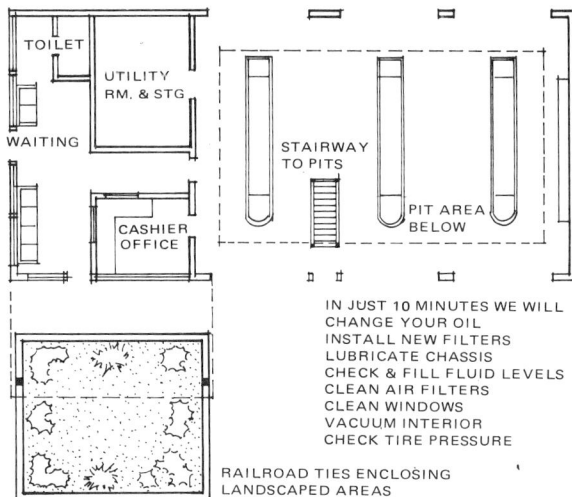

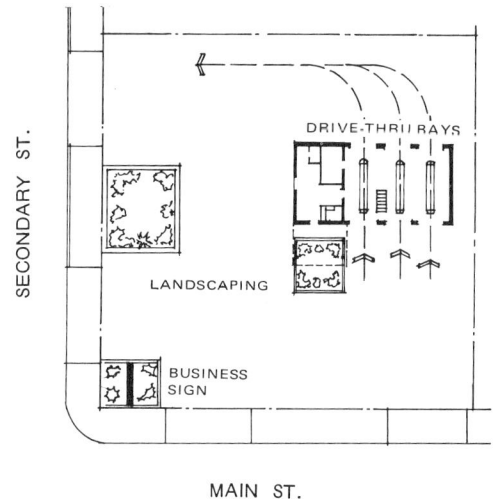

IN JUST 10 MINUTES WE WILL
CHANGE YOUR OIL
INSTALL NEW FILTERS
LUBRICATE CHASSIS
CHECK & FILL FLUID LEVELS
CLEAN AIR FILTERS
CLEAN WINDOWS
VACUUM INTERIOR
CHECK TIRE PRESSURE

RAILROAD TIES ENCLOSING
LANDSCAPED AREAS

FLOOR PLAN

PLOT PLAN

THE LUBRITORY BAYS WITH WITH A PIT AREA UNDER
THE FLOOR

FILTERS, MISCELLANEOUS ACCESSORIES AND PRICE
LIST OF SERVICES

The popularity of the ten minute quick oil change and lubrication is growing. Motorists do not care to wait long periods of time while their cars are being serviced. As full service stations continue to close, this franchise concept will flourish and grow to take its place in the automotive aftermarket. A three bay service station is perfectly designed for the large basement area that is required to provide prompt service. In addition to the lubrication and oil change, eight important engine checks are included. These inspection checks result in additional revenue for the operator.

This property is ideally situated at the intersection of a major road leading to the heart of a large city. The major expense was the installation of the pit area, together with explosion-proof lights and exhaust fan. The sales area was converted into an office and lounge. The storage area was not altered. The basement installation and sales room alterations cost $25,000, and another $2,000 was spent to install landscaping required by the city.

In 1979 the store won an aesthetic award for reconstruction from the local Chamber of Commerce.

Courtesy Grease Monkey, Denver, Colorado

DIET BAKERY AND SNACK BAR

Specialty bakeries are few and far between. When located in a densely populated area, on a heavily travelled road, they can be an instant success. Thousands of abandoned service stations are ideally situated for this business use. A two bay building, containing 1200 square feet of floor area, can be easily converted into a bakery-pastry shop with sufficient room for a snack bar.

This store specializes in sugar-free and/or salt-tree cakes, pastries, bread, pies, cookies and candy. They contain no sugar for the diabetic, no salt for the hypertensive, and low calorie content for the people with weight problems.

The parking area is important for both bakery and snack bar customers. Low calorie sandwiches and desserts would be featured at the snack bar for the transient. All of the foods would be prepared daily on the premises. Almost every favorite cake or pie would be available for purchase.

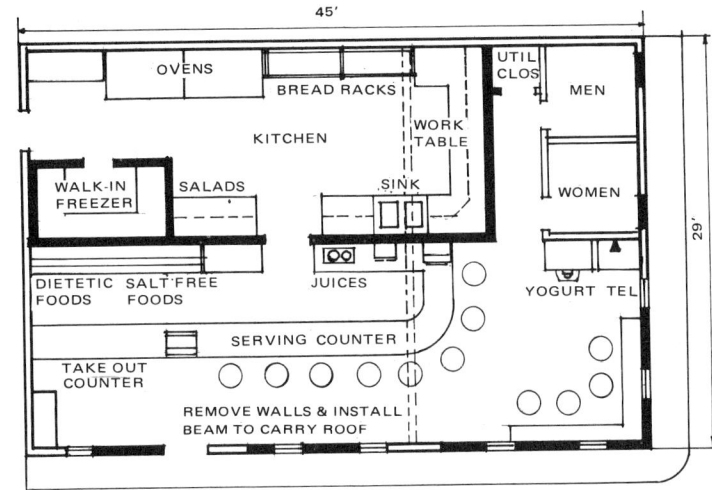

FLOOR PLAN   SCALE   0'   5'   10'   15'

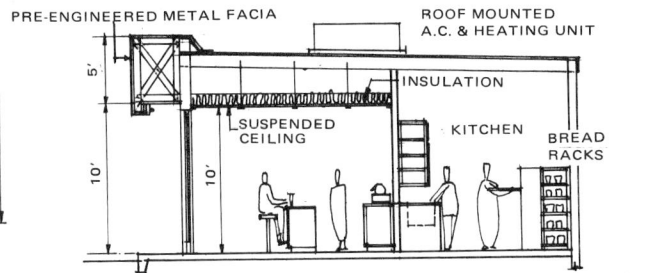

CROSS SECTION

FRONT ELEVATION

RIGHT SIDE ELEVATION

THRIFT BAKERY

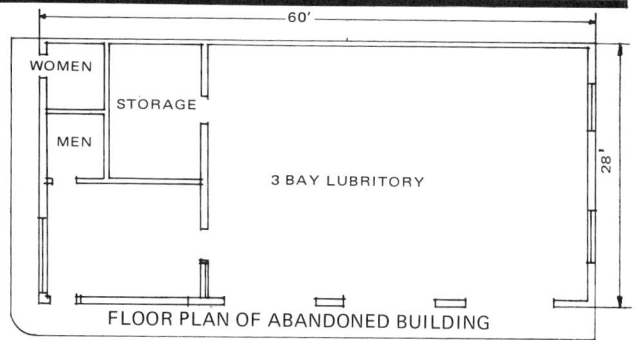

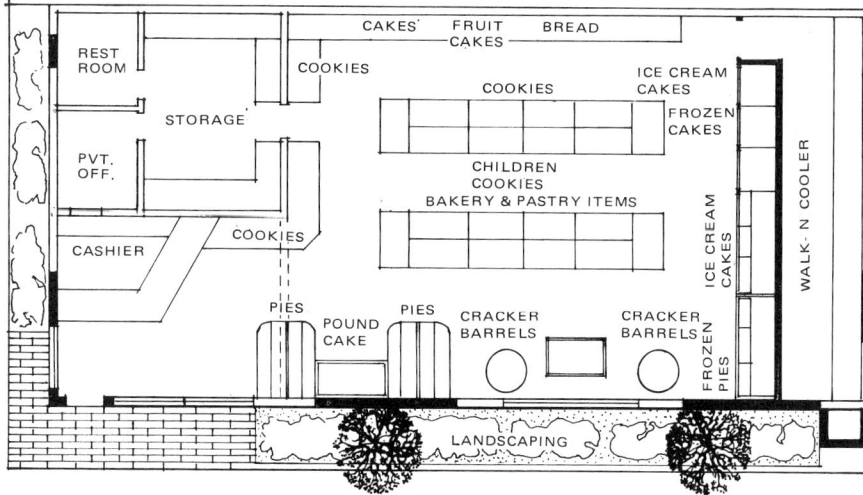

60'

WOMEN

STORAGE

MEN

28'

3 BAY LUBRITORY

FLOOR PLAN OF ABANDONED BUILDING

During periods of inflation, thrift bakery stores will experience substantial increase in business. The consumer, faced with 12% to 15% annual inflation for the next decade, will constantly search for bargains. These stores carry a full line of bread, cake and cookies at reduced prices. Abandoned service station buildings can be adapted at nominal cost. A large parking area is of major importance to accommodate peak periods of operation.

The design shown on this page can be accomplished at low cost. Minor internal changes, together with the development of a new facade, will attract the consumer.

REST ROOM

STORAGE

PVT. OFF.

CASHIER

COOKIES

CAKES FRUIT CAKES BREAD

COOKIES

COOKIES

ICE CREAM CAKES

FROZEN CAKES

CHILDREN COOKIES

BAKERY & PASTRY ITEMS

ICE CREAM CAKES

WALK - N COOLER

PIES

POUND CAKE

PIES

CRACKER BARRELS

CRACKER BARRELS

FROZEN PIES

LANDSCAPING

FLOOR PLAN

SCALE

0'   5   10'   15'

THRIFT BAKERY

DIAGONAL WOOD SIDING

WOOD SIDING

FRONT ELEVATION

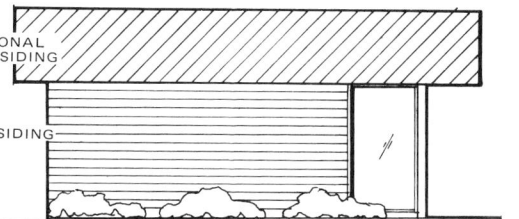

LEFT SIDE ELEVATION

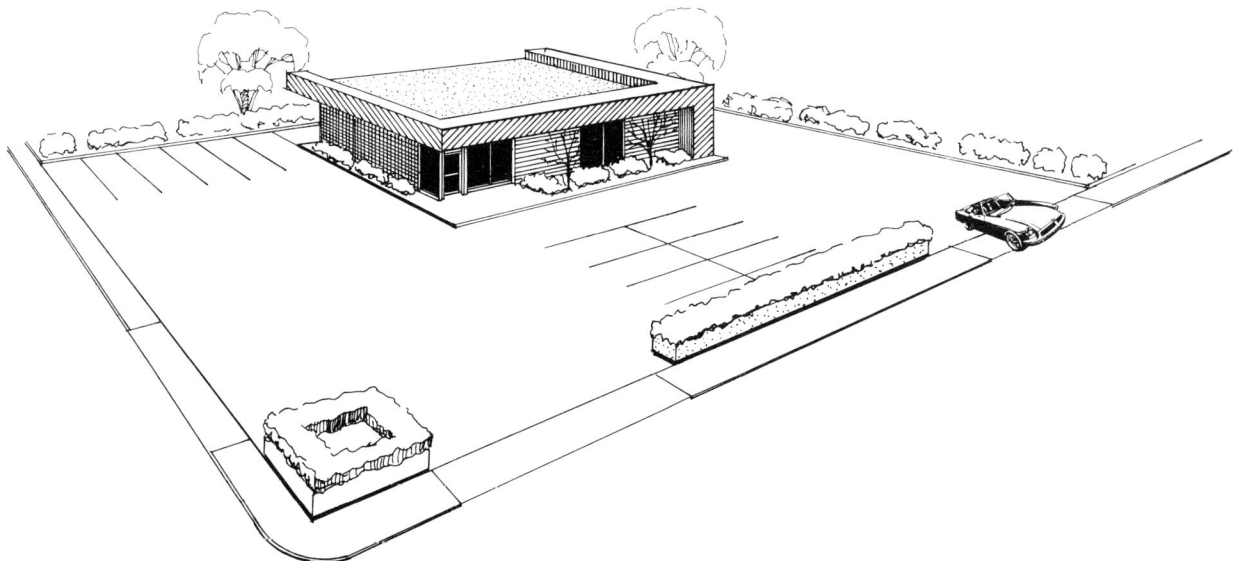

BRANCH BANK

THE CONVERSION OF A TWO BAY SERVICE STATION TO A BRANCH BANK

A CONTEMPORARY REMODELING DESIGNED BY AN ARCHITECT

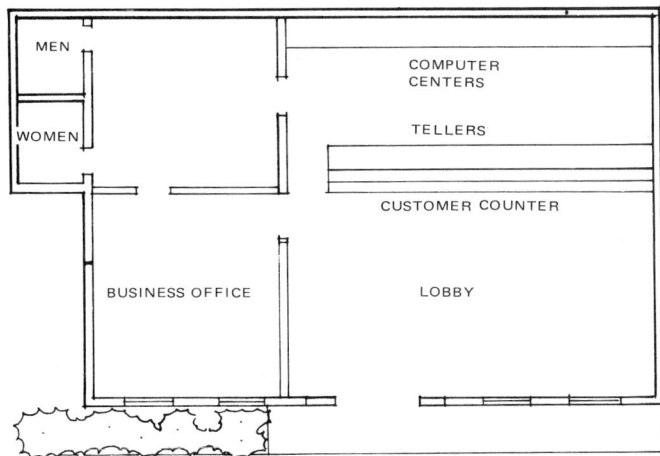

FLOOR PLAN

In 1976, a local bank was looking for a site for a new branch building. After completing a feasibility study of the area, an abandoned two bay service station was selected. The property was located on a major highway, adjacent to a busy shopping center. The decision to convert this building was made on the basis of the desirability of the site, and because there were no comparable properties in the vicinity. Various other sites were rejected because of restrictive zoning. The property was made to suit the bank's purpose, which was to established a full-service facility in a limited service area. Numerous zoning hearings were held and a subdivision was necessary. The bank utilized the service of a local architect who designed an attractive contemporary motif. The success of this branch supports the feasibility study and thorough research performed by the bank prior to the acquisition of the property. It is unusual for a bank to attain a high volume without a drive-up window area; however, this unit has succeeded.

Courtesy Red Hill Savings and Loan Association, Red Hill, Pa.

BRANCH BANK

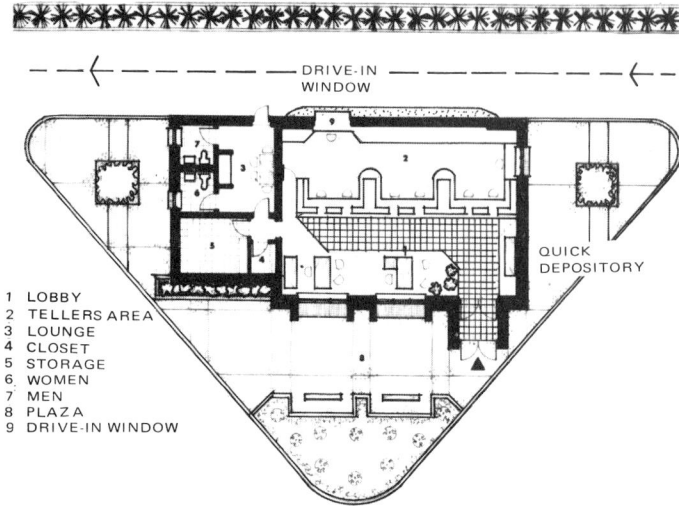

DRIVE-IN
WINDOW

QUICK
DEPOSITORY

1 LOBBY
2 TELLERS AREA
3 LOUNGE
4 CLOSET
5 STORAGE
6 WOMEN
7 MEN
8 PLAZA
9 DRIVE-IN WINDOW

FLOOR PLAN

0'   8'   12'

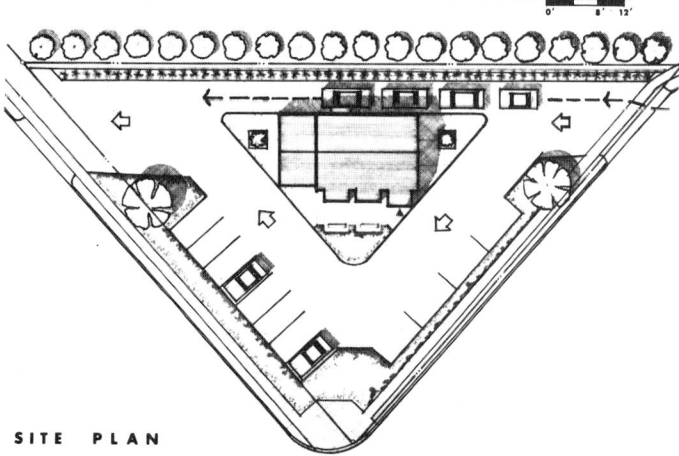

SITE PLAN

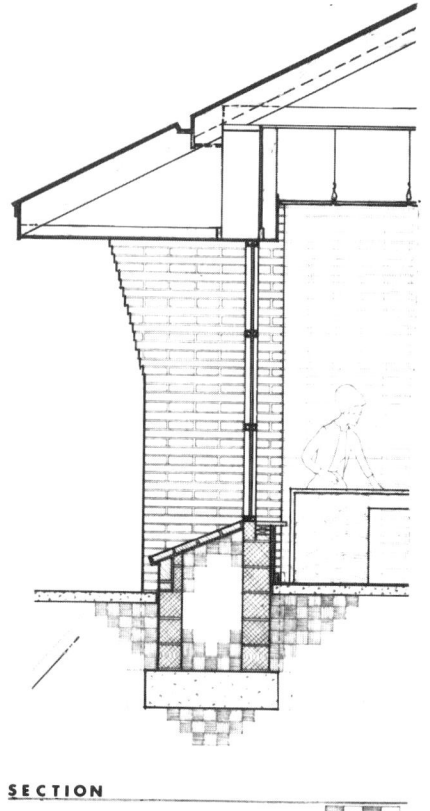

SECTION

B B Barry Bannett · Architect

Haddonfield   **New Jersey**

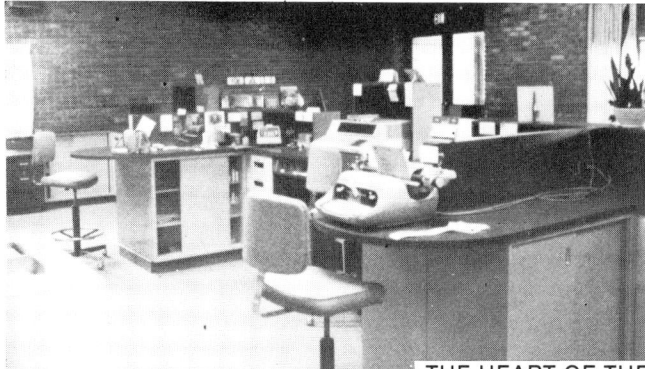

THE HEART OF THE TELLER WORK AREA

THE INTERIOR WALLS ARE FACED IN BRICK

THE TELLER COUNTERS

Drawings & Photographs Courtesy Barry Bannett, A.I.A., Architect, Haddonfield, N.J.
Courtesy Riverside Savings & Loan Assn., Riverside, N.J.

BRANCH BANK

VIEW OF THE COLONIAL SERVICE STATION

In 1974, an abandoned three bay colonial building, situated in a densely populated residential area, was purchased by a bank for $73,500. The bank retained a talented architect who provided construction management services. This permits the client to gain greater control over the construction budget and schedule while reducing costs. The brick veneered interior and exterior will require very little maintenance throughout the years. Good planning permitted the installation of a drive-in window across the rear of the building. The total remodeling was completed at a cost of $52.27 per square foot. An estimate of $104,000 had been projected. The actual cost was $100,374.

A breakdown of costs is detailed below:

Site demolition	$2,190.
Interior demolition	1,466.
Rubbish and trash	680.
Asphalt paving	2,100.
Lawn planting	3,100.
Concrete walks and curbs	8,215.
Brick veneer	10,863.
Miscellaneous metal	416.
Carpentry	3,600.
Laminated construction	8,252.
Roofing	2,545.
Aluminum entry doors and glass	4,850.
Weatherstripping	55.
Finished hardware	549.
Vinyl walls	590.
Quarry tile floor	1,350.
Acoustical tile	1,610.
Painting	600.
Toilet access	281.
Kitchen unit	817.
Bank equipment	14,504.
Carpeting	1,533.
Mats	50.
Loose furniture	3,510.
Plumbing	1,000.
Heating, vent and air conditioning	4,506.
Electrical work	9,775.
Contingency	907.
Lawn sprinkler	1,700.
Signs	6,450.
	$100,374.

Drawings & Photographs Courtesy Barry Bannett, A.I.A., Architect, Haddonfield, N.J.
Courtesy Riverside Savings & Loan Assn., Riverside, N.J.

BRANCH BANK

THE ABANDONED THREE BAY COLONIAL BUILDING

A NEW BRANCH BANK WITH DRIVE-UP TELLERS

A LARGE PARKING AREA REQUIRED FOR CUSTOMER PARKING

A VIEW OF THE ENTIRE PROPERTY

LANDSCAPING AROUND THE BUILDING AND PARKING AREA

EMPLOYEE PARKING BEHIND THE BANK

THE DRIVE-UP TELLERS

THE PROPERTY IS NOW AN ASSET TO THE COMMUNITY

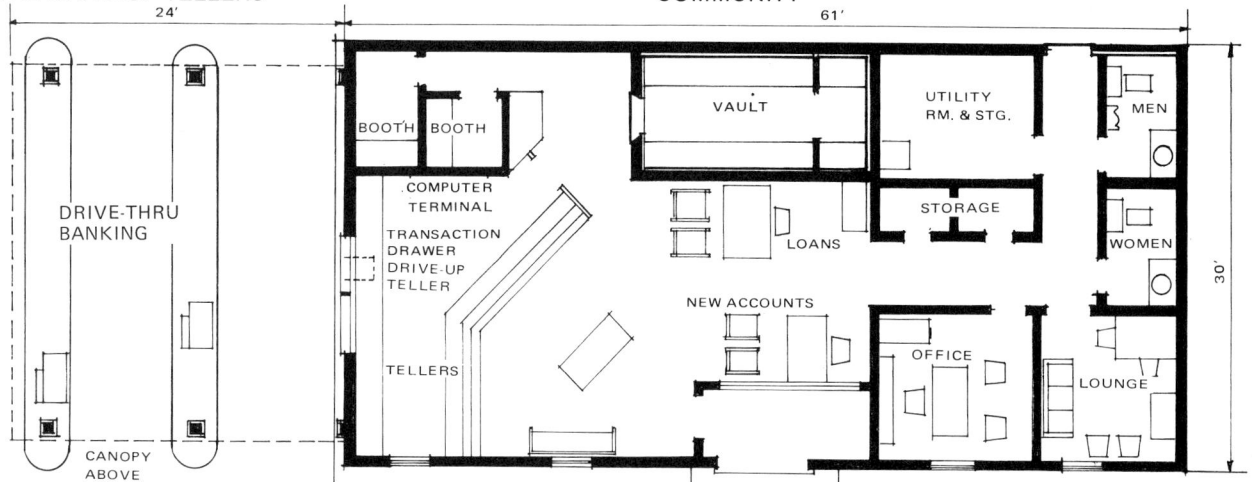

24'

DRIVE-THRU BANKING

CANOPY ABOVE

61'

BOOTH BOOTH

COMPUTER TERMINAL

TRANSACTION DRAWER DRIVE-UP TELLER

TELLERS

VAULT

LOANS

NEW ACCOUNTS

UTILITY RM. & STG.

STORAGE

OFFICE

MEN

WOMEN

LOUNGE

30'

FLOOR PLAN

Courtesy Lawrence Savings & Loan Assn., Hamilton Twp., N.J.
Drawings Courtesy Joseph D. Mason, Architect, P. A., Freehold, N. J.

BRANCH BANK

Large properties on major arteries near shopping and residential areas of a thriving community are hard to find. The President of a local savings and loan association spotted an abandoned service station. The property dimensions were 200' x 200' and the building had a deep setback, which would provide the essential off-street parking area that a bank requires.

The one acre property and building was purchased for $110,000. Drive-in banking was accommodated by adding a colonial canopy for three vehicles. The area behind the building was utilized for the traffic flow to the drive-in lanes. A ten car back-up was required by the zoning board for each drive-in customer. In addition, the board required one parking space for each ten customers. Existing yard lighting fixtures were re-used and portions of the existing paved yard were resurfaced while others were sealed. Attractive landscaping was developed, outlining the perimeter of the property and defining the various parking areas for customers and employees.

The President of the bank stated that the colonial architectural style of the abandoned building was perfect for banking purposes. The photographs illustrate the minimal interior changes to the original structure. The heating system, rest rooms, and storage area were all re-used. The lubritory area was converted to the teller and business office. A safety deposit vault with steel lining was installed to complete the conversion.

The breakdown of the construction costs is as follows:

1. Building renovation costs — $ 65,000.
2. Site work costs — 45,000.
3. Specialized bank equipment (including cagework, steel vault and equipment, remote island consoles and teller windows) — 75,000.

$185,000.

In my opinion, this is one of the finest branch bank developments I have seen, and is the result of retaining a talented architect.

THE LOBBY AND TELLER AREA

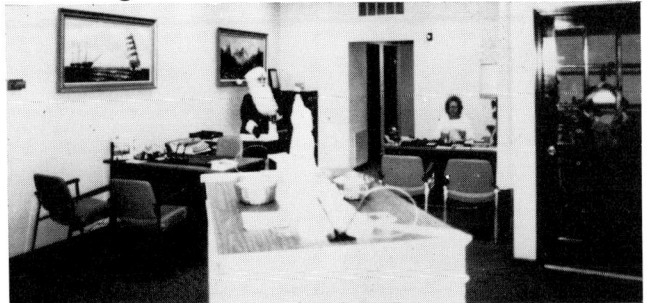

NEW ACCOUNTS AND ADMINISTRATION

THE DRIVE-UP TELLER COUNTER AND TELLER WORK AREA

INTERIOR VIEW

EMPLOYEES LOUNGE

THE INTERIOR VIEWED FROM THE ADMINISTRATION AREA

Courtesy Lawrence Savings & Loan Assn., Hamilton Twp., N.J.
Drawings Courtesy Joseph D. Mason, Architect, P. A., Freehold, N. J.

BRANCH BANK

Many banks are expanding their operations through small community branches. The conversion of abandoned service stations has been an inexpensive program for growing banks. A typical two bay colonial building contains 1260 square feet. A large property with ample room to the rear of the building is suitable for the addition of a drive-up window. The installation of dormers, and bay windows to replace the overhead doors, and double entry doors converts the former service station to an attractive colonial bank. A limited amount of interior work is required to provide a lobby, administrative area, teller stations, a lounge, two rest rooms and a vault. Parking for ten cars is an important consideration when selecting a parcel for a bank conversion. The site should be selected based on highway patterns and community growth potential.

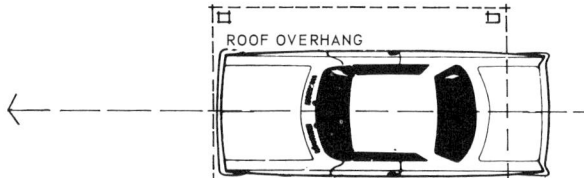

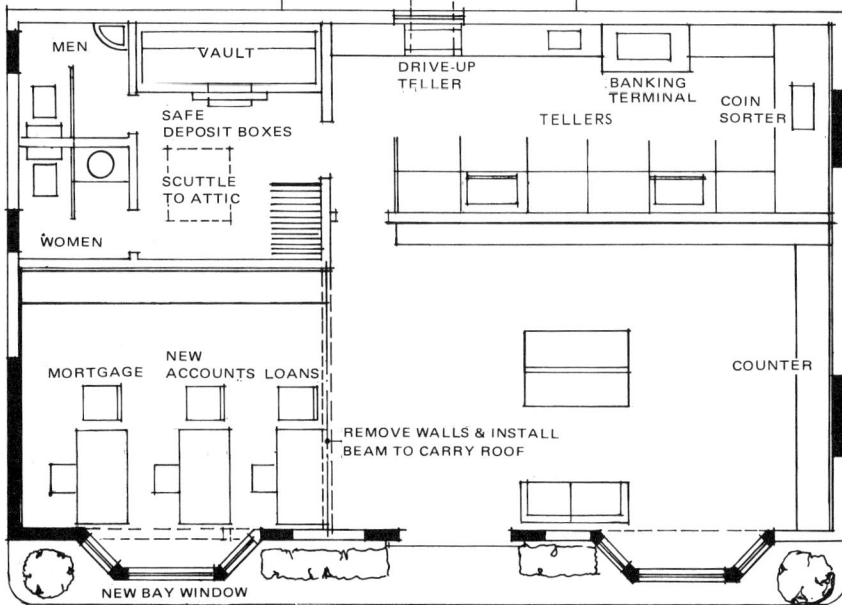

ROOF OVERHANG

MEN

VAULT

DRIVE-UP TELLER

BANKING TERMINAL

COIN SORTER

TELLERS

SAFE DEPOSIT BOXES

SCUTTLE TO ATTIC

WOMEN

MORTGAGE

NEW ACCOUNTS

LOANS

COUNTER

REMOVE WALLS & INSTALL BEAM TO CARRY ROOF

NEW BAY WINDOW

REMOVE O.H. DOORS & INSTALL NEW STORE FRONT

FLOOR PLAN

SCALE

0'   4   8'   12'

NEW DORMERS

FRONT ELEVATION

REAR ELEVATION

RIGHT SIDE ELEVATION

BRANCH BANK

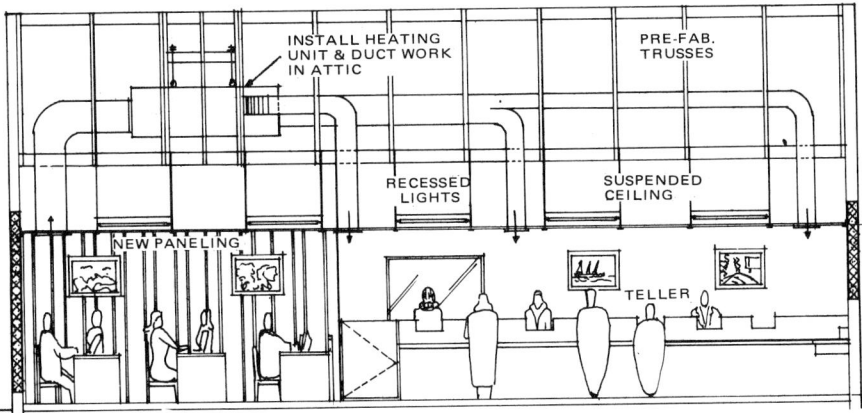

INSTALL HEATING UNIT & DUCT WORK IN ATTIC

PRE-FAB. TRUSSES

RECESSED LIGHTS

SUSPENDED CEILING

NEW PANELING

TELLER

LONGITUDINAL SECTION

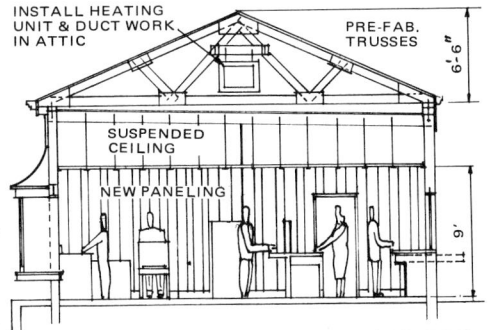

INSTALL HEATING UNIT & DUCT WORK IN ATTIC

PRE-FAB. TRUSSES

6'-6"

SUSPENDED CEILING

NEW PANELING

9'

CROSS SECTION

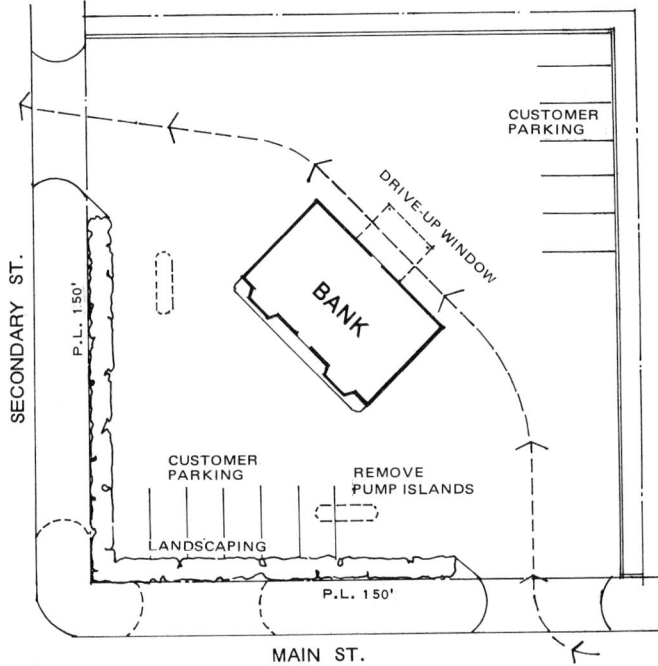

CUSTOMER PARKING

DRIVE-UP WINDOW

SECONDARY ST.

P.L. 150'

BANK

CUSTOMER PARKING

REMOVE PUMP ISLANDS

LANDSCAPING

P.L. 150'

MAIN ST.

PLOT PLAN

SCALE   0' 10' 20' 30'

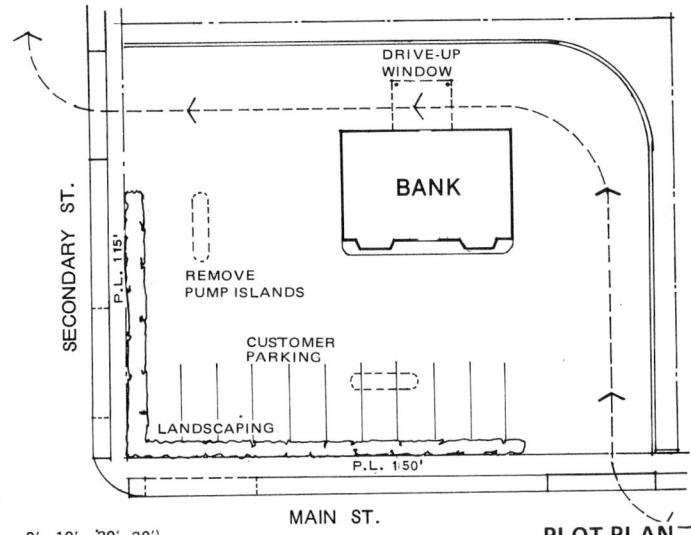

DRIVE-UP WINDOW

SECONDARY ST.

P.L. 115'

BANK

REMOVE PUMP ISLANDS

CUSTOMER PARKING

LANDSCAPING

P.L. 150'

MAIN ST.

PLOT PLAN

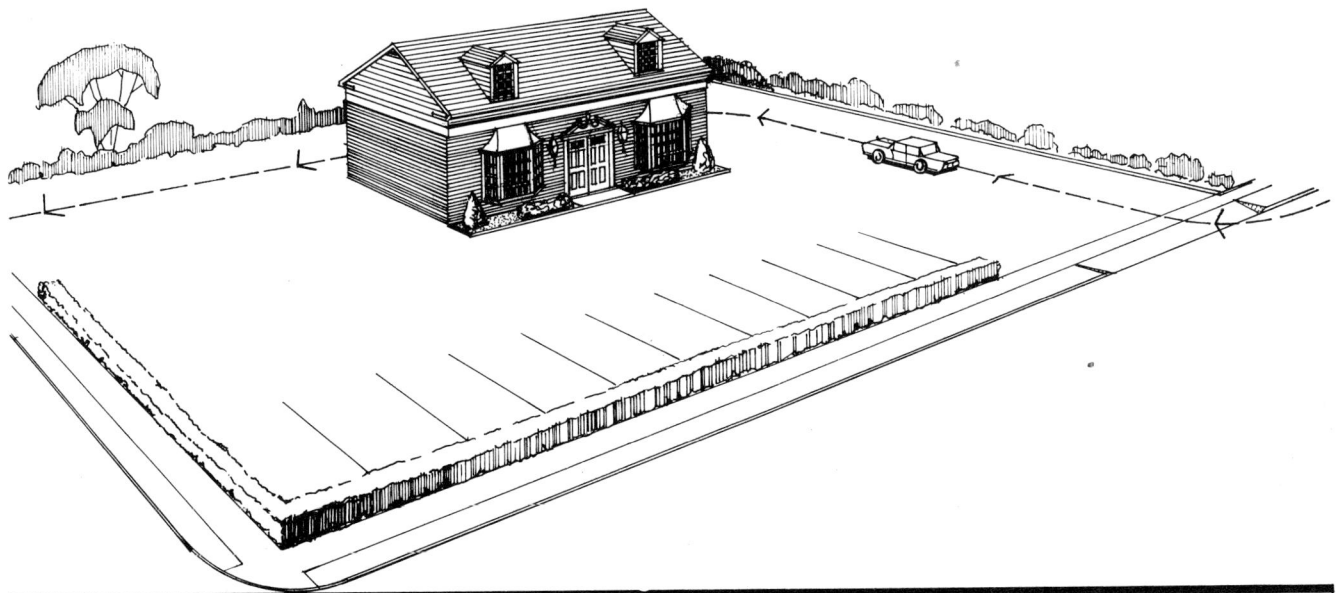

BRANCH BANK

On the West Coast, abandoned service stations dot the landscape of every community. Many intersections have two or three closed facilities. The photograph illustrates a typical abandoned three bay service station on a large parcel of land. One of the most popular successful business uses for these abandoned buildings is a branch bank.

This 1800 square foot building includes a banking area of 1500 square feet with six teller stations, a business office, four drive-up remote teller stations, employees' lounge and rest rooms. This particular service station requires extensive remodeling. Removal of the roof is suggested and a new look should be designed with a sloping contemporary roof and clerestory that will cast sunlight into the interior banking area from above. Parking for twelve to fifteen cars is required for customers. The canopy must be extended to the building to provide cover for one additional drive-up teller.

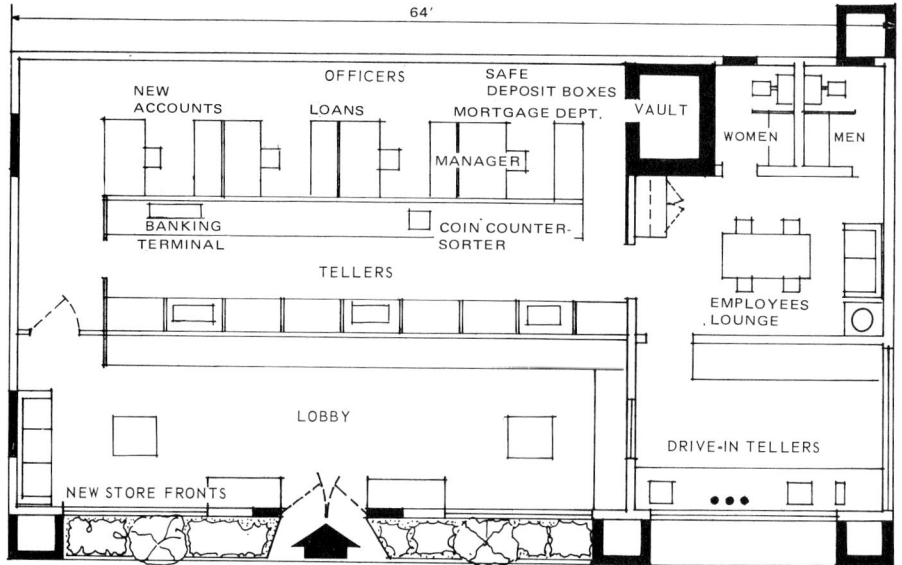

FLOOR PLAN

SCALE 0' 5' 10' 15'

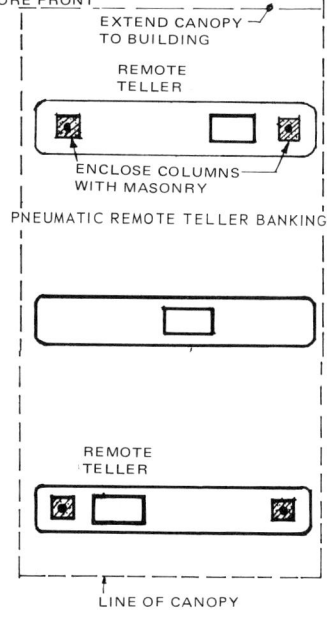

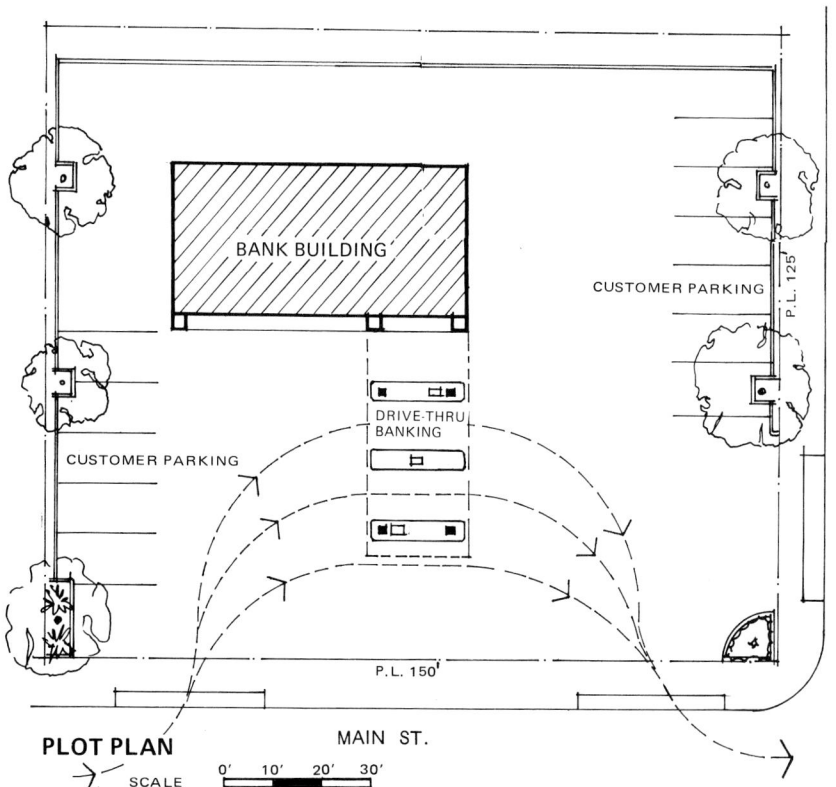

PLOT PLAN

MAIN ST.

SCALE 0' 10' 20' 30'

BRANCH BANK

FRONT ELEVATION

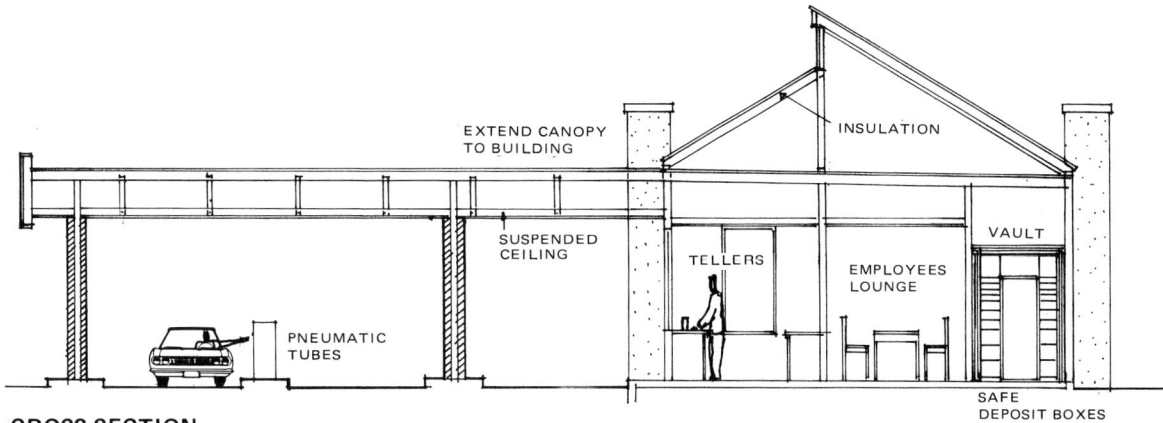

CLERESTORY

NEW ROOF

First Union Bank

TELLERS

STUCCO PILASTERS

CROSS SECTION

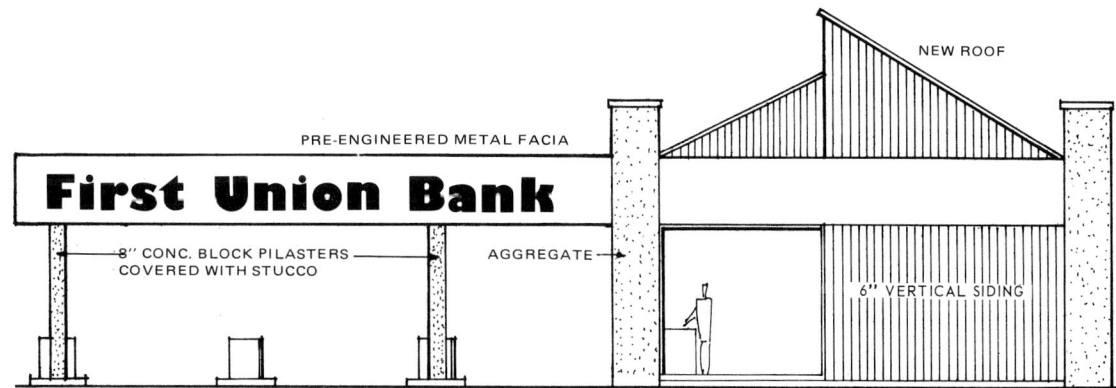

EXTEND CANOPY
TO BUILDING

INSULATION

SUSPENDED
CEILING

PNEUMATIC
TUBES

TELLERS

EMPLOYEES
LOUNGE

VAULT

SAFE
DEPOSIT BOXES

RIGHT SIDE ELEVATION

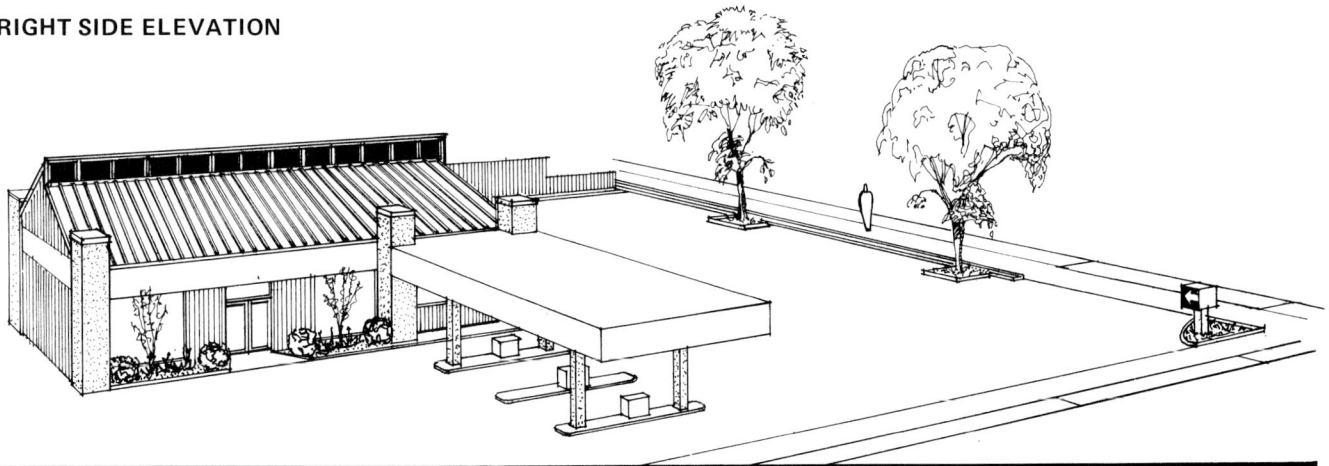

NEW ROOF

PRE-ENGINEERED METAL FACIA

First Union Bank

8" CONC. BLOCK PILASTERS
COVERED WITH STUCCO

AGGREGATE

6" VERTICAL SIDING

DRIVE-UP BANK

A number of "gas only" service stations have been abandoned but very few have been converted to alternate business uses. They may be recycled inexpensively to another drive-up use -- the drive-up branch bank that features fast, but limited, service. There is no loan service or inside banking. Three drive-up teller stations are available for fast customer service. Most "gas only" units are situated on major highways that carry considerable through and commuter traffic. A contemporary motif is achieved by installing a textured metal facade which simulates concrete. The remodeled 360 square foot office includes a self contained rest room.

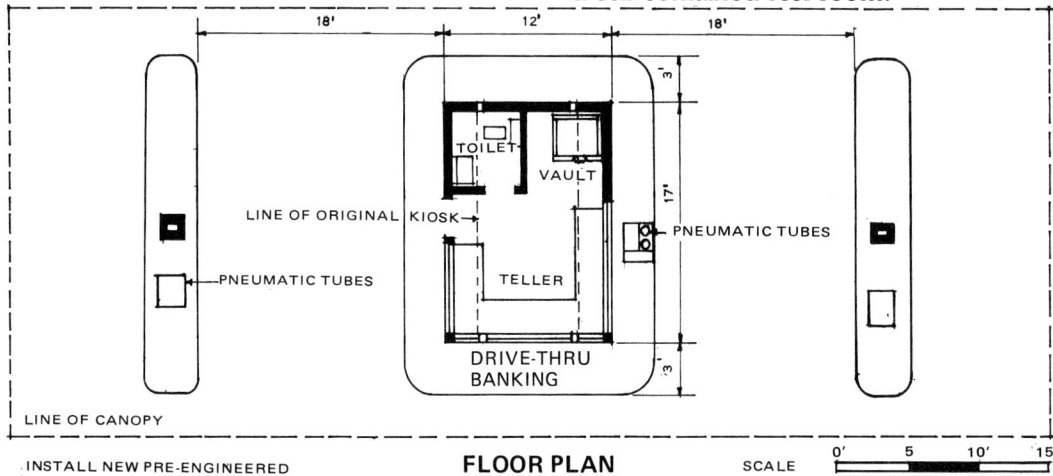

FLOOR PLAN

SCALE 0' 5 10' 15'

LINE OF CANOPY

18' 12' 18'

3' 17' 3'

TOILET

VAULT

LINE OF ORIGINAL KIOSK

PNEUMATIC TUBES

PNEUMATIC TUBES

TELLER

DRIVE-THRU BANKING

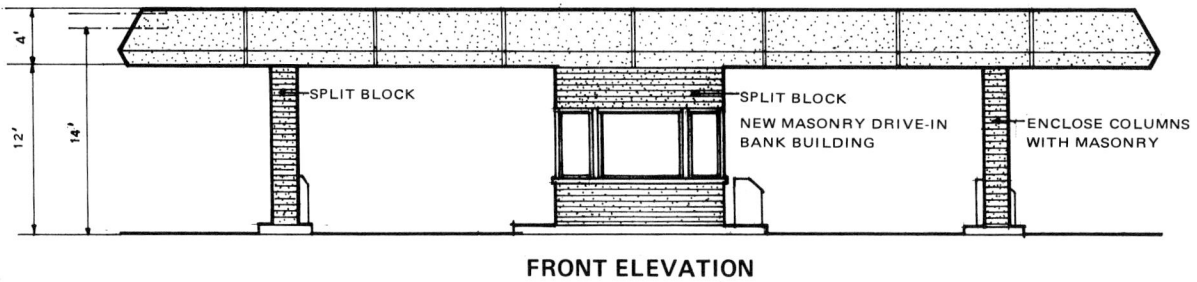

INSTALL NEW PRE-ENGINEERED FACIA OVER CANOPY

4'

12' 14'

SPLIT BLOCK

SPLIT BLOCK

NEW MASONRY DRIVE-IN BANK BUILDING

ENCLOSE COLUMNS WITH MASONRY

FRONT ELEVATION

DRIVE-UP BANK

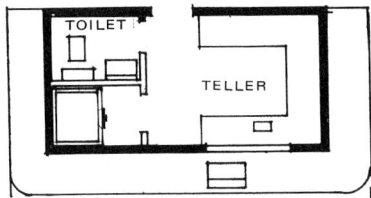

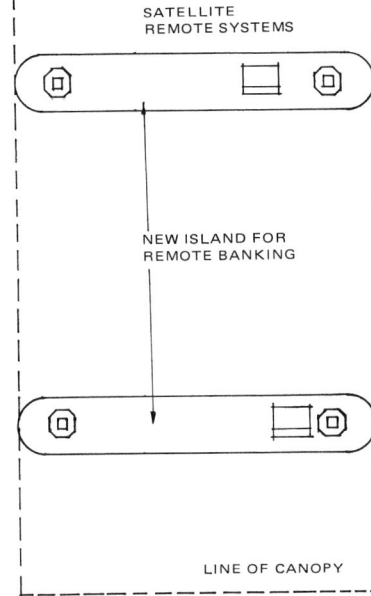

SATELLITE
REMOTE SYSTEMS

NEW ISLAND FOR
REMOTE BANKING

LINE OF CANOPY

FLOOR PLAN   SCALE   0'   5'   10'   15'

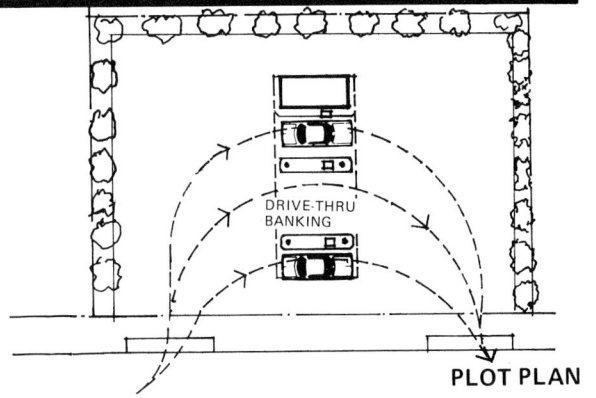

PLOT PLAN

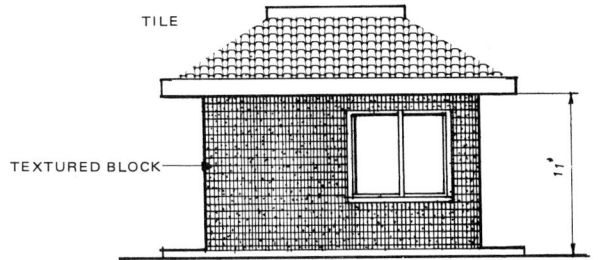

FRONT ELEVATION

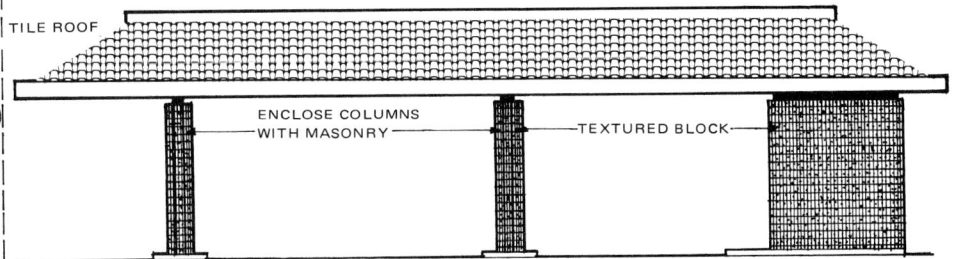

RIGHT SIDE ELEVATION

The conversion of a former self service gasoline facility, with a ten or eleven foot high canopy, can be accomplished by using a pre-engineered roofing systems. The cross section through the canopy illustrates the erection of lightweight structural elements, and the installation of Spanish roofing tile that creates a distinctive new look for the drive-up branch bank. The renovation of the kiosk into a self-contained bank teller office can be performed inexpensively. The major portion of the work will be the installation of appropriate security measures. Three remote teller stations are outlined on the plan.

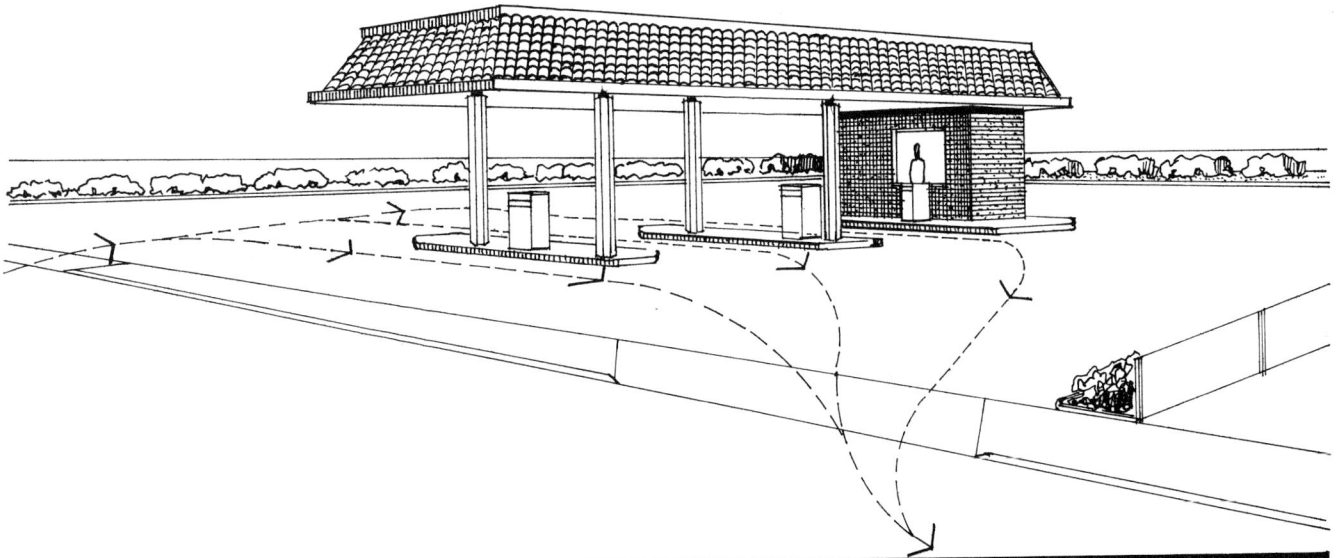

BICYCLES

In 1974, a young-at-heart retiree had plans for opening a new bicycle store in an affluent Pennsylvania city of 110,000. He properly researched approximately forty bicycle stores. Where should he locate? He contacted the local Planning Commission to determine their projected growth pattern for the suburbs. He examined five or six surveys of the projected growth for housing. After determining the general area, he searched for a site with easy access to a highway and large enough to permit customer test riding of new bicycles--an important feature of the "Pedaler". He came upon an abandoned two bay service station situated on a 300' x 190' parcel. The entrepreneur leased the facility for three years at $450 per month and proceeded to convert the building and erected a 28' x 30' addition for a repair shop and service department and cycle storage.

Lifts were removed, floor drains filled with concrete and the two overhead doors were relocated to become the overhead doors of the new addition. A glass partition, separating the old sales room from the lubritory, was removed and installed as a store front in one of the former overhead door openings. A matching window was installed in the other door opening. Inexpensive paneling was used to cover the interior walls of the lube and sales areas. Indoor-out carpeting was laid over the entire floor. The present heating system was adequate for the new use. The old storeroom was converted to an exercise cycle room. Cycle accessories are stored in large cabinetry along the walls, and bicycles are displayed throughout the old lubritory area. At each window, bicycles are suspended from the ceiling in an eyecatching manner. Steel supports were installed in the new storage addition and plywood, which formerly was used to board

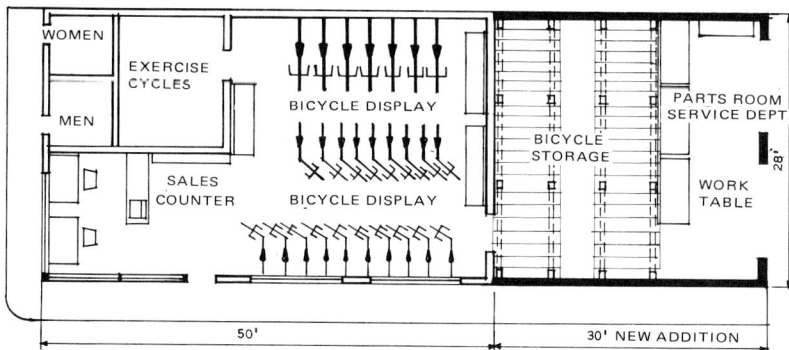

FLOOR PLAN

THE INSTALLATION OF AN AWNING PROTECTS THE OPEN BICYCLE DISPLAY

EXERCISE CYCLES DISPLAYED IN FORMER STORE ROOM

SERVICE AND REPAIR DEPARTMENT AND BICYCLE STORAGE

DURING SUMMER MONTHS THE OVERHEAD DOORS ARE OPENED

Courtesy Village Pedaler, Whitehall, Pa.

BICYCLES

up the building, served as a platform for bicycles stored in cartons. The remodeling cost the owner approximately $12,000 and the service department and storeroom addition cost $30,000. The total floor area is 2100 square feet, considered minimal for a successful bicycle store.

At the expiration of the three year lease, the businessman purchased the property for $160,000 and supplemented his income by adding another profit center, the sale of wood frame storage buildings and dog houses. In a recent year, approximately 2200 bicycles were sold. During the winter months, two bicycle repairmen are retained to service customers, and in the summer as many as ten may be hired to make repairs.

An important feature for all cycle shops is sufficient cycle storage facilities, particularly during the Christmas season when it is not uncommon to sell seventy-five bicycles during a weekend. An ample inventory becomes a tremendous asset.

Recently the owner leased a portion of the property to a convenience store chain. These two uses will complement each other. The convenience store anticipates 5,000 customers per month. These additional customers will have a positive effect on the continued success of the Village Pedaler.

A SUCCESSFUL BUSINESS OCCUPIES A FORMER ABANDONED BUILDING

BICYCLE SALES DISPLAYS

Courtesy Village Pedaler, Whitehall, Pa.

BICYCLE AND MOPED STORE

THE ABANDONED SERVICE STATION

The rapid increase in gasoline prices has accelerated bicycle sales throughout the country. This is the cheapest form of transportation for the commuter who lives within three miles of a railroad station. The cost is approximately one-quarter the price of an old used car and a bicycle runs on leg power not horse power. The repairs are less expensive and costly parking space is not necessary. The bicycle provides an excellent form of exercise for all age groups. Peak sales can be anticipated for three and ten speed bikes in affluent communities.

40'

WORK BENCH

PARTS DEPT.

REPAIRS & SERVICE 5 SERVICEMEN

15'

STORAGE

STORAGE, 3 TIERS FOR 360 BIKES

57'

ACCESSORIES, CHAINS, LOCKS, WHEELS, SEATS, PEDDLES, BASKETS & HELMETS

WOOD COUNTER WITH CABINETS FOR ACCESSORY SALES

REMOVE WALLS & INSTALL BEAM TO CARRY ROOF

3 SPEED BICYCLES

10 SPEED BICYCLES

ROLLER SKATES & SKATE BOARDS

TOILET

ACCESSORIES

PVT. OFF.

28'

DISPLAY OF USED BICYCLES OR MOPEDS

BICYCLE DISPLAY 10 SPEED BICYCLES

REMOVE WALLS

SALE & DISPLAY COUNTER

REMOVE O.H. DOORS & INSTALL NEW STORE FRONT

EXERCISE CYCLES

BICYCLE DISPLAY 10 SPEED BICYCLES

ENCLOSE CANOPY AREA WITH NEW MASONRY WALLS

47'

40'

57'

ADDITION

ADDITION

BICYCLE & MOPED STORE

CUSTOMER PARKING

45'

P.L. 125'

SECONDARY ST.

CUSTOMER PARKING

P.L. 160'

PLOT PLAN

MAIN ST.

SCALE 0' 5' 10' 15'

FLOOR PLAN

SCALE 0' 5' 10' 15'

BICYCLE AND MOPED STORE

Income opportunities are excellent from bicycle sales, sales of special accessories and repair services. In 1976, more than ten million bicycles were sold and more than one hundred million people in the United States are currently riding two-wheelers.

Accessory equipment ranges from CB radios to electric motors. These items can be easily installed on any bicycle.

A college town is an exceptionally good location for bicycle sales and repairs. Many students cannot afford a car; however they still need some type of transportation and a bicycle is an excellent choice.

The peak periods for selling bikes are during the spring, summer, and at Christmas time. It is important to have a paved area where the cycles may be test ridden. This is a big plus for selecting an abandoned service station site. Storage and warehouse facilities are essential to maintain a high sales volume.

A well situated bicycle store can generate gross sales ranging from $150,000 to $300,000 annually. Income from repairs can range from $50,000 to $250,000 annually, depending on the size of the service department.

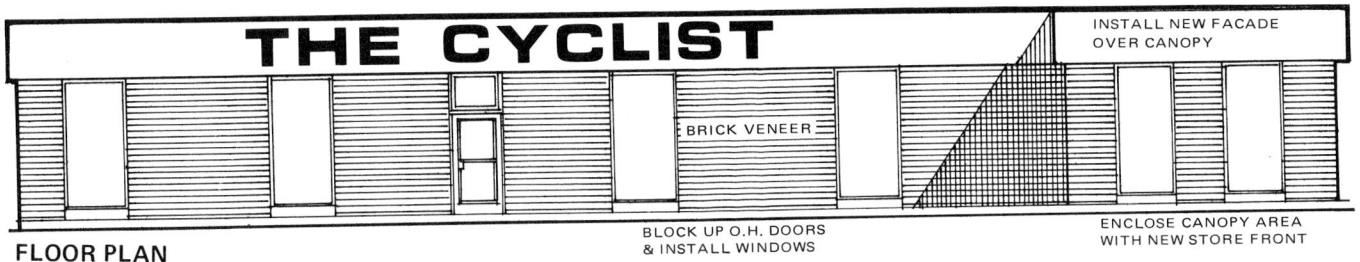

THE CYCLIST

INSTALL NEW FACADE OVER CANOPY

BRICK VENEER

FLOOR PLAN

BLOCK UP O.H. DOORS & INSTALL WINDOWS

ENCLOSE CANOPY AREA WITH NEW STORE FRONT

SUSPENDED CEILING

INSULATION

ACCESSORIES, CHAINS, LOCKS, WHEELS, SEATS, PEDDLES, BASKETS & HELMETS

SERVICE DEPT.

BICYCLE DISPLAY

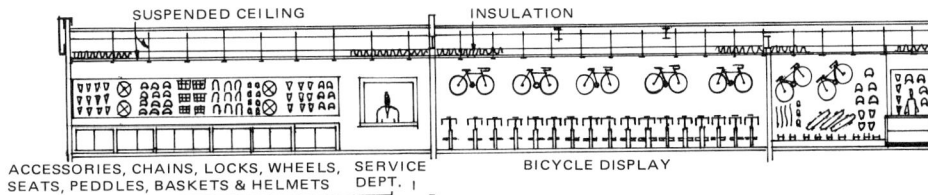

LONGITUDINAL SECTION

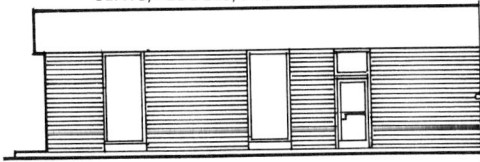

RIGHT SIDE ELEVATION

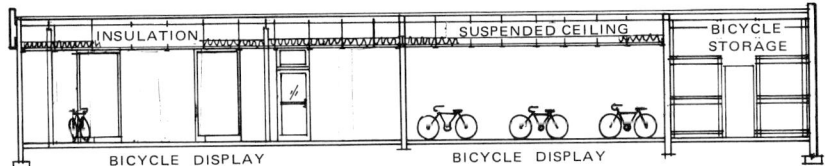

INSULATION

SUSPENDED CEILING

BICYCLE STORAGE

BICYCLE DISPLAY

BICYCLE DISPLAY

CROSS SECTION

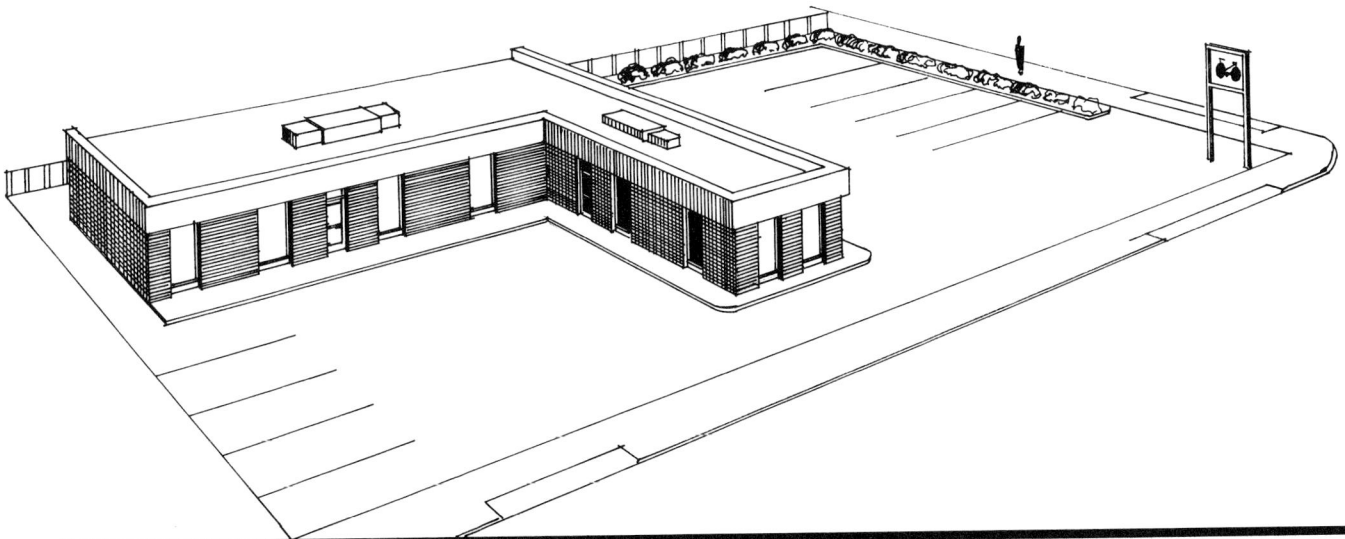

BICYCLE STORE

The businessman who intends to open a new bicycle store on a limited budget should consider the expansion of the business in four stages. The major factors to be considered for the success of the business are: location, display of merchandise, storage area and the repair and service department. The drawings illustrate the growth of a business as follows:

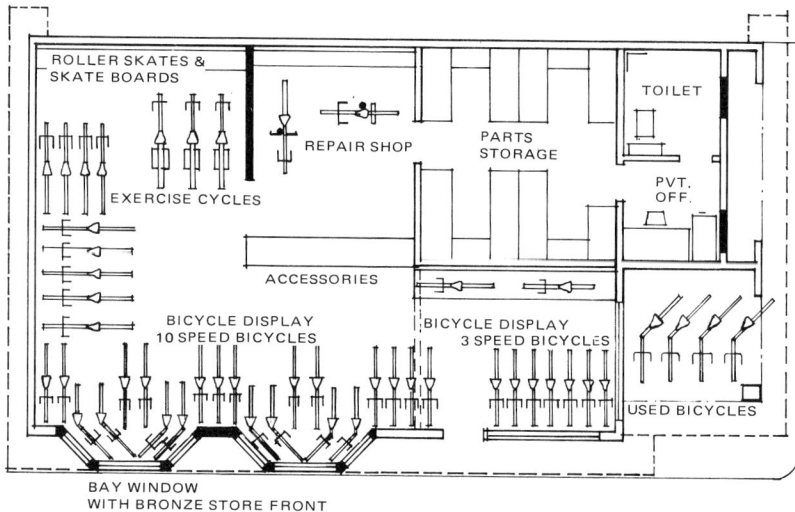

ROLLER SKATES & SKATE BOARDS

REPAIR SHOP

PARTS STORAGE

TOILET

PVT. OFF.

EXERCISE CYCLES

ACCESSORIES

BICYCLE DISPLAY 10 SPEED BICYCLES

BICYCLE DISPLAY 3 SPEED BICYCLES

USED BICYCLES

BAY WINDOW WITH BRONZE STORE FRONT

FLOOR PLAN   SCALE   0'   5'   10'   15'

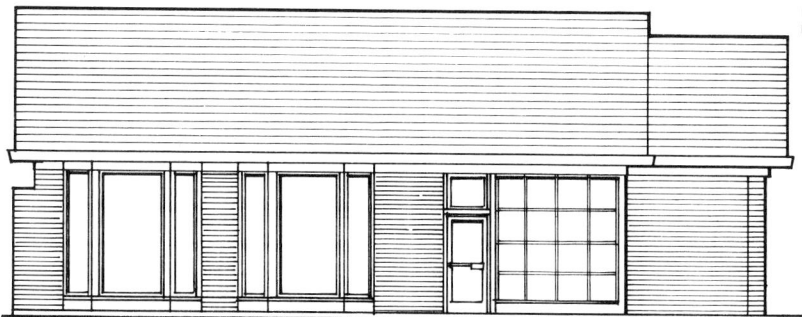

NEW BAY WINDOWS

FRONT ELEVATION

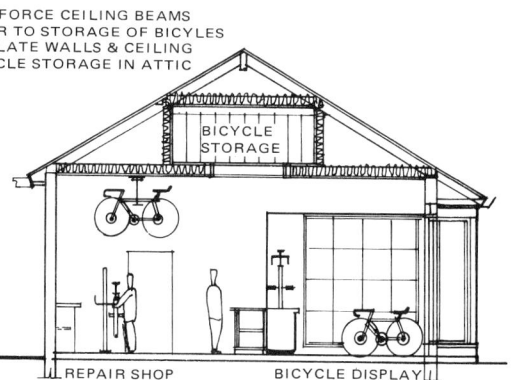

REINFORCE CEILING BEAMS PRIOR TO STORAGE OF BICYLES
INSULATE WALLS & CEILING
BICYCLE STORAGE IN ATTIC

BICYCLE STORAGE

REPAIR SHOP   BICYCLE DISPLAY

CROSS SECTION

Stage 1: The selection of a location in a growing but populated area, situated on a heavily travelled main artery that leads to a large city. A large property near a growth area is recommended. The drawings are based on acquiring a 125' x 125' corner parcel with an abandoned two bay colonial service station containing 1200 square feet of floor area.

The drawings outline the limited work that is required to convert the original building into a small bicycle store. The following scope of work should be performed: removal of overhead doors and installation of a store front for display purposes; closing of exterior access to rest rooms, the rearrangement of plumbing fixtures for one rest room and the remodel-ing of the women's room into a private office; the installation of an access door and disappearing stairs to the attic area, the installation of a plywood floor and a frame partition and the insulation of the attic area for bicycle storage and the erection of a partition and counter to create a sales and service department.

Stage 2: The erection of a 20' x 30' freestanding pre-engineered building for a sales display showroom.

Stage 3: The construction of a 17' x 54 storeroom addition for bicycle inventory.

Stage 4: The construction of a 17' x 45' addition for a large service and repair department, parts room, and enlargement of the sales display showroom.

If finances permit, several stages may be completed at one time.

BICYCLE STORE

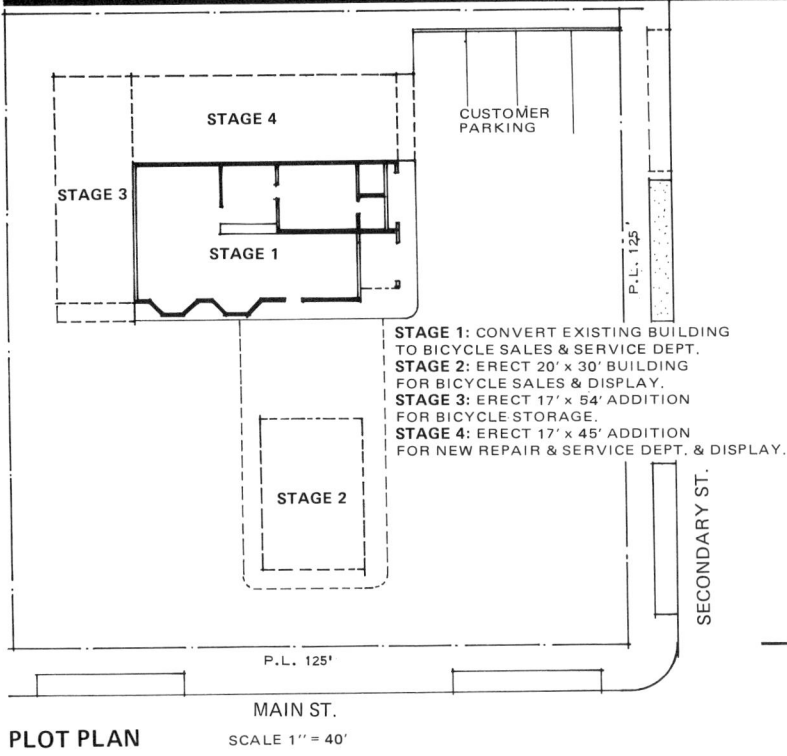

STAGE 4

STAGE 3

STAGE 1

CUSTOMER PARKING

P.L. 125'

STAGE 1: CONVERT EXISTING BUILDING TO BICYCLE SALES & SERVICE DEPT.
STAGE 2: ERECT 20' x 30' BUILDING FOR BICYCLE SALES & DISPLAY.
STAGE 3: ERECT 17' x 54' ADDITION FOR BICYCLE STORAGE.
STAGE 4: ERECT 17' x 45' ADDITION FOR NEW REPAIR & SERVICE DEPT. & DISPLAY.

STAGE 2

SECONDARY ST.

P.L. 125'

MAIN ST.

PLOT PLAN   SCALE 1'' = 40'

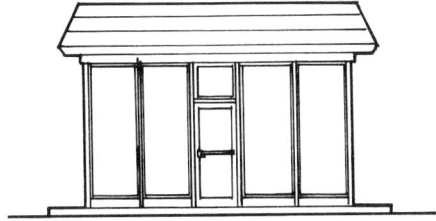

ELEVATION

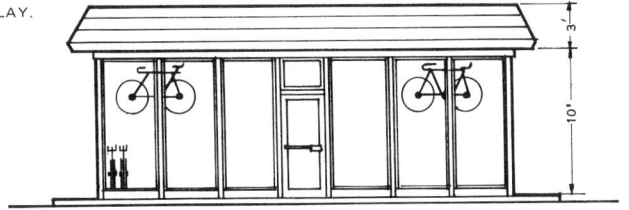

3'

10'

SIDE ELEVATION

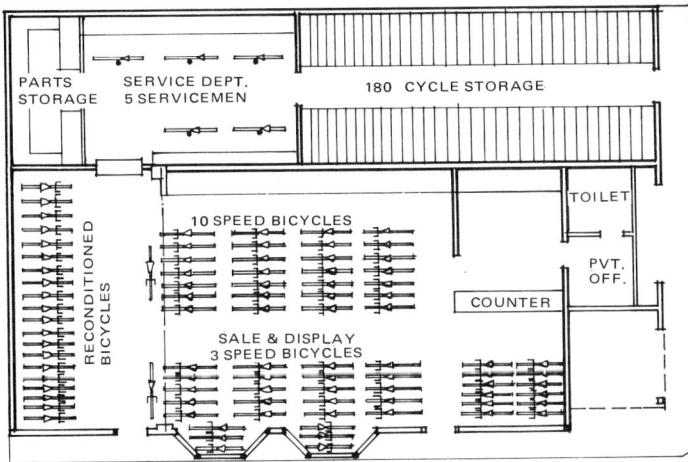

PARTS STORAGE

SERVICE DEPT. 5 SERVICEMEN

180 CYCLE STORAGE

TOILET

10 SPEED BICYCLES

PVT. OFF.

RECONDITIONED BICYCLES

COUNTER

SALE & DISPLAY 3 SPEED BICYCLES

FLOOR PLAN OF PROPOSED IMPROVEMENTS

SCALE 0' 5' 10' 15'

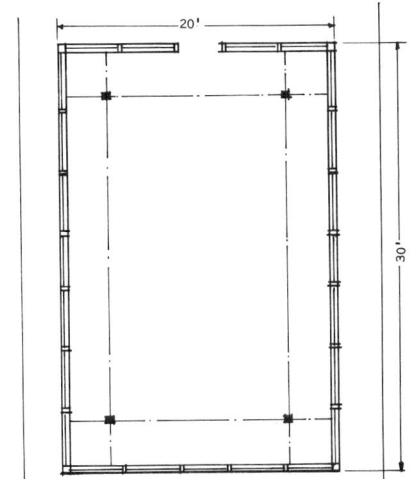

20'

30'

STAGE 2: ERECT 20' x 30' BUILDING FOR BICYCLE SALES & DISPLAY.

FLOOR PLAN

BICYCLE & MOPED STORE

THE IMPORTANCE OF SHOW WINDOWS

A VIEW OF THE 100' x 50' PLOT AND BUILDING

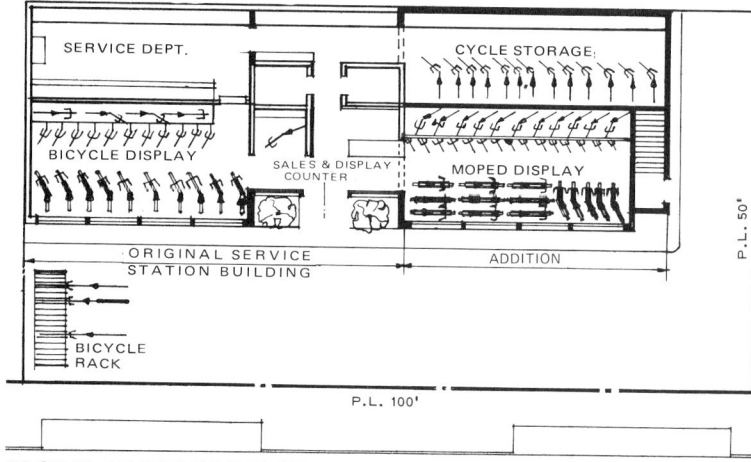

PLOT PLAN & FLOOR PLAN

This bicycle and moped store is located in an historical university town. For more than twenty years it operated from a store in the heart of the shopping district with very limited storage area for cycles. In 1976, a two bay service station in a superior location on Main Street was abandoned. It was purchased for $95,000 by the cycle shop owner. It was necessary to secure the approval of the zoning board to change the use of the premises and erect an addition. Since the service station had existed prior to the adoption of the present zoning law, the variance was ultimately approved.

The building was gutted, leaving just the shell, and a 28' x 27' wing was erected. The conversion included a new roof, suspended ceiling, paneled walls, tile and carpeted floor, recessed lighting and a large floor to ceiling store front. The new addition and remodeling were completed at a cost of approximately $65,000.

THE SALES COUNTER AND DISPLAY AREA FOR ACCESSORIES

Courtesy Jay's Cycles, Princeton, N.J.

BICYCLE & MOPED STORE

PARTS AND BICYCLES STORAGE

The floor plan provides two large sales areas for bicycles and mopeds which are most popular with students and faculty, a service department, and a parts and accessories storage area. It is quite evident, from the amount of business being conducted, that this is not only an attractive conversion but a successful business venture as well.

SEASONAL SPORTS EQUIPMENT

MOPED SALES DISPLAY

MOPED AND BICYCLE DISPLAYS

SERVICE AND REPAIR DEPARTMENT

BICYCLE SALES DISPLAYS

BICYCLE SALES DISPLAYS

Courtesy Jay's Cycles, Princeton, N.J.

BOOK STORE

A SIMPLE AND INEXPENSIVE CONVERSION

VIEW OF PROPERTY FROM THE BUSY INTERSECTION

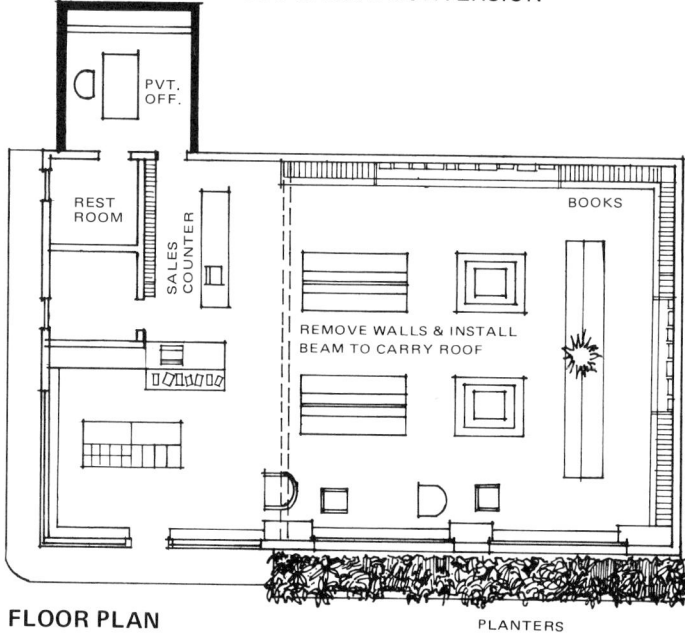

FLOOR PLAN

In 1973, the enterprising owner of an urban bookstore paid $35,000 for a two bay service station that had been abandoned for more than a year. A ten foot road widening and new thirty foot radius at the corner had rendered the service station useless. The new owner removed the interior walls which had formed the sales room and storage area and a steel beam was installed to replace a masonry bearing wall. The overhead doors were retained in a locked position and a large planter with attractive shrubbery was carried across the front of the bay area. The walls were panelled and a suspended acoustical ceiling and carpeting were installed. The interior of this bookstore is not cluttered as many are. The owner has thoughtfully created a small reading area with comfortable chairs where a customer can leisurely browse through books. The exterior remodeling included a mansard roof, natural wood cedar shingles and avocado painted porcelain walls which has resulted in an attractive conversion completed in the $15,000 to $17,000 range.

A SUSPENDED CEILING WITH ABOVE AVERAGE ILLUMINATION FOR READING

CHAIRS AND STOOLS FOR CUSTOMER COMFORT

UNCONGESTED LAYOUT OF BOOK DISPLAYS

THE UNCLUTTERED LOOK OF GOOD PLANNING

Courtesy Asheville Bookstore, Asheville, N. C.

RELIGIOUS BOOK STORE

A REJUVENATED 40 YEAR OLD SERVICE STATION

STUCCO AND HALF TIMBER EXTERIOR

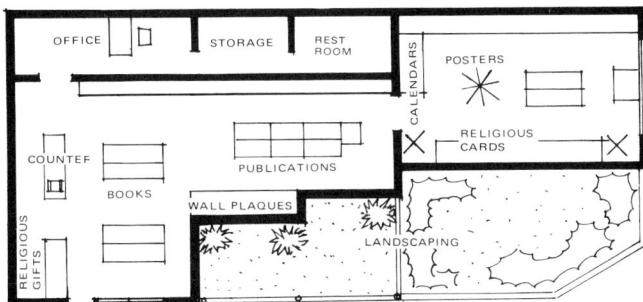

FLOOR PLAN

A nondescript service station that was erected more than forty years ago is now the home for an attractive Christian Book Store. It is situated on a major New Jersey highway. This irregular shaped building was purchased several years ago for $65,000. At the time of purchase, the building had dirt piled high in the bay area, no heating system, no electrical wiring and two wood barn doors, which did not close properly. The new owner inspected the interior and the exterior of the building and decided to do the work in two stages. The roof, which had four different levels, was realigned with a mansard treatment and the exterior walls were resurfaced with stucco and half timber. New floors, a heating system, electrical work and paneling completed the final stage. It cost the contractor-owner approximately $20,000 to accomplish the work.

You may wonder why anyone would buy an abandoned service station requiring so much work. There is only one reason, and that is LOCATION. The owner stated: "The key to a successful operation is highway exposure".

RELIGIOUS CARDS AND POSTERS

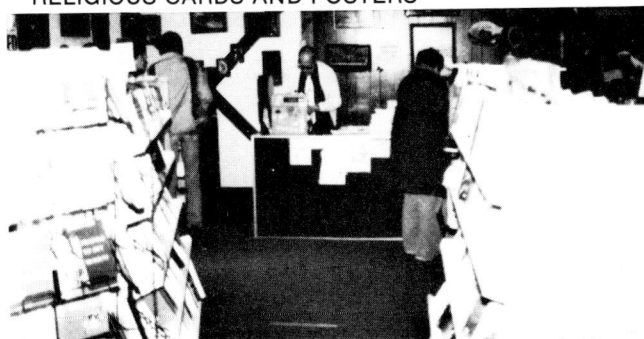

LITERATURE, BOOKS AND POSTERS

Courtesy Jesus Book & Gift Store, Islien, N.J.

THE ABANDONED SERVICE STATION

A closed service station can be converted into a bus terminal without difficulty. The property should contain at least 15,000 square feet with at least 150′ of frontage. A three bay building has sufficient area to accommodate a large waiting room with a seating capacity for forty-nine passengers, a baggage section, ticket counter, a vending area, rest rooms and a drivers' room. A canopy is essential to provide cover while luggage is being loaded and buses are being boarded by passengers. Abandoned service station sites in the inner cities are generally limited to 100′ x 100′ or less. These properties require study to determine if they can be developed with an acceptable traffic flow.

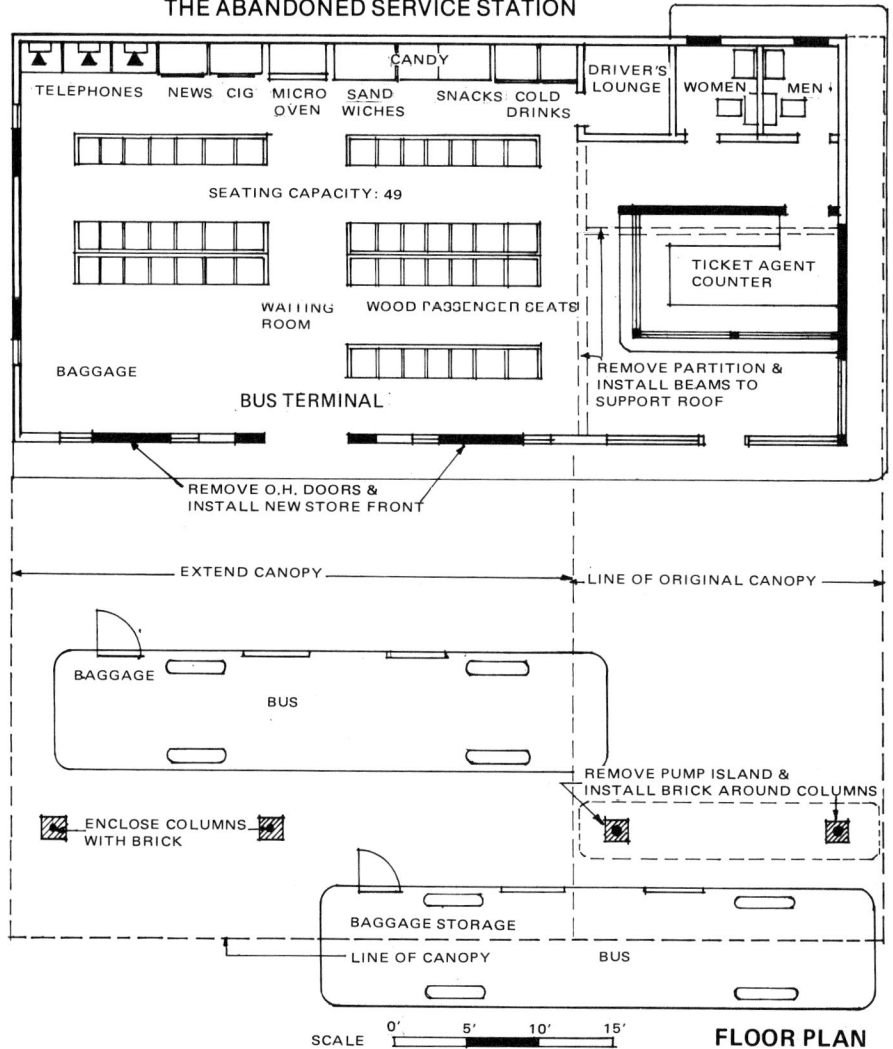

TELEPHONES NEWS CIG MICRO OVEN SAND WICHES CANDY SNACKS COLD DRINKS DRIVER'S LOUNGE WOMEN MEN

SEATING CAPACITY: 49

WAITING ROOM WOOD PASSENGER SEATS

BAGGAGE

TICKET AGENT COUNTER

REMOVE PARTITION & INSTALL BEAMS TO SUPPORT ROOF

BUS TERMINAL

REMOVE O.H. DOORS & INSTALL NEW STORE FRONT

EXTEND CANOPY LINE OF ORIGINAL CANOPY

BAGGAGE BUS

REMOVE PUMP ISLAND & INSTALL BRICK AROUND COLUMNS

ENCLOSE COLUMNS WITH BRICK

BAGGAGE STORAGE BUS

LINE OF CANOPY

SCALE 0′ 5′ 10′ 15′

FLOOR PLAN

INSTALL NEW PRE-ENGINEERED FACIA OVER CANOPY & MARQUEE

ABC Bus Terminal

BRICK VENEER

REMOVE O.H. DOORS & INSTALL NEW STORE FRONT

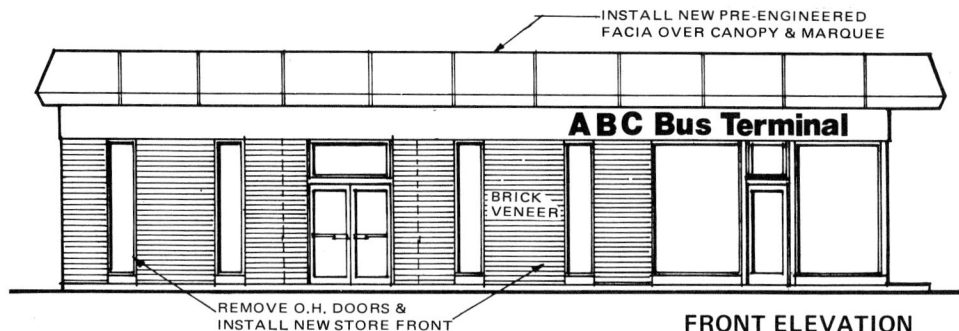

FRONT ELEVATION

BUS TERMINAL

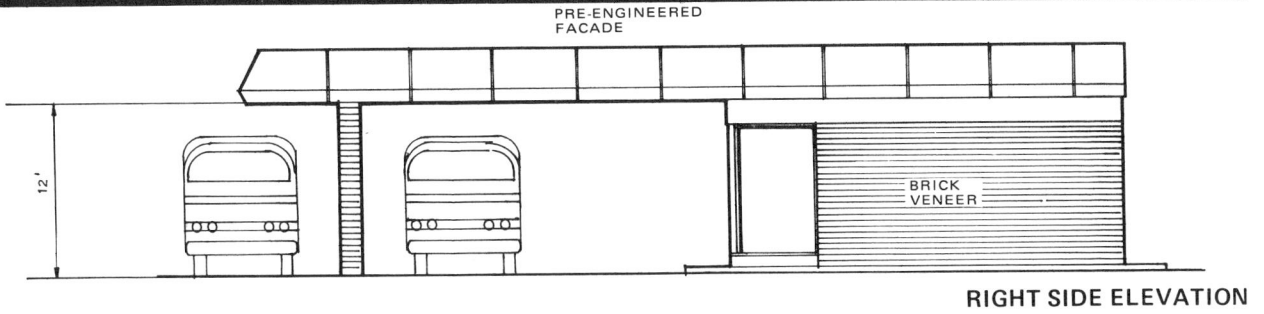

PRE-ENGINEERED
FACADE

12'

BRICK
VENEER

RIGHT SIDE ELEVATION

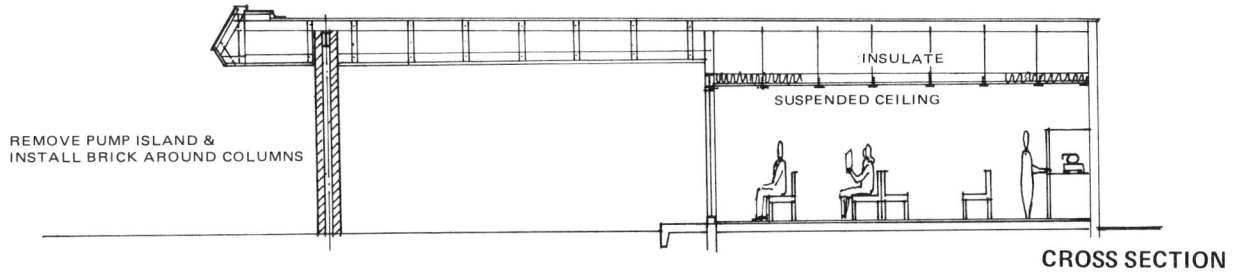

INSULATE

SUSPENDED CEILING

REMOVE PUMP ISLAND &
INSTALL BRICK AROUND COLUMNS

CROSS SECTION

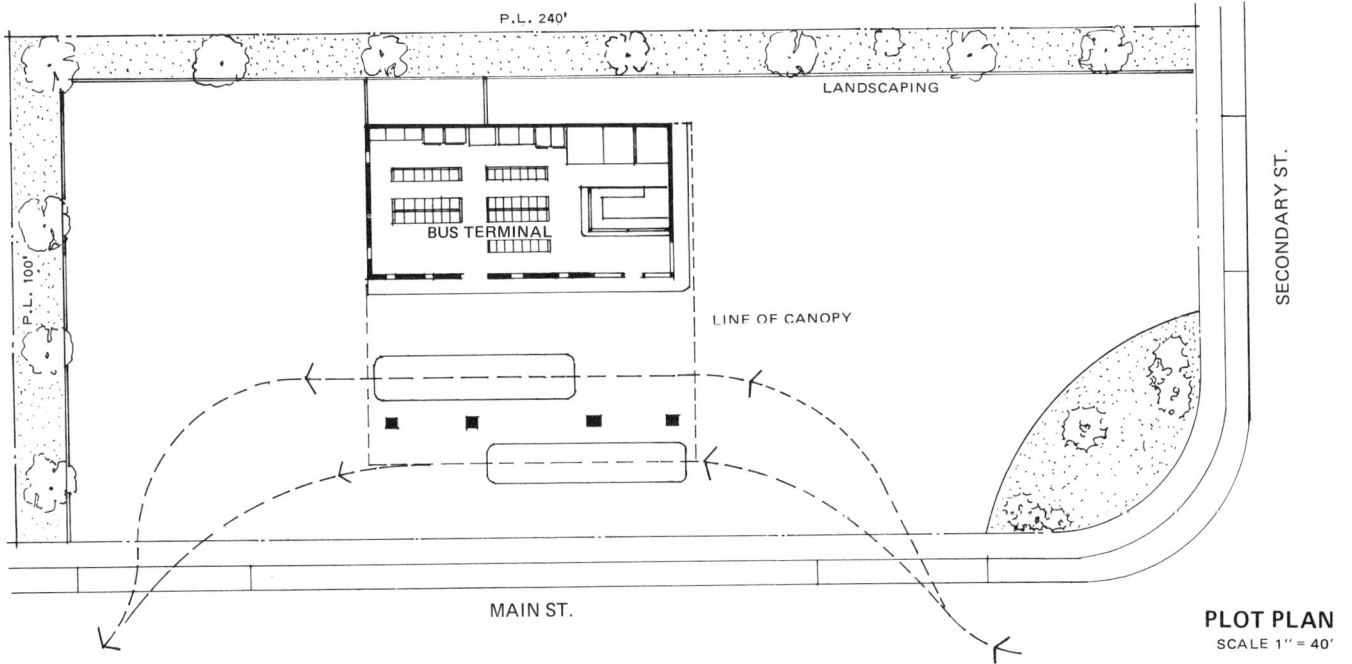

P.L. 240'

LANDSCAPING

SECONDARY ST.

P.L. 100'

BUS TERMINAL

LINE OF CANOPY

MAIN ST.

PLOT PLAN
SCALE 1" = 40'

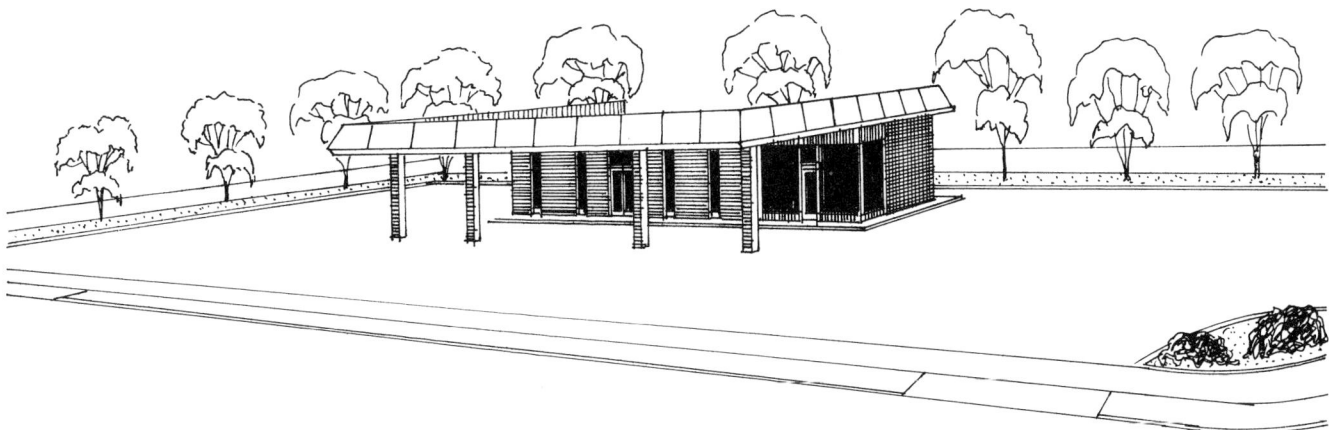

BUSINESS MACHINES, SALES & SERVICE

ABANDONED 3 BAY SERVICE STATION WITH DOUBLE CANOPY.

REMODELED SALES AND SERVICE FACILITY FOR ELECTRONIC CASH REGISTERS

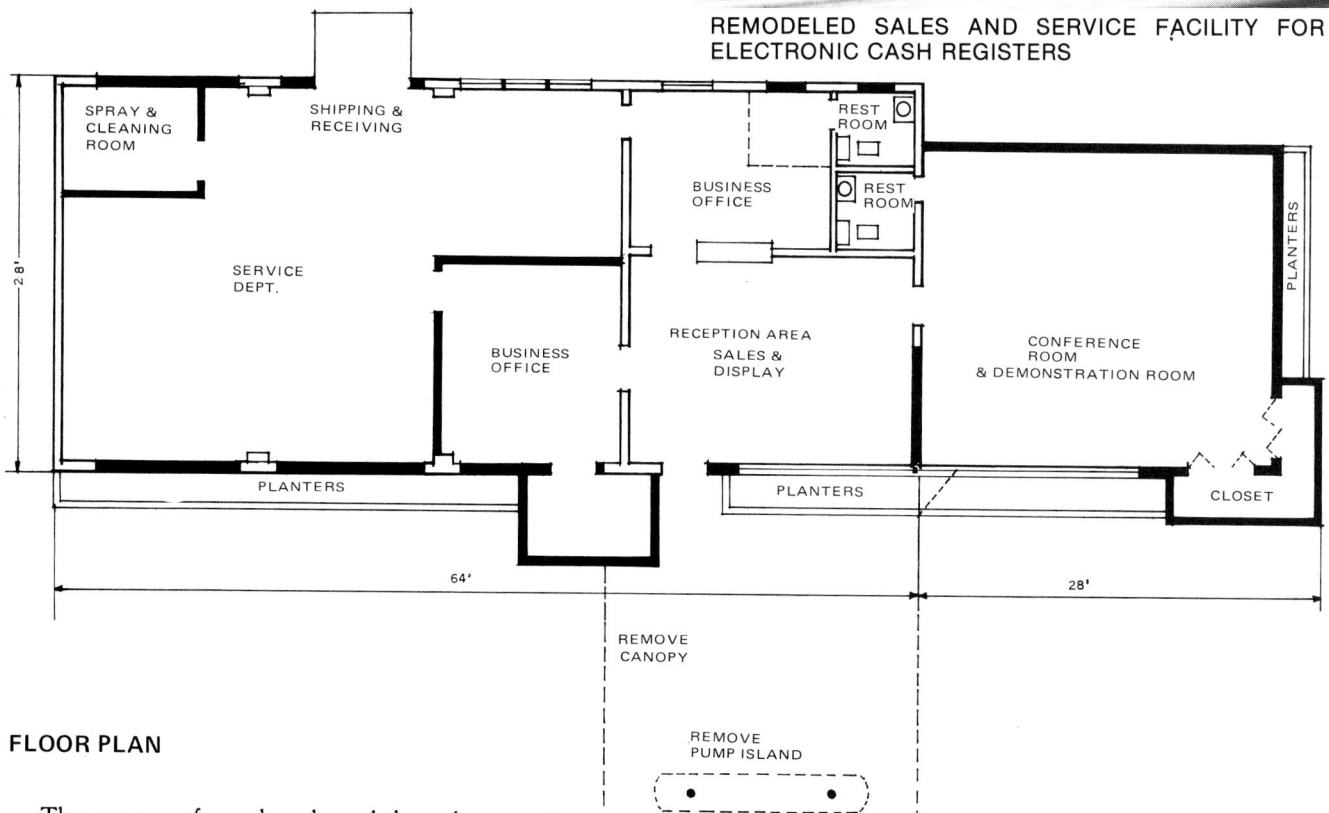

FLOOR PLAN

The owner of an abandoned three bay service station in Florida put down the author's first book without finding exactly what he wanted. It did, however, give him the incentive to search out the final results obtained. He then retained the services of a talented local architect to prepare plans that met his requirements. No problems were encountered with local governing authorities and the conversion commenced.

A 24′ x 32′ addition was erected on the north end of the building and the lubritory bays were converted into a storage and repair center for electronic cash registers. The former service station sales and store room have been converted into a display center and business office. One rest room door was relocated to provide interior access. An attractive facade, using random stone piers, a mansard with hand split wood shingles, and a glass store front, combined with a planter area completed the transformation. There is no evidence that a service station ever occupied the property. The city's Chamber of Commerce presented the owner with an award for the "most improved appearance". The owner gives credit for the fine conversion to the architect. This is another example of how an architect has contributed to the successful conversion of an abandoned service station.

Courtesy R. & S. Supply, Ft. Pierce, Florida
Drawing Courtesy, Stebbins & Scott Architects, Ft. Pierce, Florida

BUSINESS MACHINES, SALES & SERVICE

DEMONSTRATION ROOM

SERVICE DEPARTMENT

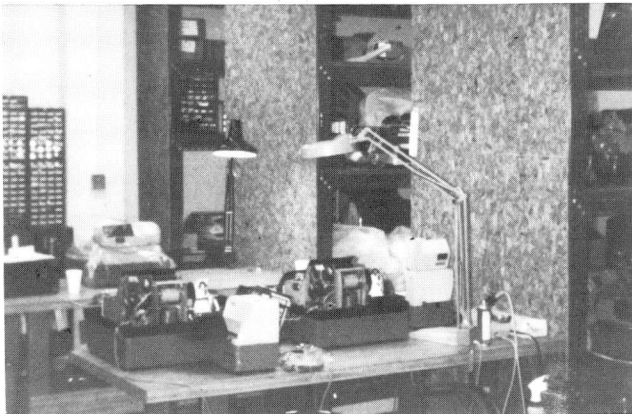

SERVICE DEPARTMENT

SERVICE DEPARTMENT

DEMONSTRATION ROOM

DISPLAY AREA FOR ELECTRONIC CASH REGISTERS

SERVICE DEPARTMENT

SALES DISPLAYS

Courtesy R. & S. Supply, Ft. Pierce, Florida
Drawing Courtesy, Stebbins & Scott Architects, Ft. Pierce, Florida

CAMERA STORE

48'

38'

STORAGE

WOMEN

MEN

3 BAY LUBRITORY

SALES

FLOOR PLAN OF ABANDONED BUILDING

PAPER DEVELOPER

SINK

BLOCK UP O.H. DOORS & INSTALL WINDOWS

DARK ROOM

TRIPODS

BINOCULARS

STORAGE SHELVING

TELESCOPES

CAMERA CASES

LENSES

PHOTOBOOKS

FILM

SALES COUNTER

PROJECTORS

MOVIE EQUIPMENT

SCREENS

REMOVE WALLS & INSTALL BEAM TO CARRY ROOF

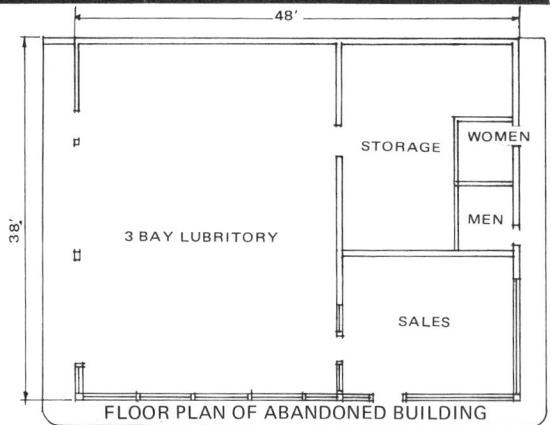

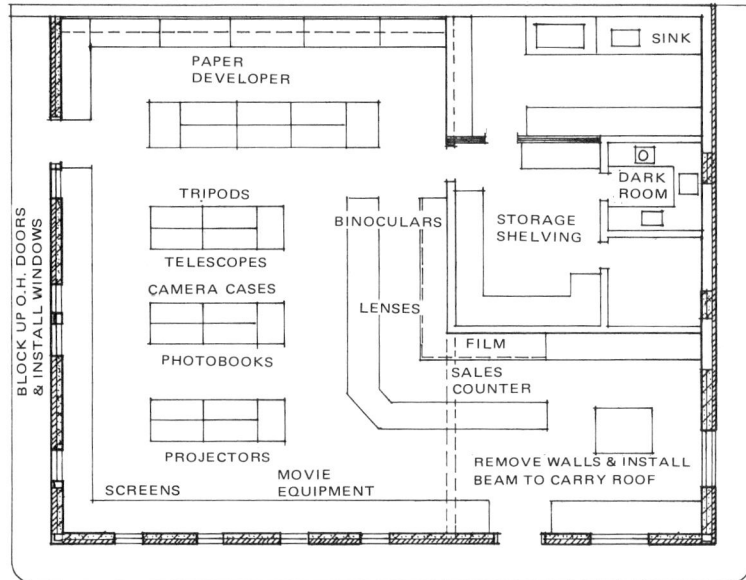

FLOOR PLAN   SCALE   0'   5   10'   15'

The success of a camera shop depends on locating a site on a primary artery in a heavily populated area. The business section of town is the most important consideration. A successful camera store must include a "film developing drop" that specializes in processing customer film in eight hours. Customers should be able to leave their order on the way to work and pick up the completed photos on the way home. This may require on premises developing and printing.

Provisions must be made for a display area for convenience items such as film, flashbulbs, low and moderate priced cameras, gadget bags, viewers, slide projectors and accessories. A storage area is required for rental equipment for hotels or motels.

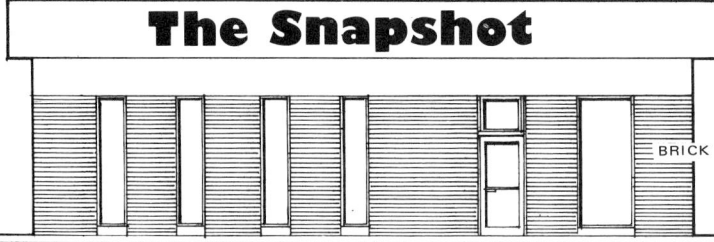

The Snapshot

PRE-ENGINEERED FACADE

BRICK VENEER

FRONT ELEVATION

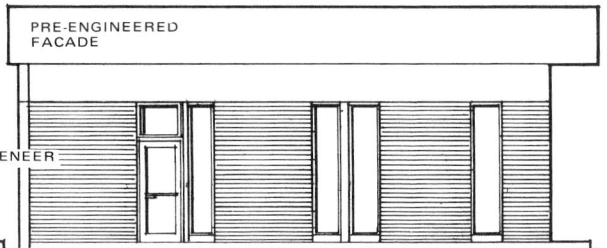

LEFT SIDE ELEVATION

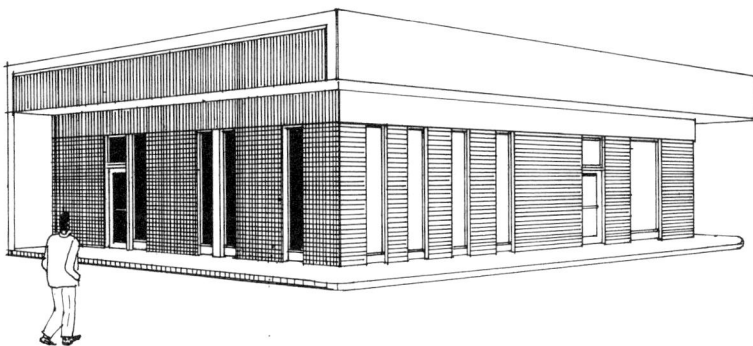

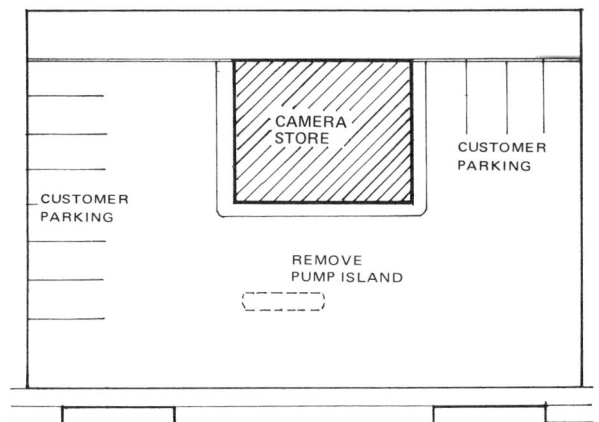

CUSTOMER PARKING

CAMERA STORE

CUSTOMER PARKING

REMOVE PUMP ISLAND

MAIN ST.

PLOT PLAN

CANDY AND CONFECTIONERS STORE

THE STORE IS ADJACENT TO THE RAILROAD STATION

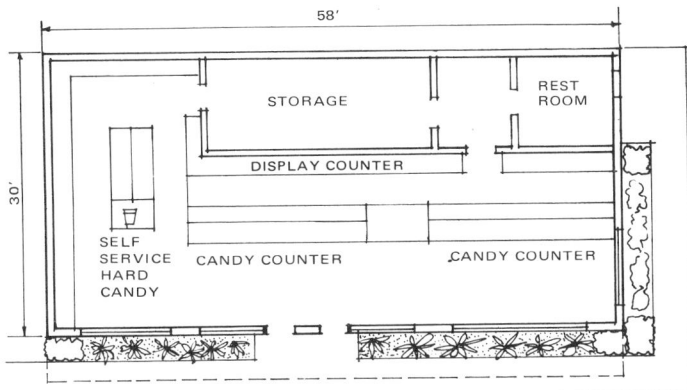

FLOOR PLAN

On the outskirts of an attractively restored New Jersey town, situated close to the railroad station, is a unique candy store.

Located at the entrance is a 45′ display case, where holiday specials are displayed, as well as an assortment of chocolates and fudge.

The interior is spacious and immaculate with a self-service area for hard candies. The overhead doors were removed and the openings were converted into a glass storefront. Awnings were installed on the southern exposure to shield the candy from the hot summer sun. During the winter months, when fuel conservation is important, they are not used.

Well placed landscaping provides additional color and beauty. This is an outstanding conversion in an excellent location.

THE LONG CHOCOLATE COUNTER DISPLAY

CHOCOLATE DISPLAYS AT CASHIER'S COUNTER

THE SELF SERVICE HARD CANDIES

Courtesy Fannie May Candies, Cornwell Heights, Pa.

CONVERSION OF AN ABANDONED THREE BAY SERVICE STATION TO A SELF-SERVICE COIN-OPERATED WAND CAR WASH

The wave of the future in car washing is the growth of the self-serve coin-operated high pressure wand car wash. This is one of the most rapidly growing businesses throughout the country, with the exception of the Northeastern States. Abandoned service stations offer excellent opportunities for the investor or businessman. If permits are available to convert the lubritory bays for self-serve car washing, the prospective buyer may wish to consider the addition of two or three bays to improve the income. A six bay car wash located in a good neighborhood will yield a good income. The self wash consists of a hot soapy wash, hot rinse, and a hot wax application. Successful car wash operators recommend equipment including engine degreaser, white wall cleaner, and, the latest innovation, a super foam brush. In most cases the latter will increase gross income from 30% to 40%. In order that the reader may evaluate this business concept for abandoned service stations, a budget estimate has been included. The cost will vary depending on the manufacturer, any discount he may grant, the type of equipment, size of pumps, the range of construction costs in a particular area, city and state taxes, local requirements, fire insurance and climate. The estimate is based on furnishing and installing equipment in an abandoned three bay service station: (1981 costs)

1. Furnish and install three bay, self-service car wash equipment, including super foam brush, engine degreaser, whitewall cleaner and three vacuums.

$27,000

2. Remodel the existing building, install three overhead doors in rear of building, 3' x 4' x 5' deep floor drains in bays, partitions separating bays, waterproof lights, paving the rear of the property estimated at 5000 square feet.

$45,000

Total $72,000

The cost to add a 45' x 28' three bay addition with a heated floor and overhead doors, front and rear is estimated at $45,000 to $50,000. (The weather in the northern states requires this provision. In southern and milder climates heated floors and overhead doors may be eliminated which reduces costs substantially.)

An income projection is included for the reader based on a three bay car wash. The gross income is based on the owner/buyer performing his own maintenance.

No. Cars Washed Per Bay Per Month	75¢ Wash, Wax, Rinse Average Income Per Bay Per Month	Total Income 3 Bays Month	Vacuums and Vending 20%	Gross Income	30% Expenses	Net Profit Per Month
500	$375	$1125	$225	$1350	$405	$ 945
600	450	1350	270	1620	485	1135
700	525	1575	315	1900	570	1330
800	600	1800	360	2160	650	1510
900	675	2025	405	2455	735	1720
1000	750	2250	450	2700	810	1890
1100	825	2475	495	2970	890	2080
1200	900	2700	540	3240	970	2270
1300	975	2920	585	3505	1050	2455
1400	1050	3150	630	3780	1135	2645
1500	1125	3375	675	4050	1215	2835
1600	1200	3600	720	4320	1295	3025
1700	1275	3825	775	4600	1380	3220

The Gross Income represents 100% wash, plus 20% vacuums and vending.

The car wash cycle is 75¢ for 5 minutes and includes the super foam brush, a wash, rinse and wax application. (In many states the basic self-service car wash is now $1.00 for four minutes)

The vacuum cycle is 25¢ for four minutes and 50¢ for towels and air fresheners. (In many states the vacuum charge is 50¢ for five minutes. A number of car washes include a tire cleaner and/or engine degreaser at an additional cost of 75¢).

SELF SERVICE WAND CAR WASH

The Income Projection outlined above is a conservative estimate prepared by the author. Many three bay self-service car washes situated in densely populated areas adjacent to both business and residential communities will gross between $40,000 and $55,000 annually. Location and size of the property are essential to the success of the business.

The operating expenses of a typical self-service car wash are calculated as follows:

	% of Gross Income
Utilities (Fuel Oil-Add 5% for gas)	8.0%
Water and Sewer	6.0%
Chemicals (Detergents & Wax)	4.5%
Sand Trap Cleanout & Trash Pickup	1.5%
Equipment Maintenance (Based on new equipment)	6.0%
Merchandise purchased for resale - vending	4.0%
Total Expenses	30.0%

SITE SELECTION GUIDELINES FOR A SELF SERVICE WAND CAR WASH

These guidelines are very similar to those used for the selection of service stations or retail business sites, and as outlined on page 12.

1. Neighborhood: Upper middle income area with 12,000 to 18,000 families in a three mile radius consisting of one-family residences, multi-story and garden apartments and shopping center.

2. Traffic Count: A steady heavy volume in excess of 18,000 cars per day.

3. Property: Visual inspection will reveal whether a potential ingress or egress problem exists or can develop for car wash customers. Example: A near corner site at a busy intersection may have a right turn lane which could create an egress problem for customers. Is there a road divider? Will it adversely affect business? Are there traffic signals that will cause vehicle traffic to back up and block ramps?

4. Visibility: It is very important that the car wash sign can be seen at a great distance by the motorist. Visibility of the building is desirable but not essential.

5. Layout: The position of the building on the property is very important for a drive-thru car wash. A 30' minimum rear yard behind the building is required for a turning radius exit. 40' is preferred where egress to the right is required.

6. Competition: The most important consideration for the success of the new business use is its relationship to competition within a three mile radius of the site.

SITE AND BUILDING INVESTIGATION: After having selected the location, investigation of the abandoned service station is required. Many of the existing utilities and services will be adequate, thereby keeping startup costs to a minimum. A thorough investigation is necessary to determine the following:

1. Are city or municipal sewers available? A septic tank disposal system is not adequate for a car wash.

2. Is the electric service adequate? Is 200 amp. single phase service available? This will have a bearing on the type of motors ordered for the pumps.

3. What is the size of the water main and how much pressure is there? A main with 40 p.s.i. will generally be adequate for a six bay car wash.

4. Is the floor area sloped to the drain in the center of the lubritory bay or is it level? A level floor will have to be replaced and properly sloped to a drain that must be connected to the sewer.

5. Is the storeroom large enough for the installation of the car wash equipment?

6. What type of heating system exists? Heated bays are not required in the sunbelt areas, or where the winter temperatures do not generally fall below 20°. In colder northern regions, floor heat will be required.

7. Is there sufficient room on the property for the installation of the vacuum islands and space for "detailing" cars?

SELF SERVICE WAND CAR WASH

STREET VIEW OF CAR WASH. THIS IS THE EXIT SIDE. THE FORMER PUMP ISLAND AND ISLAND LIGHT STILL REMAINS.

THE VACUUM ISLANDS ARE PROPERLY POSITIONED AT THE ENTRY TO THE WASH BAYS

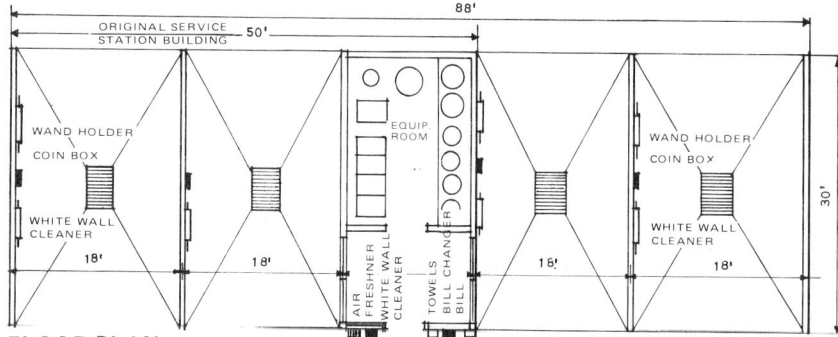

FLOOR PLAN

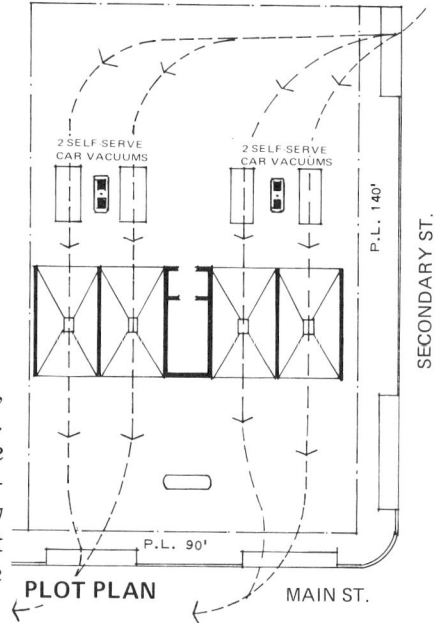

PLOT PLAN

This former three bay service station, (with one bay an exterior wash), was converted to a four bay self service wand car wash in the early 1970s. The old sales section was converted to an equipment room and storage area for chemical tanks. Access to the car wash is gained from the secondary street. Customers vacuum their cars, then enter the bays where they wash their cars with safety trigger wands. Customers apply softened hot soapy water, hot rinse and hot wax at a pressure of 1,100 P.S.I. Tire cleaner and degreaser are also available at the meter.

This is one of the finest recession-proof business ventures for the small businessman, particularly for abandoned service stations with sufficient property to install a drive-thru facility. At the time the photographs were taken of this location, the car wash costs were 50¢ for five minutes and the vacuum was 25¢ for five minutes. These costs have been increased respectively to $1.00 for four minutes and 50¢ for a four minute vacuum as a result of increased operating expenses. The owners have also improved the quality of the car wash equipment by installing trigger wands. The operators of this location also operate three additional properties with thirteen wash bays situated within one mile of this facility and a seven bay in the south area of Aurora.

Notice the excellent traffic flow. Entry to the car wash is from the secondary street, where there is sufficient land to stack up eleven cars on the premises. The recommended position for the vacuum island is 20' from entrance wall of the building to the vacuum pad.

A WELL DESIGNED VACUUM ISLAND, AT LEAST 24" HIGH TO PROTECT THE EQUIPMENT FROM ERRANT DRIVERS. TWO 35 GALLON BARRELS ARE USED AS TRASH RECEPTACLES.

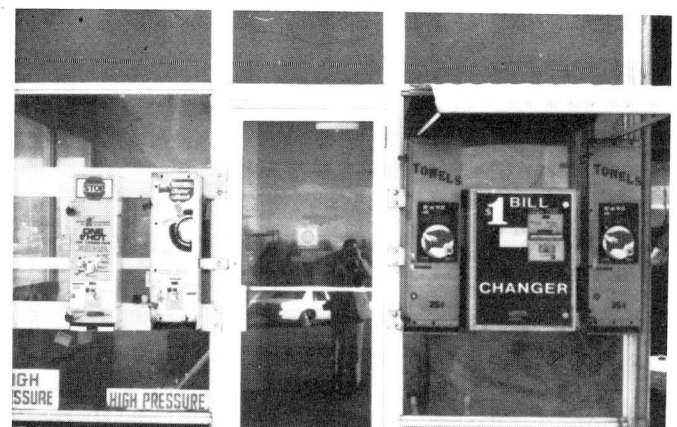

THE VENDING AREA FOR THE DOLLAR BILL CHANGER, TOWELS, WHITEWALL CLEANER AND AIR FRESHENERS

Courtesy Herlyn Enterprises, Aurora, Colorado

SELF SERVICE WAND CAR WASH

THE WAND FOR CLEANING ENGINES AND WHITEWALL TIRES IS ON THE LEFT. THE WAND FOR CAR WASHING IS POSITIONED TO THE RIGHT. THE MAT HOLDERS ARE ON THE FAR RIGHT.

A TYPICAL ESTIMATE OF THE MONTHLY INCOME and EXPENSE STATEMENT FOR A HIGH PRESSURE SELF SERVICE WAND CAR WASH WITH 4 BAYS

INCOME

Car washes: 1,108 per bay per month @ $1.00		$4,432.00
Vacuum (4): 1600 per month @ 50¢		800.00
Vending: 8% of the gross monthly income		250.00
Total Gross Monthly Income		$5,482.00

OPERATING EXPENSES

Utilities	9%	$ 493.00
Water and sewer	2%	109.00
Chemicals and detergents	5%	274.00
Sand trap cleanout and trash	2%	109.00
Equipment maintenance	5%	274.00
Vending merchandise (for resale)	3%	164.00
Total Monthly Operating Expense	26%	$1,423.00
Gross Margin (for expenses listed below)	74%	$4,059.00

OTHER EXPENSES (Approximately 10% of Gross Income)

Insurance

Office, legal and bookkeeping

Rent and/or land building payments and equipment

Taxes

Wages

Courtesy Heryln Enterprises, Aurora, Colorado

CARPET STORE

THE MODEST CONVERSION OF A FORMER TWO BAY SERVICE STATION

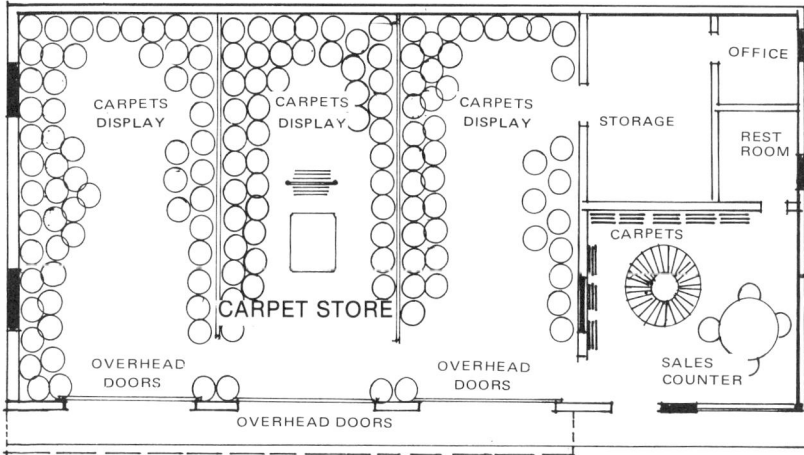

FLOOR PLAN

In August 1975, this three bay abandoned service station, situated on a major artery with a very high traffic count, was leased and converted into a carpet store. The store manager, an excellent merchandiser, has generated "impulse" sales by utilizing the former pump island, the front of the building and a delivery truck, parked adjacent to the sidewalk, for the display of carpeting. The bays are used as a show room for approximately 250 remnants. The high lubritory ceiling provides an excellent area to stack 12′ rugs. Additional samples of remnants line the wall of the sales room. The cost to convert the building was $8000.

CARPET REMNANTS STACKED IN THE LUBRITORY RAYS

SAMPLES OF CARPET REMNANTS DISPLAYED IN THE SALES ROOM

Courtesy Furhman's Carpet Store, Levittown, N.Y.

CHEESE AND WINE STORE

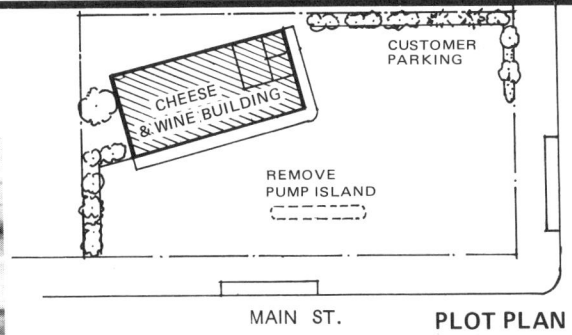

CUSTOMER PARKING

CHEESE & WINE BUILDING

REMOVE PUMP ISLAND

MAIN ST.

PLOT PLAN

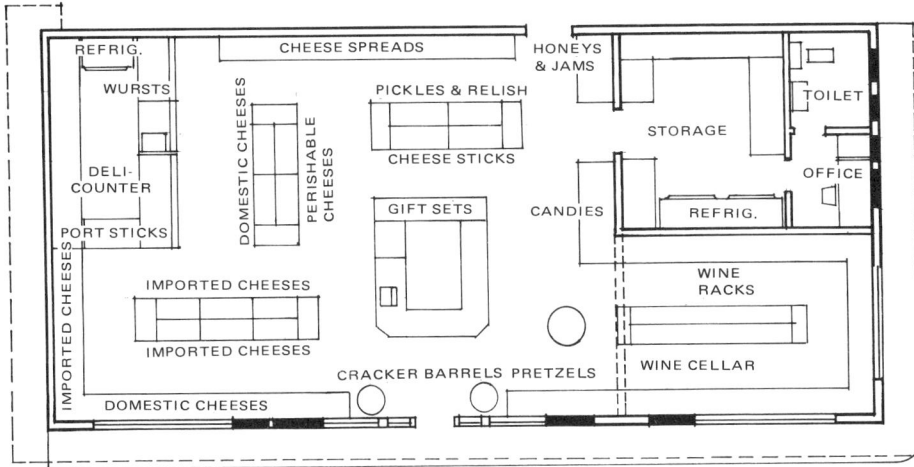

REFRIG.

WURSTS

DELI-COUNTER

PORT STICKS

IMPORTED CHEESES

DOMESTIC CHEESES

PERISHABLE CHEESES

CHEESE SPREADS

PICKLES & RELISH

CHEESE STICKS

GIFT SETS

IMPORTED CHEESES

IMPORTED CHEESES

CRACKER BARRELS   PRETZELS

DOMESTIC CHEESES

HONEYS & JAMS

CANDIES

STORAGE

TOILET

OFFICE

REFRIG.

WINE RACKS

WINE CELLAR

FLOOR PLAN   SCALE   0'   5'   10'   15'

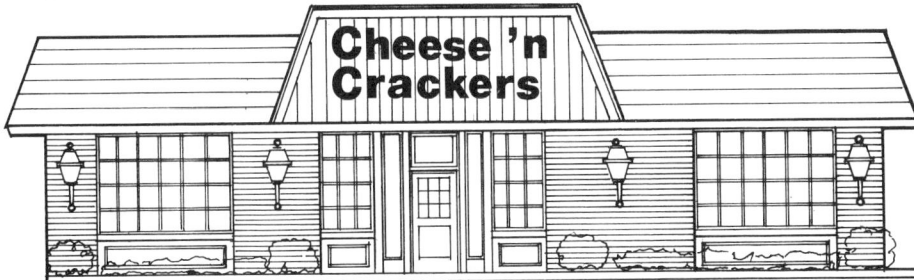

FRONT ELEVATION

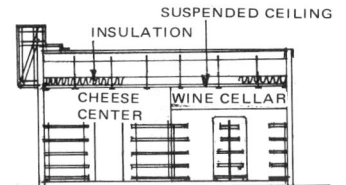

SUSPENDED CEILING

INSULATION

CHEESE CENTER   WINE CELLAR

CROSS SECTION

Don't overlook the abandoned service station in the suburbs of large cities or inner cities. These sites are generally situated in areas where vacant land is at a premium. Many of them are suitable for conversion to retail business uses. One of these retail uses is a unique store specializing in domestic and imported cheese, gourmet meats, sandwiches, crackers, honey, jams, a broad line of fine tea and coffees, candies and many other gourmet food items.

The success of the store will depend largely upon the selection of a site with population density, in a middle to upper middle income area, with adjacent stores and a good traffic flow.

A three bay building provides ample floor space for a quality sandwich counter, display cases, coolers, and shelving for the wide range of cheese and snack items. An interior panelled in rich redwood creates a warm inviting atmosphere for customers.

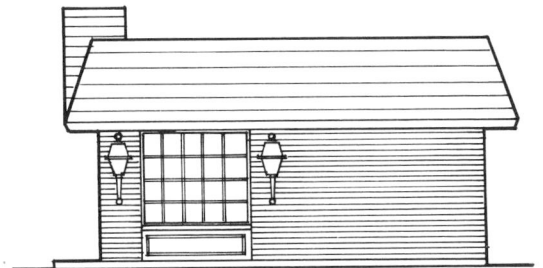

RIGHT SIDE ELEVATION

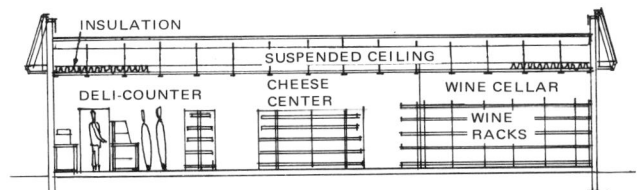

INSULATION

SUSPENDED CEILING

DELI-COUNTER   CHEESE CENTER   WINE CELLAR

WINE RACKS

LONGITUDINAL SECTION

CLOTHING STORE

A COLORFUL FACADE ATTRACTS MOTORIST

A TWO BAY WING OF THE FORMER SERVICE STATION

The initial conversion of this nondescript service station to a beer and soda distributorship was unsuccessful. A local resident, with twenty years' experience in the clothing business, "took a flyer" and rented the well situated service station and garage in 1973. He knew the residents in the community were "style conscious". Since the clothing business can fluctuate rapidly due to style changes, one must be extremely careful not to get caught with a large inventory.

The interior is carpeted and shelves are installed to a height of twelve feet because of the limited storage area. Sections of the building are devoted to sneakers, jackets, jeans, T-shirts, press-on letters, sweat shirts, and other accessories.

In 1977, after evaluating his successful venture, the owner purchased the adjacent two story building and both garages and expanded his operations to include winter sports clothes and ski clothing. On Saturdays during peak periods he has ten to twelve young people working on sales and scampering up and down ladders to accommodate customers.

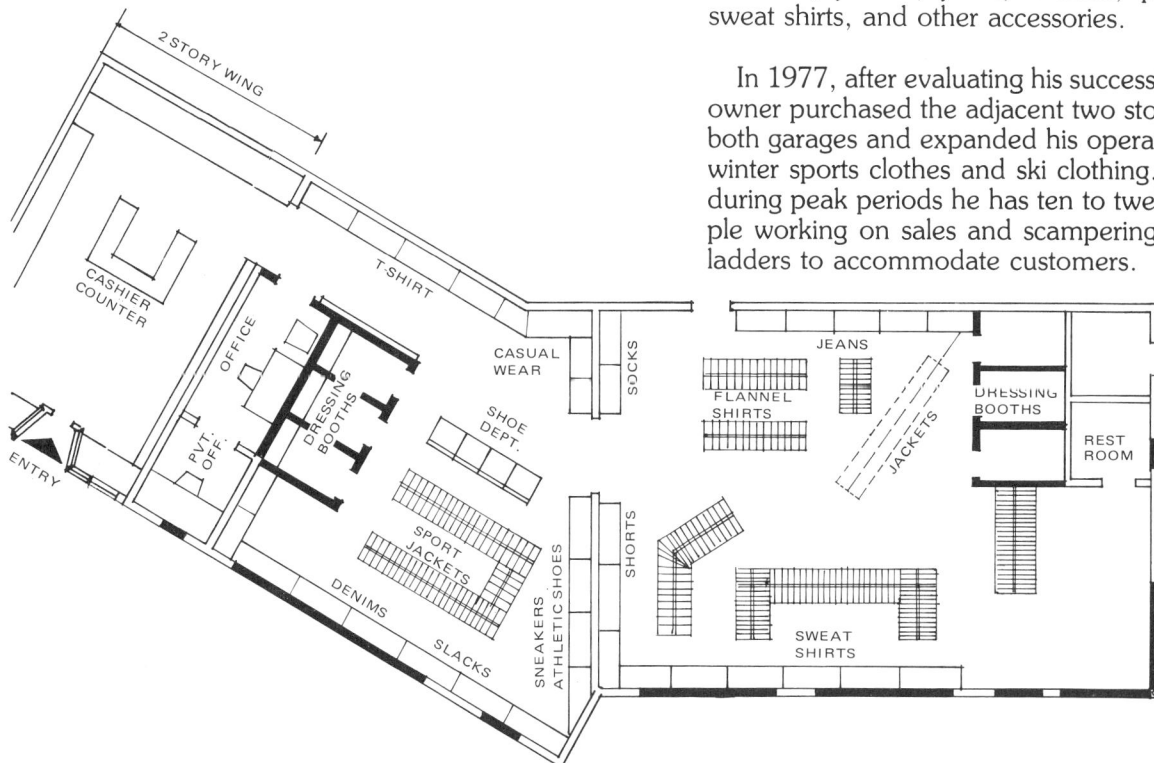

FLOOR PLAN

Courtesy Kemps Korner, Merrick, N.Y.

CLOTHING STORE

THE SALES COUNTER

FLANNEL SHIRTS

CHANGE BOOTHS

SLACKS, CASUAL WEAR AND DRESSING BOOTHS

SWEAT SHIRTS, T-SHIRTS, PANTS AND DENIMS

JEANS, SOCKS, SHORTS, AND FLANNEL SHIRTS

Courtesy Kemps Korner, Merrick, N.Y.

CLOTHING STORE

THE ABANDONED 3 BAY SERVICE STATION

THE CONVERSION TO A CLOTHING STORE

THE STORE ROOM ADDITION AT THE REAR OF THE BUILDING

SIDE VIEW OF BUILDING

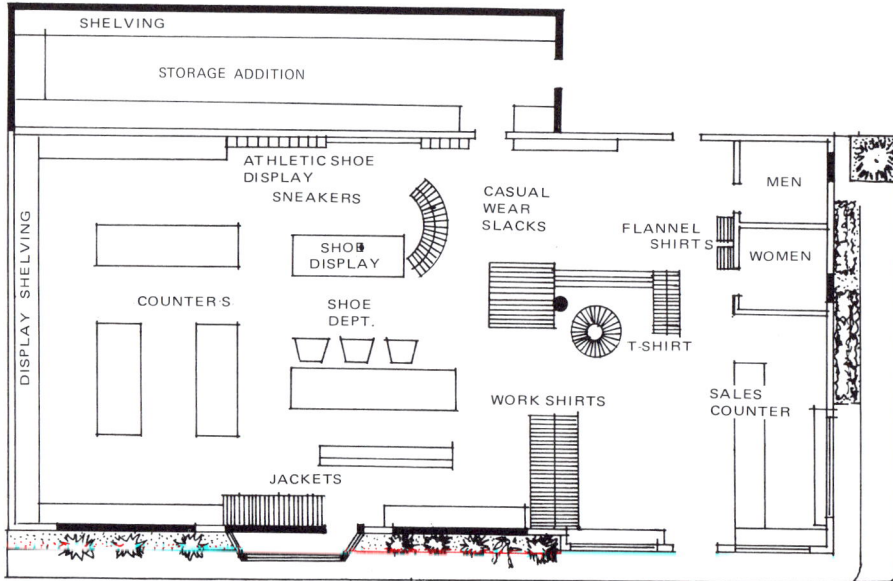

FLOOR PLAN

THE CASHIER'S COUNTER

THE NEW STORE ROOM AND SHELVING

Courtesy Chamber's Army-Navy Store, Ridgefield, Conn.

CLOTHING STORE

INTERIOR OF THE STORE

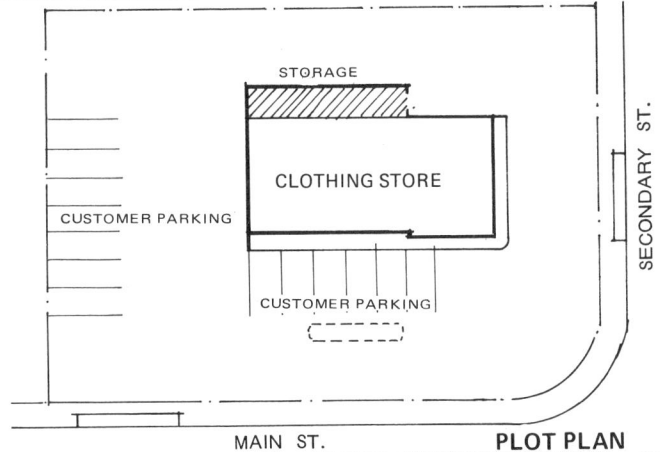
PLOT PLAN

STORAGE

CLOTHING STORE

CUSTOMER PARKING

CUSTOMER PARKING

SECONDARY ST.

MAIN ST.

WALL AND GONDOLA DISPLAYS

DISPLAY COUNTERS

JOGGERS ATHLETIC SHOES

SHOE DEPARTMENT

In 1980 an enterprising businessman purchased an existing twenty-five year lease, with an option to buy, on an abandoned four-bay colonial service station. This facility is situated on a large parcel of land on the main intersection of an affluent country community. The entrepreneur appeared before the local zoning board and was informed that there would be strict enforcement of the parking requirements for customers. A new sidewalk, curbing at the intersection, and the elimination of one ramp from each street became mandatory, together with the installation of landscaping. The owner obtained approval to erect a new storage addition and was directed to install sand fill in seven gasoline tanks.

The business owner spent $60,000 to install a new flat roof, new lighting fixtures and wiring, blacktop, water system and insulation. The rest rooms were remodeled and a new attic was constructed over the sales wing. The interior was furnished with attractive paneling. The variety of clothing for all ages and the finest corner location in this community should make this a successful business venture.

Courtesy Chamber's Army-Navy Store, Ridgefield, Conn.

CLOTHING STORE

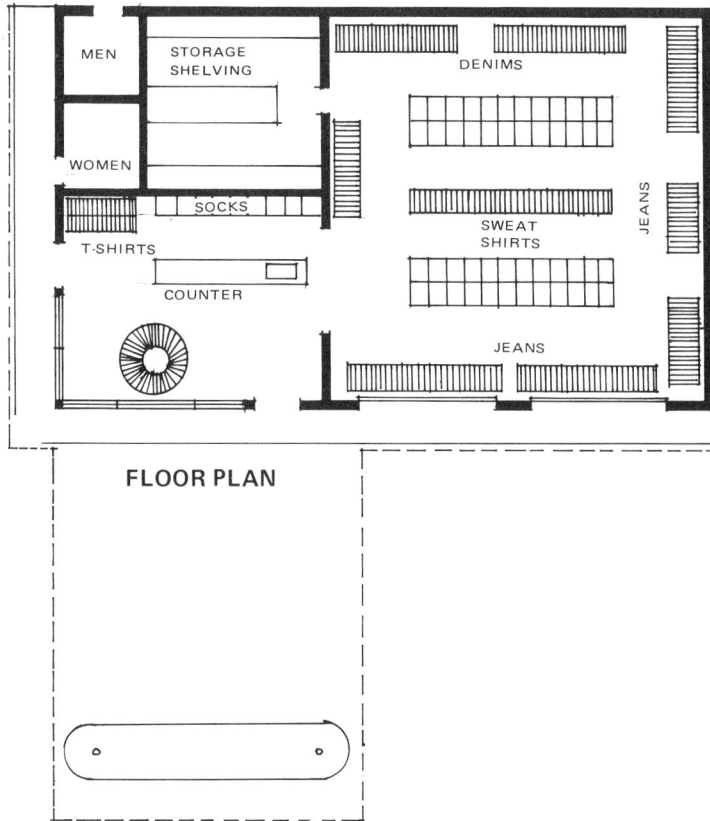

FLOOR PLAN

In 1977, the author passed a brightly painted former two bay canopy service station that had been converted into a T-shirt and jeans store. The interior was decorated inexpensively but effectively. The walls were painted, a suspended acoustical ceiling was installed in the lubritory, and the sales room was brightly painted with graphics in the standard color of the owner's stores. The overhead doors were locked into position and the upper panels were painted. The exterior signs were painted on the porcelain panels, and a three sided wood sign was installed on the canopy. The site is at one of the heaviest travelled intersections in Columbia, South Carolina. Eighteen months after these photos were taken, the store was relocated to a shopping center in the city. Business was so good at the old service station that the owner had decided to expand to several shopping center sites in the suburbs. This is one of a number of stores the owner operates in two states.

AN INEXPENSIVE CONVERSION ATTRACTS ATTENTION

CHEAP JOE'S CLOTHING

THE REMODELED SALES ROOM

A VIEW OF THE CLOTHING DISPLAYS IN THE FORMER LUBRITORY

Courtesy Cheap Joe's Jean Corp., Lexington, N.C.

SITE SELECTION GUIDELINES FOR A CONVENIENCE FOOD STORE

Convenience food store firms are interested in sites that conform with their analysis evaluations. The following site evaluation may be of assistance to the reader.

Major factors to be considered:

1. Competitive number of locations within a three mile radius.

2. Population density within a three mile radius, type and number of housing units. Aerial photos are important to verify evaluation.

3. The average expenditures for food and the average income. (Determine these statistics from a governmental agency.)

4. Obtain growth data from the local planning group. Many times a planning commission will have a completed study of growth patterns and potential. Building Departments have knowledge of currrent patterns and issuance of building permits.

5. Economic levels of the community should be determined.

6. Check with the State and County Highway Departments to determine if there are any future road widenings contemplated that will adversely affect the location. Radius corner ac-quisitions to improve right turns are commonplace today. Traffic Departments can assist in this regard.

7. Obtain traffic data from State or County Highway Departments, namely the most recent traffic counts. Determine location of traffic lights; is there a divided highway or one-way traffic? Check visibility distance of the position of an identity sign. What is the visibility distance of the building?

8. Prepare a site plan listing all competitive locations, grocery stores, service stations, (they can be converted to convenience stores) shopping centers, schools.

9. Site diagram showing adjacent business, retail developments, schools and shopping centers.

10. Verify the number and location of ingress and egress ramps. This data can be obtained from the State or County Highway Departments.

11. Check with local zoning boards to determine if there are any regulations that may bar the conversion of the service station building to a convenience food store. If you intend to market gasoline, there may be a local problem.

12. Sites located near a military base and/or a college campus have excellent potential sales.

CONVENIENCE FOOD STORE

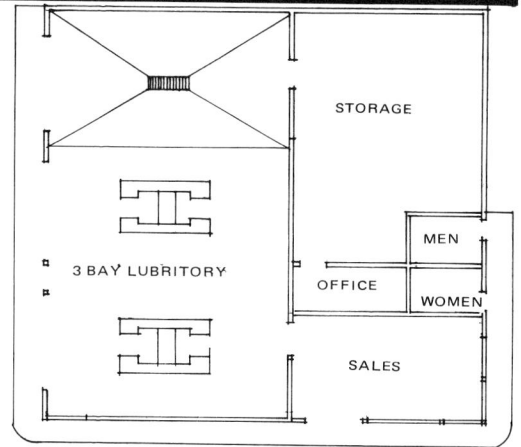

FLOOR PLAN ABANDONED BUILDING

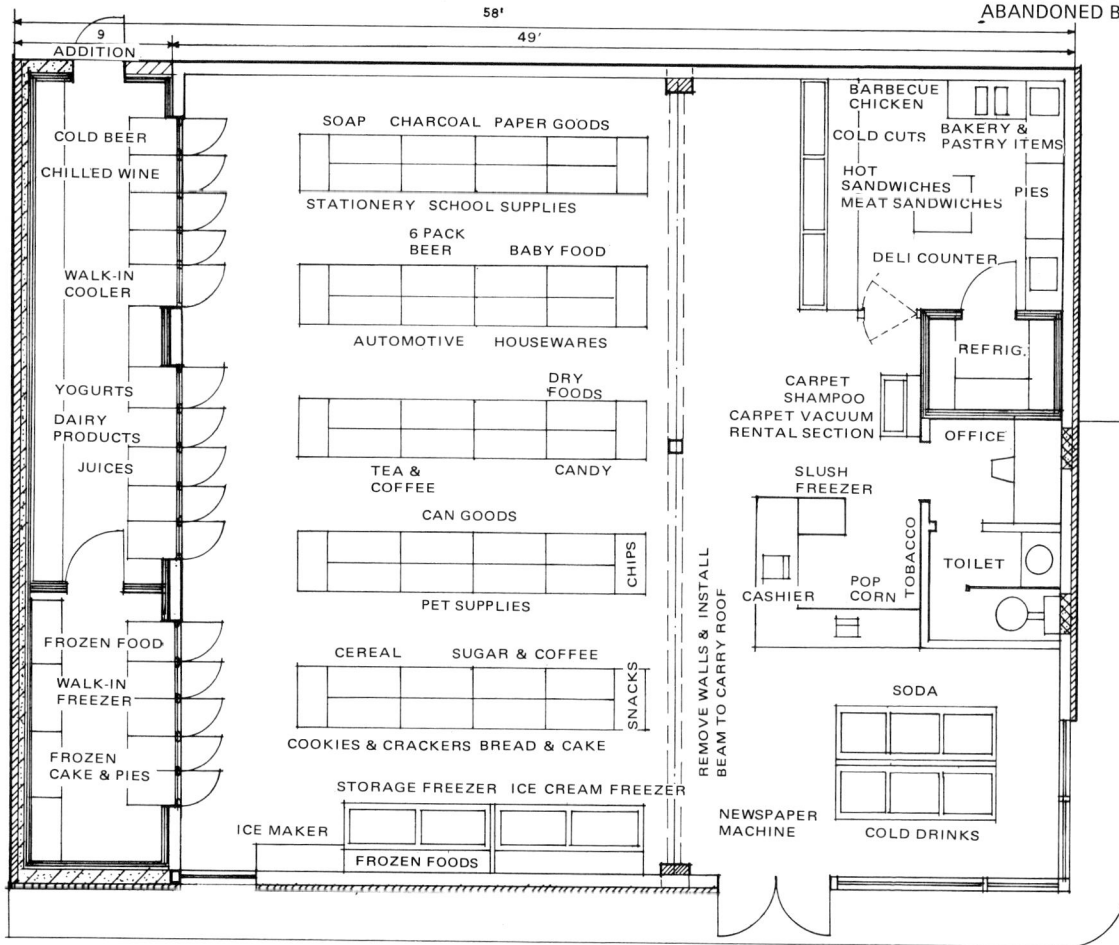

STORAGE

3 BAY LUBRITORY

OFFICE

MEN

WOMEN

SALES

FLOOR PLAN

9 ADDITION

58'

49'

COLD BEER

CHILLED WINE

WALK-IN COOLER

YOGURTS

DAIRY PRODUCTS

JUICES

FROZEN FOOD

WALK-IN FREEZER

FROZEN CAKE & PIES

SOAP   CHARCOAL PAPER GOODS

STATIONERY SCHOOL SUPPLIES

6 PACK BEER   BABY FOOD

AUTOMOTIVE   HOUSEWARES

DRY FOODS

TEA & COFFEE   CANDY

CAN GOODS

PET SUPPLIES

CEREAL   SUGAR & COFFEE

COOKIES & CRACKERS BREAD & CAKE

STORAGE FREEZER   ICE CREAM FREEZER

ICE MAKER

FROZEN FOODS

CHIPS

SNACKS

REMOVE WALLS & INSTALL BEAM TO CARRY ROOF

BARBECUE CHICKEN

COLD CUTS

BAKERY & PASTRY ITEMS

HOT SANDWICHES
MEAT SANDWICHES

PIES

DELI COUNTER

REFRIG.

CARPET SHAMPOO
CARPET VACUUM
RENTAL SECTION

OFFICE

SLUSH FREEZER

CASHIER

POP CORN

TOBACCO

TOILET

SODA

NEWSPAPER MACHINE

COLD DRINKS

SCALE   0'   4'   8'   12'

PRE-ENGINEERED FACADE

FOOD MART

5'

3'

10'

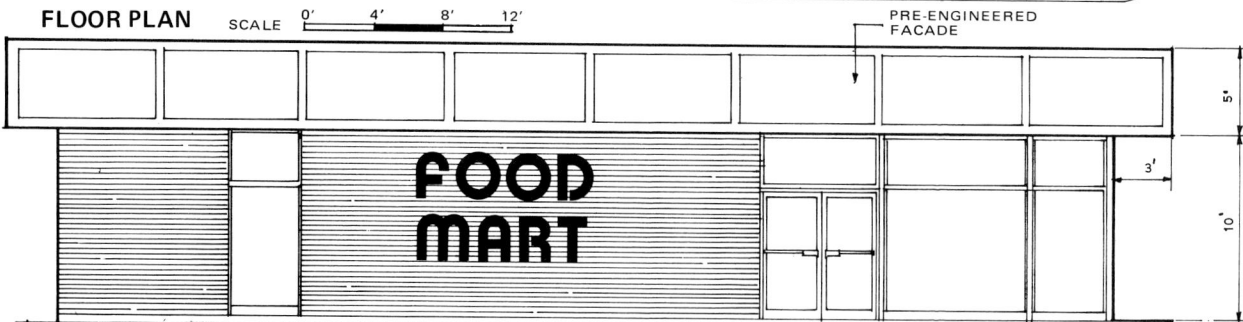

FRONT ELEVATION

CONVENIENCE FOOD STORE

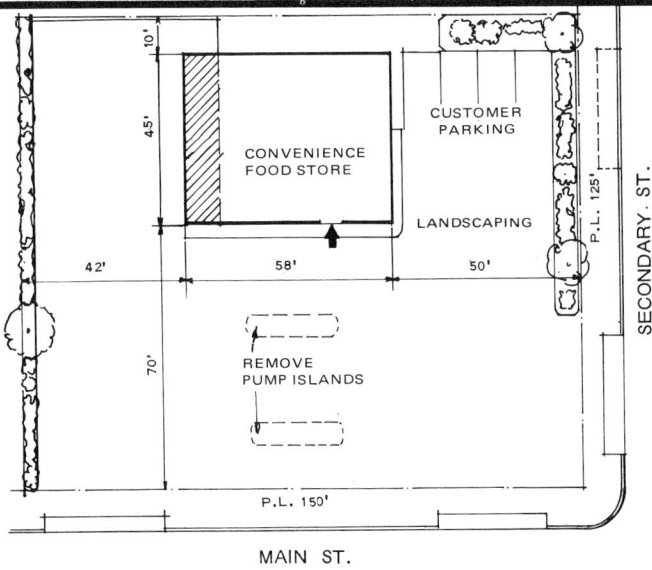

PLOT PLAN

SCALE 0' 10' 20' 30'

Abandoned side-entry bay buildings contain more than 1800 square feet of floor area that can be converted to a convenience store at a reasonable cost. The removal of interior walls permits a layout easily controlled from the cashier's counter. This convenience store features a deli-counter, with fried or barbecued chicken; made-to-go sandwiches; a fast food center; five aisles of products; a ten door cooler; a five door freezer; and a soda cooler and ice cream freezer flanking the entrance door.

The 9' x 45' addition for the walk-in cooler and freezer is essential for a spacious layout to serve a large community. The present store front should be retained and the exterior of the building faced with brick. The installation of prefinished 5' high fascia panels attached to the overhang completes an economic and attractive conversion that requires minimum external maintenance. The mechanical equipment to provide air conditioning and heating is mounted on the roof.

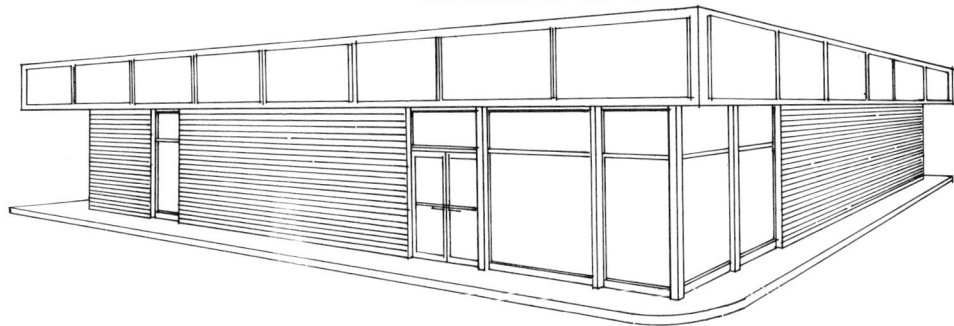

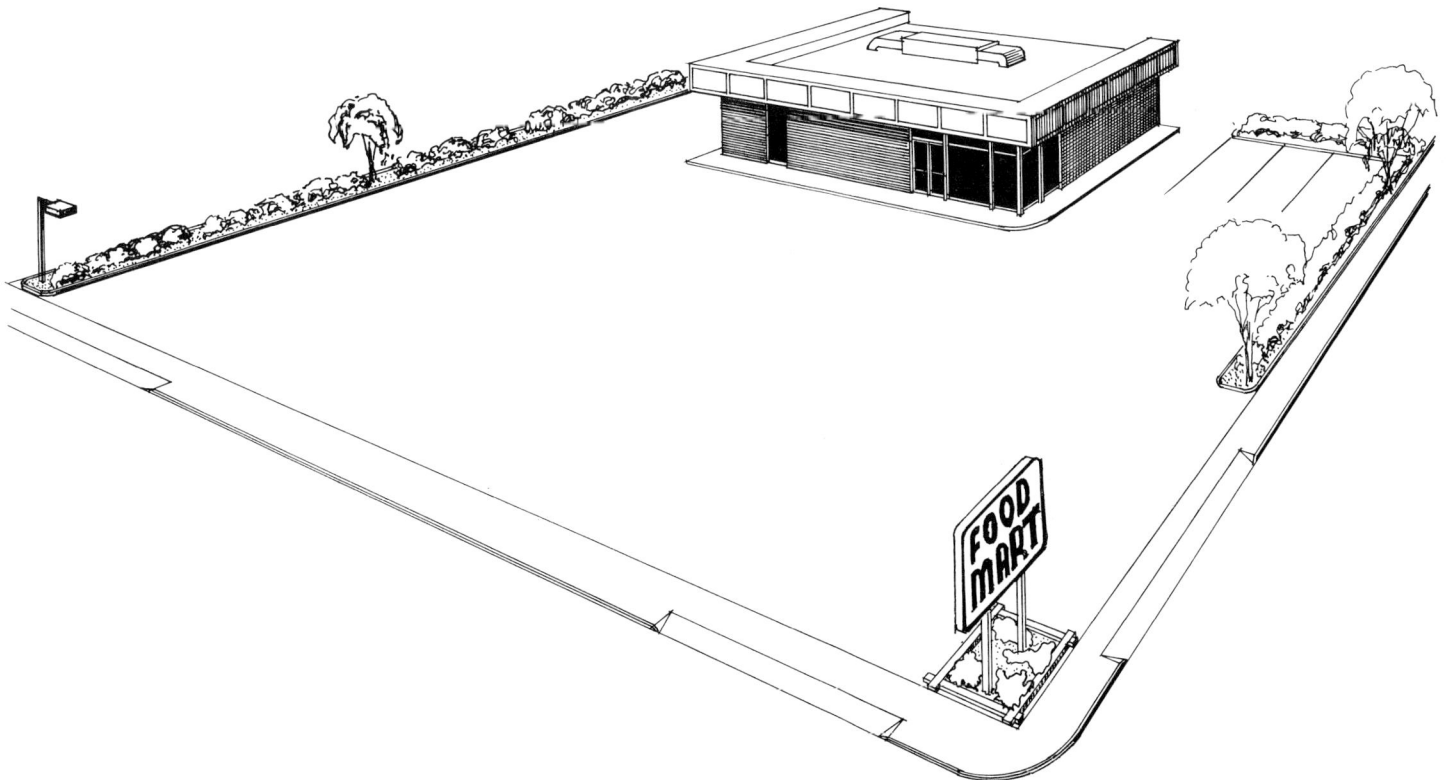

CONVENIENCE FOOD STORE

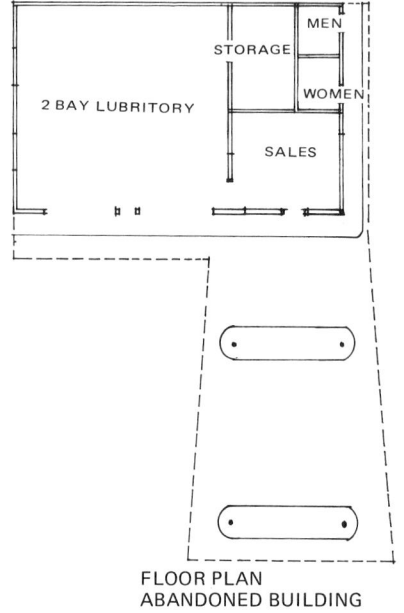

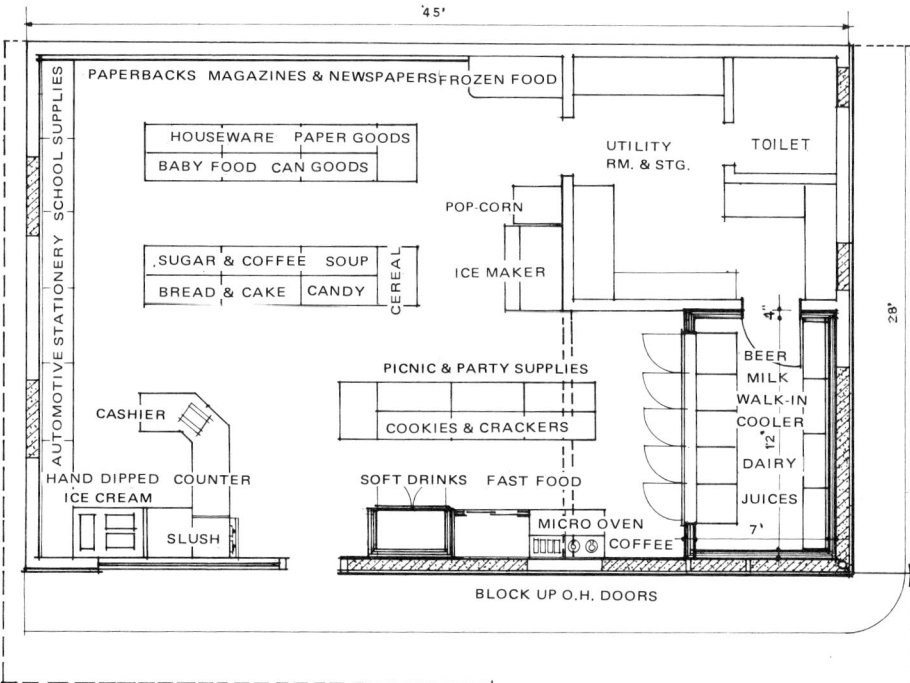

**FLOOR PLAN
ABANDONED BUILDING**

2 BAY LUBRITORY — STORAGE — MEN — WOMEN — SALES

FLOOR PLAN

SCALE 0' 5 10 15'

Floor plan labels: PAPERBACKS MAGAZINES & NEWSPAPERS / FROZEN FOOD / SCHOOL SUPPLIES / STATIONERY / AUTOMOTIVE / HOUSEWARE / PAPER GOODS / BABY FOOD / CAN GOODS / UTILITY RM. & STG. / TOILET / POP-CORN / ICE MAKER / SUGAR & COFFEE / SOUP / BREAD & CAKE / CANDY / CEREAL / BEER / MILK / WALK-IN COOLER / DAIRY / JUICES / PICNIC & PARTY SUPPLIES / COOKIES & CRACKERS / CASHIER / HAND DIPPED ICE CREAM / COUNTER / SOFT DRINKS / FAST FOOD / SLUSH / MICRO OVEN / COFFEE / BLOCK UP O.H. DOORS / 45' / 28' / 4' / 12' / 7' / 15' / 12'

VENDING ISLAND

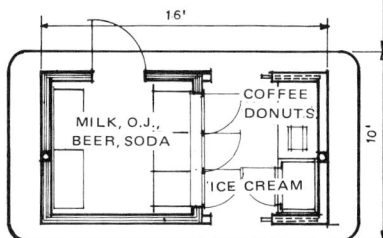

VENDING REFRESHMENTS / CIG / ICE CREAM / COLD DRINKS / POP CORN / SNACKS CHIPS / MILK / ICE MAKER / COFFEE HOT DRINKS / 8'

QUICK-PICK-UP

MILK, O.J., BEER, SODA / COFFEE DONUTS / ICE CREAM / 16' / 10'

This closed canopy service station offers many opportunties for the convenience store industry. In addition to a 1200 square foot building, the two pump island areas can be converted to a drive-up refreshment vending unit and a quick-pick-up drive-up building for the customer looking for limited dairy items. The interior layout contains a complete line of products, a fast food section, loose ice cream, a walk-in cooler and a cold drink case. Only two walls were partially removed and door openings were blocked up and covered with vertical siding. Energy conservation is an important factor in the financial success of a convenience food store.

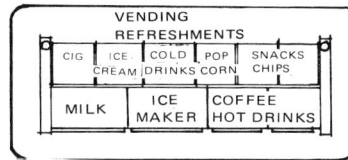

CONVENIENCE FOOD STORE

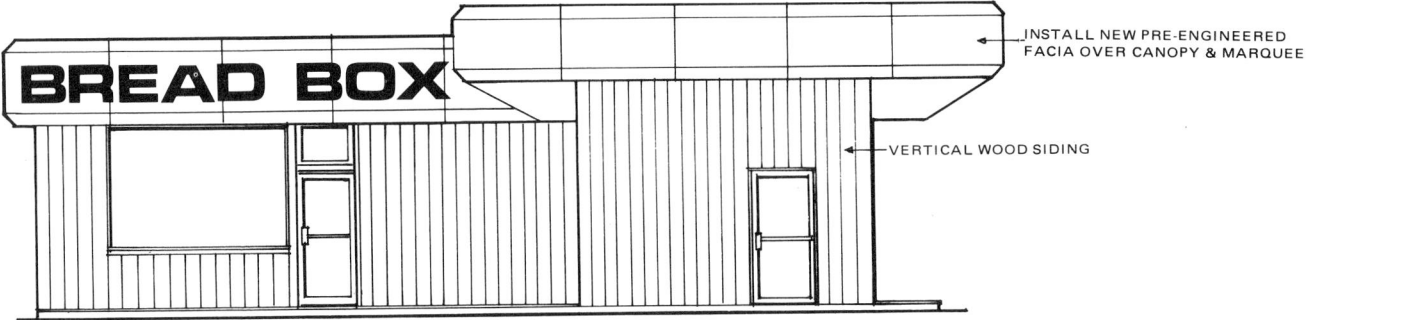

BREAD BOX

INSTALL NEW PRE-ENGINEERED
FACIA OVER CANOPY & MARQUEE

VERTICAL WOOD SIDING

FRONT ELEVATION

REMOVE O.H. DOORS &
INSTALL NEW STORE FRONT

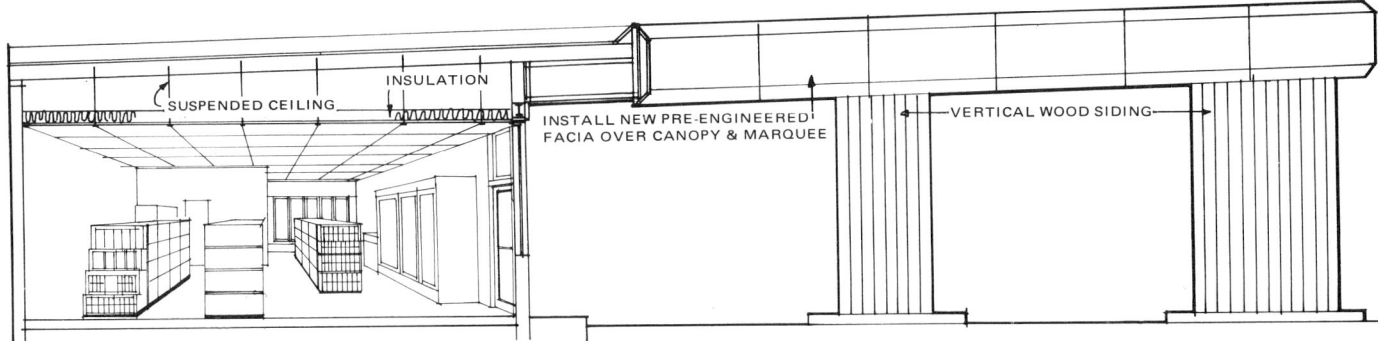

SUSPENDED CEILING

INSULATION

INSTALL NEW PRE-ENGINEERED
FACIA OVER CANOPY & MARQUEE

VERTICAL WOOD SIDING

CROSS SECTION

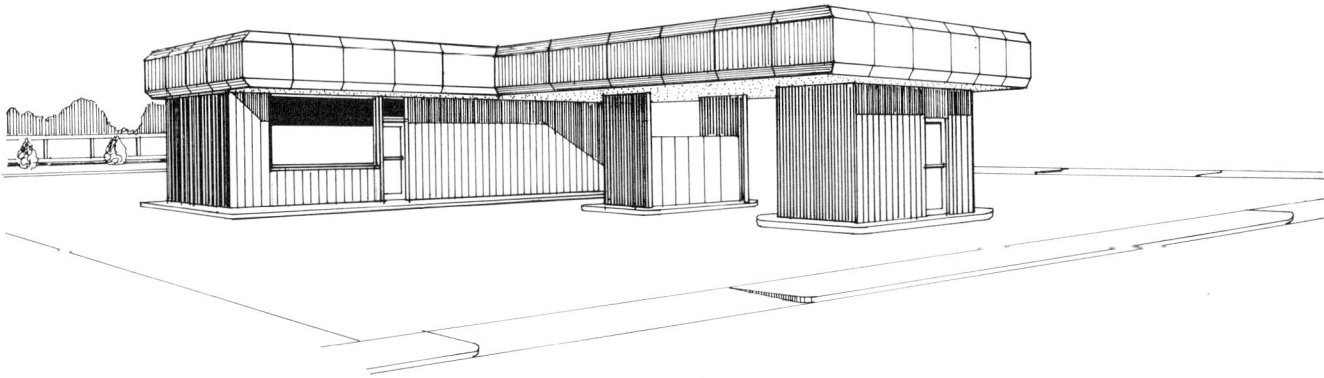

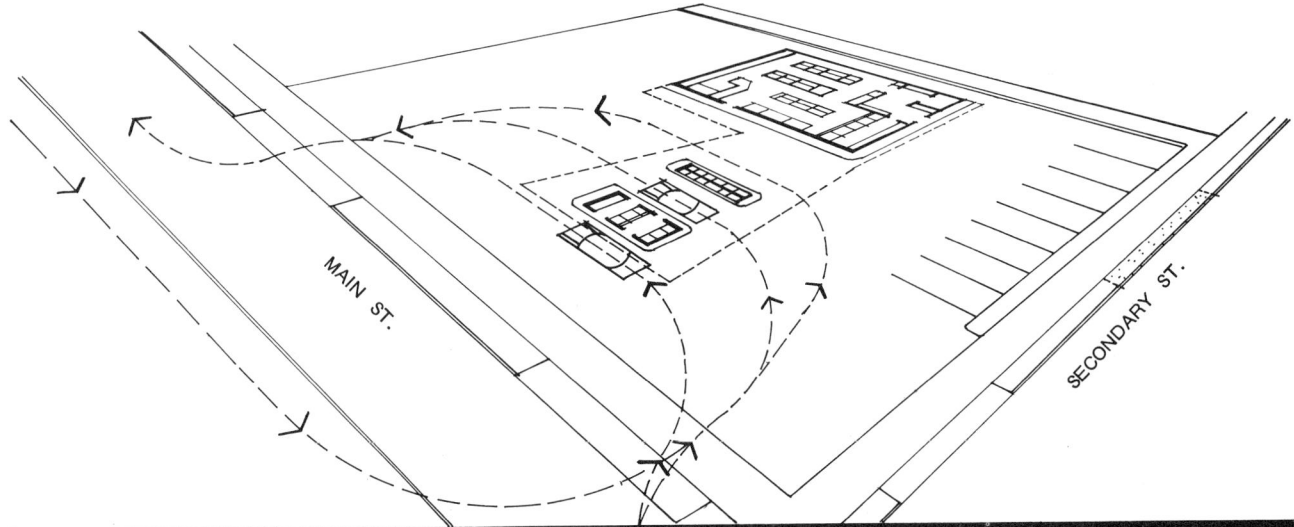

MAIN ST.

SECONDARY ST.

CONVENIENCE FOOD STORE AND AUTO ACCESSORIES

THE ABANDONED SERVICE STATION

As service stations close and self-service expands, automotive accessories and motor oil are being sold at convenience stores, drug stores, supermarkets, and large chain stores in shopping centers. As our economy becomes sluggish, the "do-it-yourself" consumers, who change their own motor oil and perform minor repairs and replacements, become a potentially large market for automotive accessories and motor oil sales.

The concept of combining the convenience store with automotive accessories has been successfully tested in many of the sunbelt states. The major problem that the convenience store marketer faces is that he is unfamiliar with the hundreds of automotive items the store would have to carry. This also presents a problem of inventory control. The approach recommended in these drawings is to carry a limited number of items from each business which constitute the high volume of sales at the average stores.

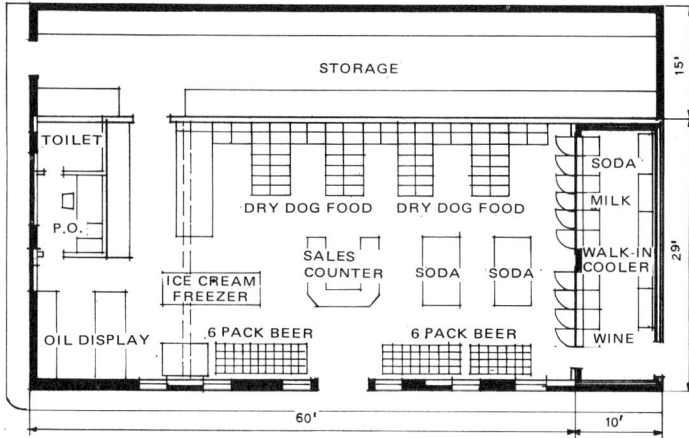

FLOOR PLAN

SCALE 0' 5' 10 15'

FRONT ELEVATION

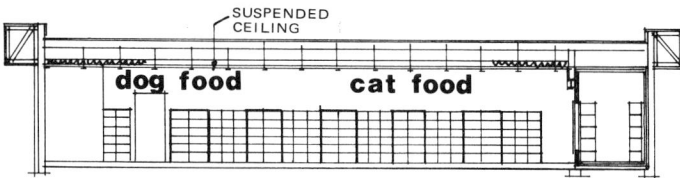

LONGITUDINAL SECTION

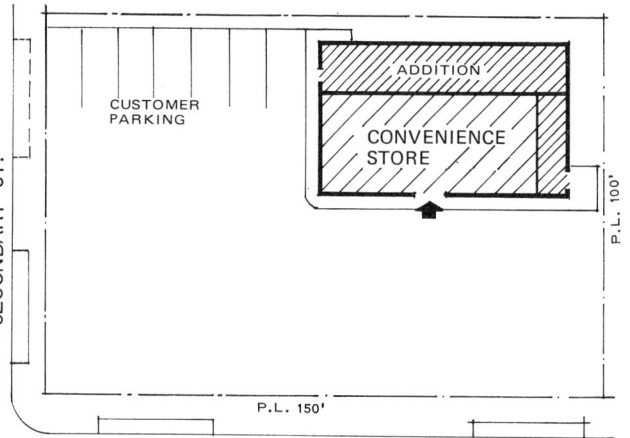

PLOT PLAN

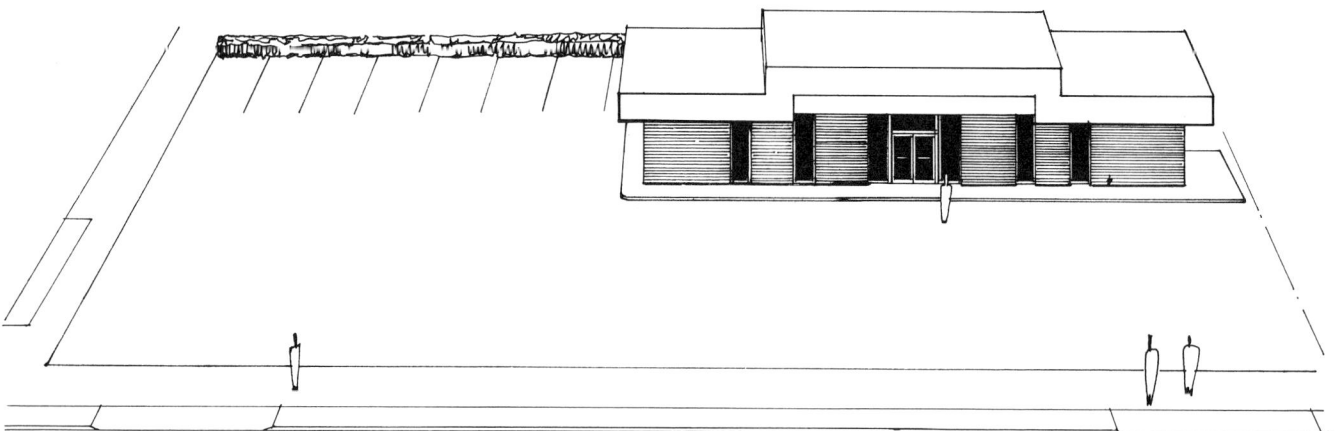

CONVENIENCE FOOD STORE

The convenience food industry carefully screens the lists of abandoned service stations throughout the country for new store acquisitions. High gasoline prices have increased the local convenience store sales, particularly on weekends and holidays. This typical 1200 square foot two bay building has been remodeled into the basic store at minimum cost. A walk-in cooler and fast food section represents the major expenditure. This store should carry only the fast turnover items and limit its inventory. An excellent profit maker for the small store is a slush machine and take-out service for ice cream. This is an energy-conscious conversion with only two small windows and a door remaining. Inexpensive diagonal wood siding covers the exterior. The mansard would be painted with a color that is compatible with the siding. Parking space for at least eight customers is desirable.

CONVENIENCE FOOD STORE STATISTICS:

According to the Department of Commerce, convenience food store growth has increased at a rate of slightly over 6% during each of the past three years. (1979, 1980 and 1981). In 1979, there were 14,660 stores with sales of approximately $6.13 billion. A 6% increase was projected for 1980 of 15,543 stores with total sales of $6.9 billion and a 6.2% increase was projected for 1981 of 16,509 stores with total sales of $7.6 billion.

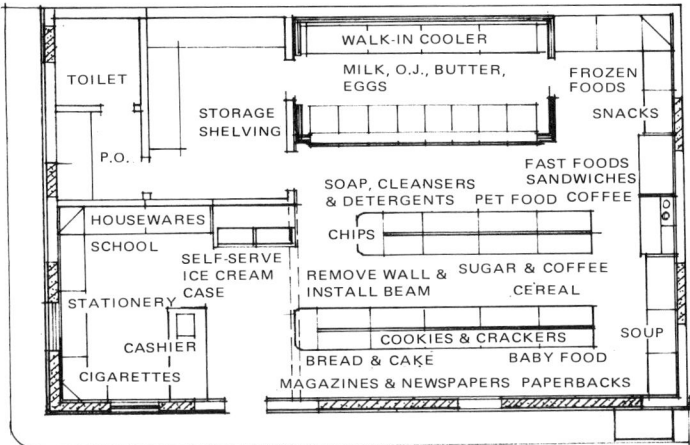
FLOOR PLAN

FRONT ELEVATION

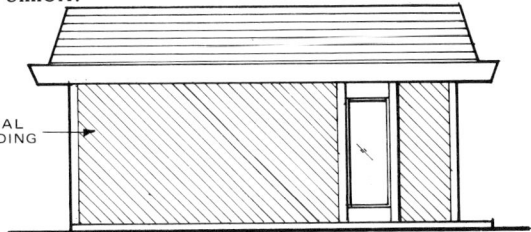
LEFT SIDE ELEVATION

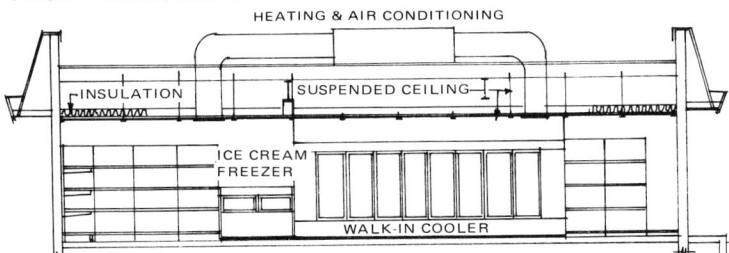
LONGITUDINAL SECTION

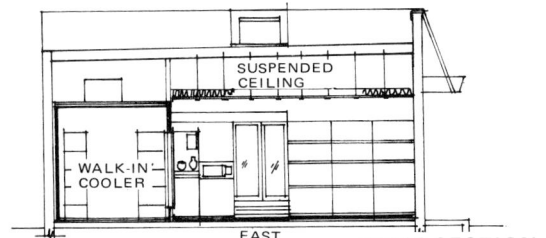
CROSS SECTION

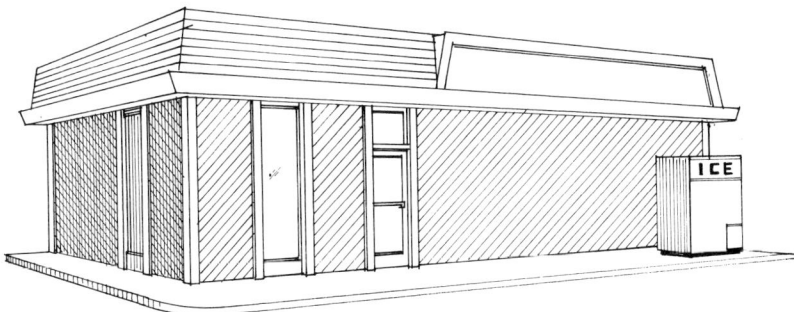

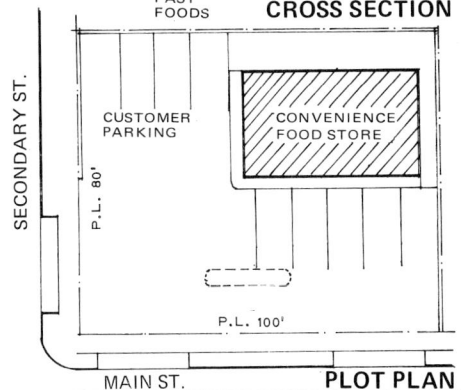
PLOT PLAN

CONVENIENCE FOOD STORE AND DONUT SHOP

Two excellent business uses that complement each other are the convenience food store and donut shop. A three bay service station has ample floor area for this purpose. The donut shop should have a seating capacity for at least twelve customers. A donut display case for take-out sales is placed adjacent to the cashier. The small convenience store requires an addition for the freezer and walk-in cooler. This permits sufficient room for gondola display. A self-service ice merchandiser is placed on an exterior platform.

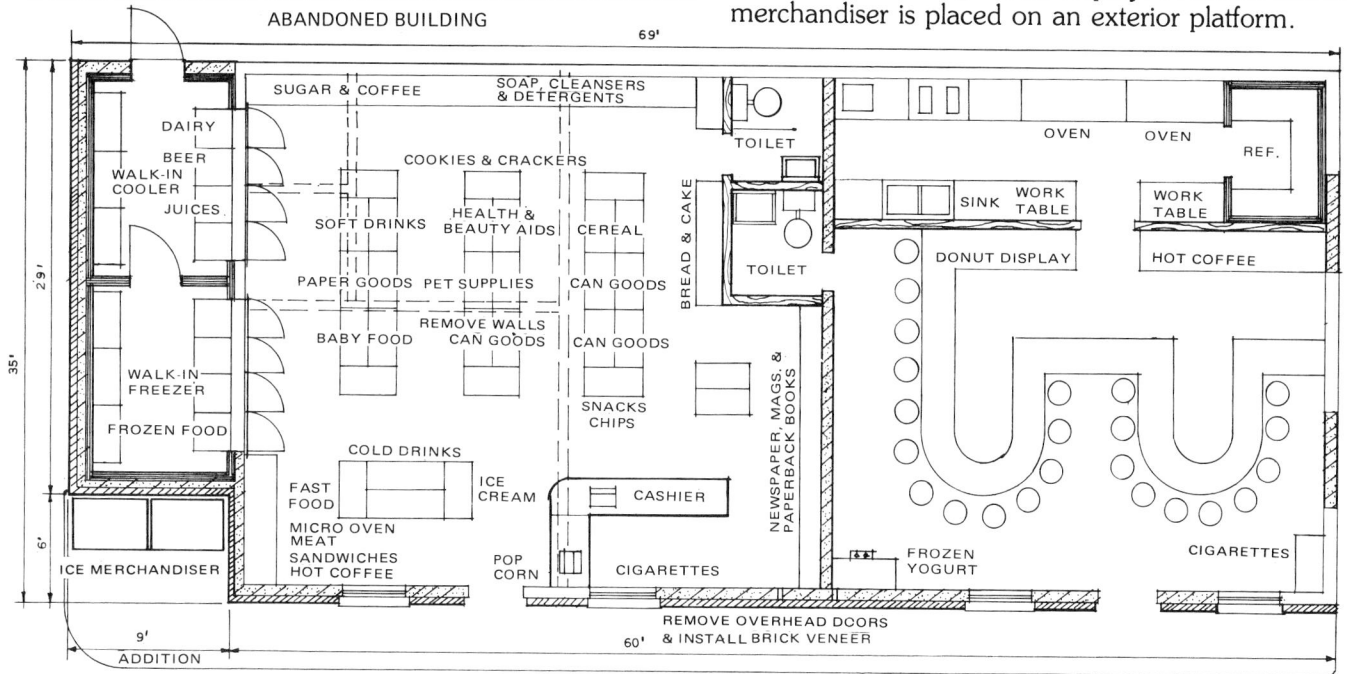

ABANDONED BUILDING

FLOOR PLAN

69'

SUGAR & COFFEE

SOAP, CLEANSERS & DETERGENTS

DAIRY

BEER

WALK-IN COOLER

JUICES

COOKIES & CRACKERS

SOFT DRINKS

HEALTH & BEAUTY AIDS

CEREAL

PAPER GOODS   PET SUPPLIES

CAN GOODS

BABY FOOD

REMOVE WALLS
CAN GOODS

CAN GOODS

WALK-IN FREEZER

FROZEN FOOD

SNACKS CHIPS

COLD DRINKS

FAST FOOD

ICE CREAM

CASHIER

MICRO OVEN
MEAT
SANDWICHES
HOT COFFEE

POP CORN

CIGARETTES

ICE MERCHANDISER

9'
ADDITION

29'

35'

6'

TOILET

TOILET

BREAD & CAKE

NEWSPAPER, MAGS, & PAPERBACK BOOKS

OVEN   OVEN   REF.

SINK   WORK TABLE

WORK TABLE

DONUT DISPLAY

HOT COFFEE

FROZEN YOGURT

CIGARETTES

REMOVE OVERHEAD DOORS
& INSTALL BRICK VENEER

60'

SCALE 3/32 = 1'-0"

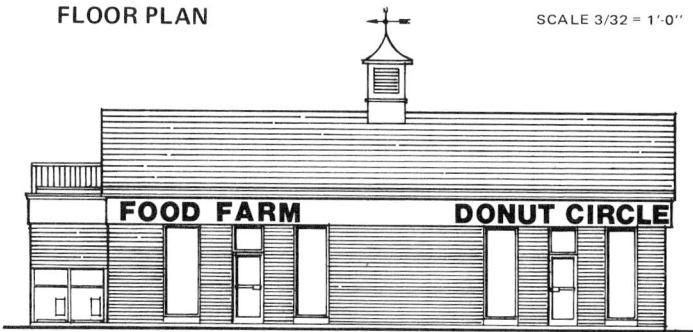

FRONT ELEVATION

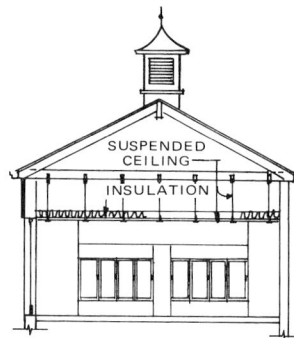

FOOD FARM   DONUT CIRCLE

CROSS SECTION

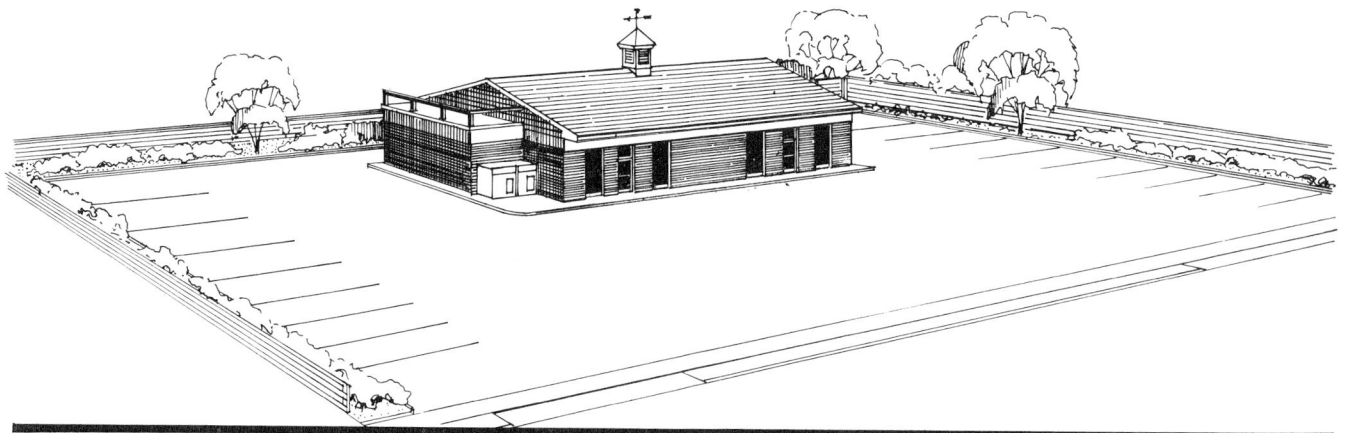

SUSPENDED CEILING
INSULATION

CONVENIENCE FOOD STORE AND SELF SERVICE WAND CAR WASH

The self-service wand car wash is a profit center that can contribute 50% of the income required to support the purchase and conversion of an abandoned service station to a convenience food store. The drive-thru wash bays can contribute $20,000 to $30,000 annual net profit before taxes. The size of the property is an important factor for the success of this business venture.

The conversion of the service station includes an addition for a walk-in cooler and freezer. This increases the floor area of the store from 1200 to 1425 square feet and allows for the installation of a deli-counter and fast food area. The car wash bays and equipment room are constructed of jumbo brick. The roof structure should be pre-engineered with pre-finished ceiling and fascia panels. In cold northern climates, the floors must be heated. It is important to lay out facilities to assure that there is no conflict between the convenience store customers and those desiring a car wash.

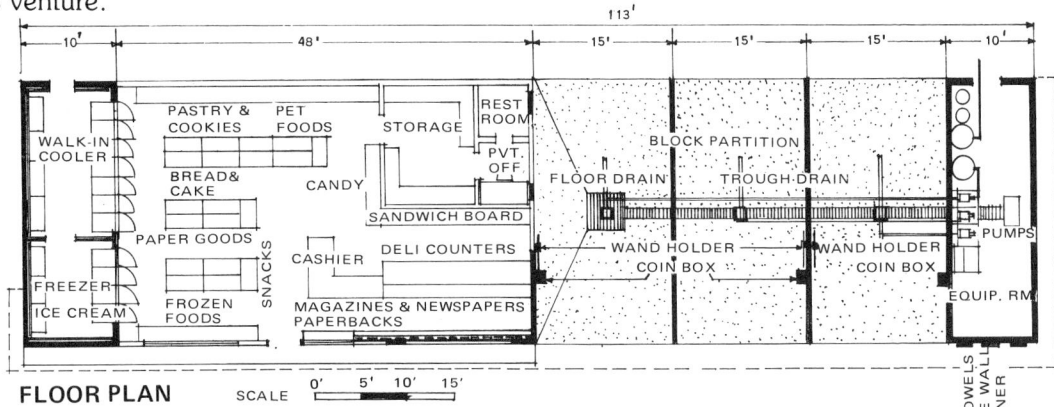

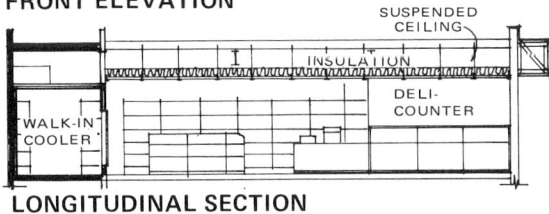

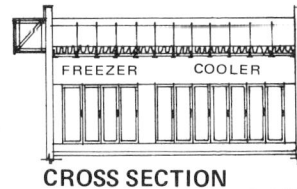

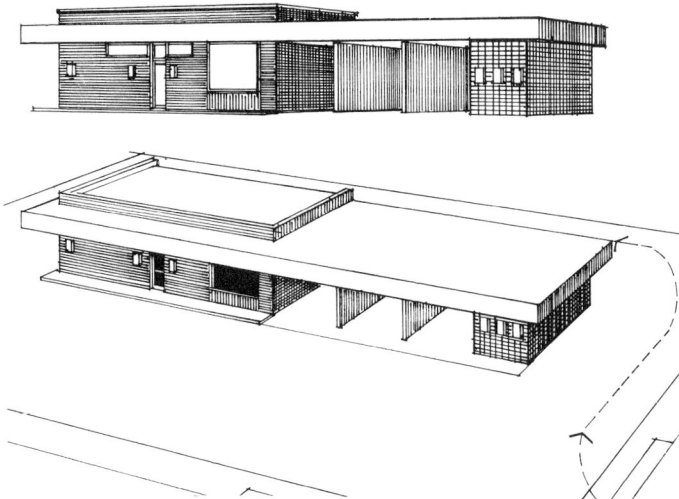

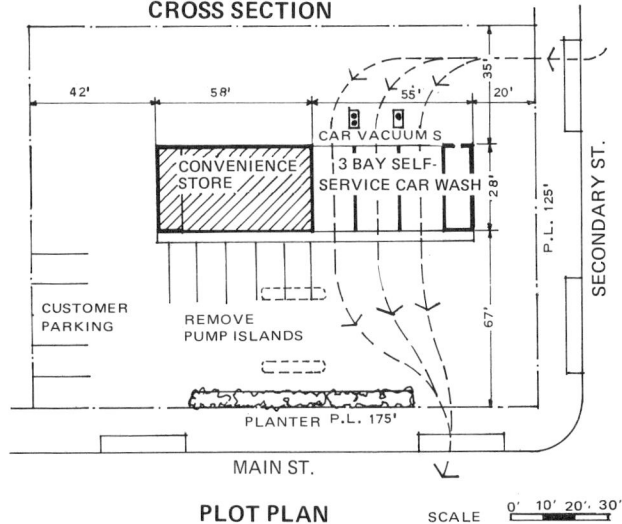

FLOOR PLAN SCALE 0' 5' 10' 15'

FRONT ELEVATION

LONGITUDINAL SECTION

CROSS SECTION

PLOT PLAN SCALE 0' 10' 20' 30'

CONVENIENCE FOOD STORE AND SELF SERVICE WAND CAR WASH

THE ABANDONED SERVICE STATION

A vacant two bay service station with a canopy, situated on a relatively large parcel, can be converted into a quick serve mini-mart and a two bay self-service wand car wash. These two uses complement each other. This unit can be operated by two employees. The quick serve unit sells eight to ten limited items, such as cigarettes, coffee, soda, beer and dairy products. The main store sells the above items, plus certain canned goods, cake, frozen foods, crackers, and pet food. It also features a fast food center.

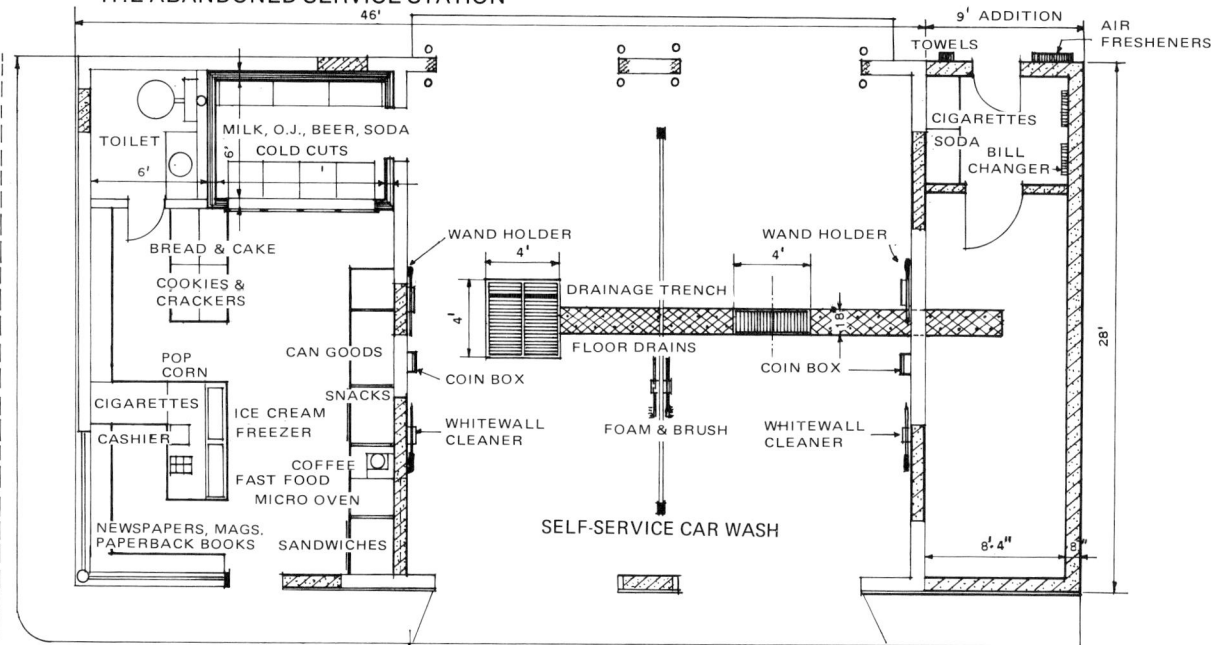

FLOOR PLAN

SCALE 3/32'' = 1'-0''

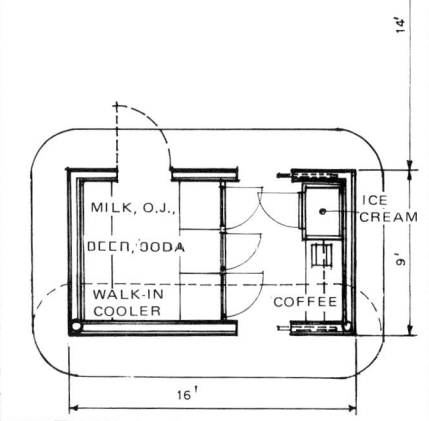

QUICK-PICK-UP

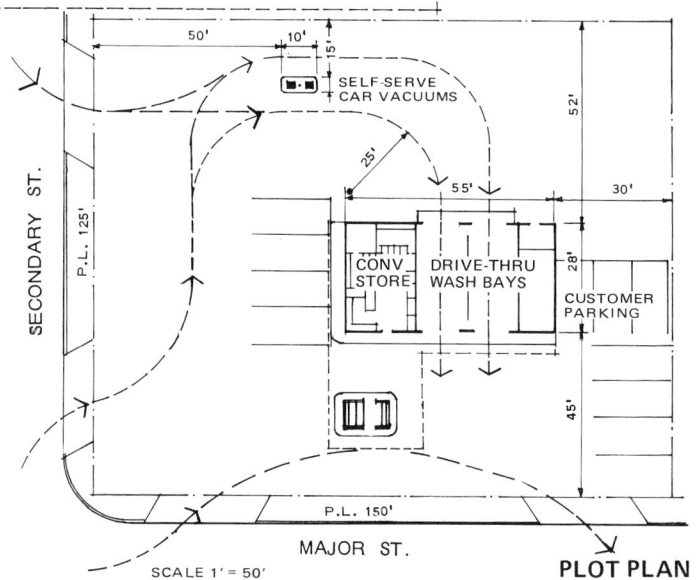

PLOT PLAN

SCALE 1' = 50'

The vacuum island and a rug shampoo unit are additional profit centers. Each customer will spend an average of $1.50 to $2.00 per car wash, far less than a tunnel wash costs. An addition is required for the equipment and vending room for towels, white wall and vinyl cleaners, and air fresheners. Equipment must deliver 1,000 p.s.i. at the wand nozzle. A super foam brush wand in each bay will insure a high return on investment.

CONVENIENCE FOOD STORE AND SELF SERVICE WAND CAR WASH

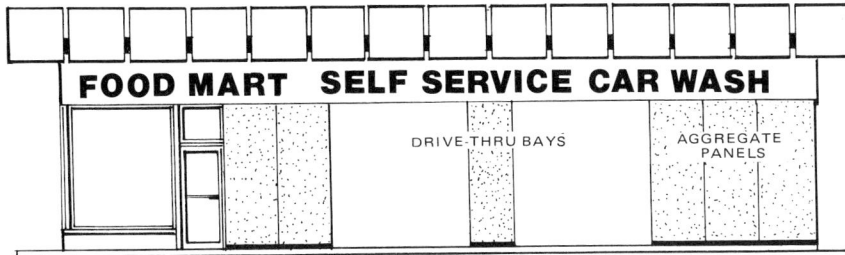

FOOD MART SELF SERVICE CAR WASH

DRIVE-THRU BAYS

AGGREGATE PANELS

FRONT ELEVATION

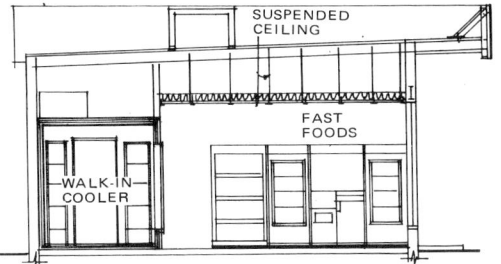

SUSPENDED CEILING

FAST FOODS

WALK-IN COOLER

CROSS SECTION

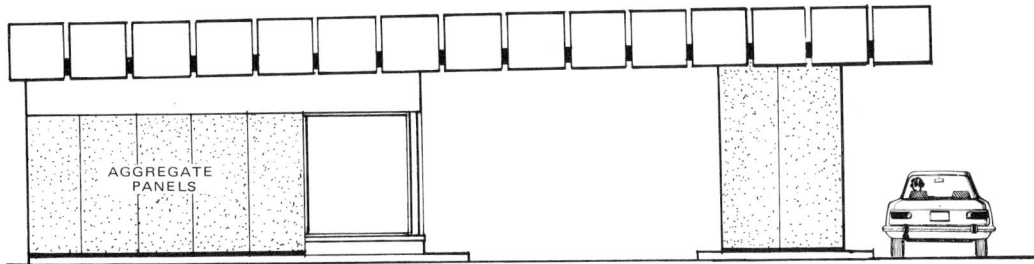

AGGREGATE PANELS

LEFT SIDE ELEVATION

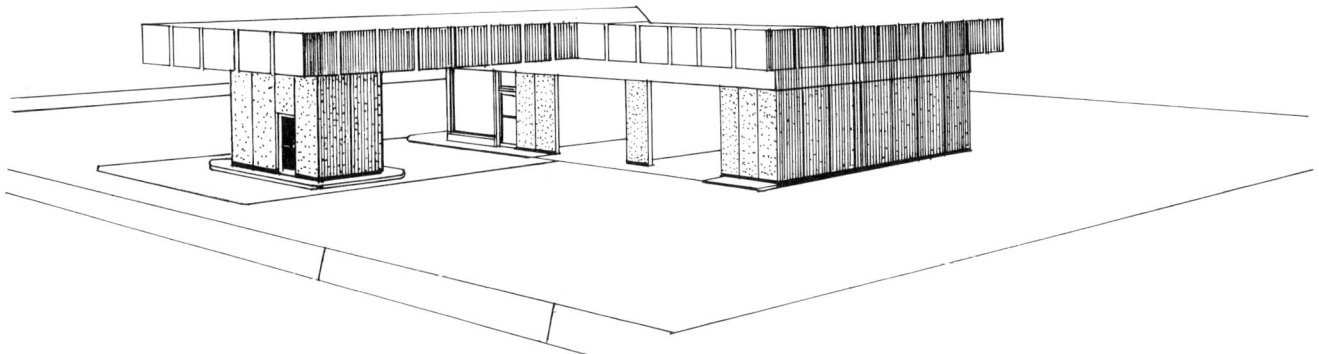

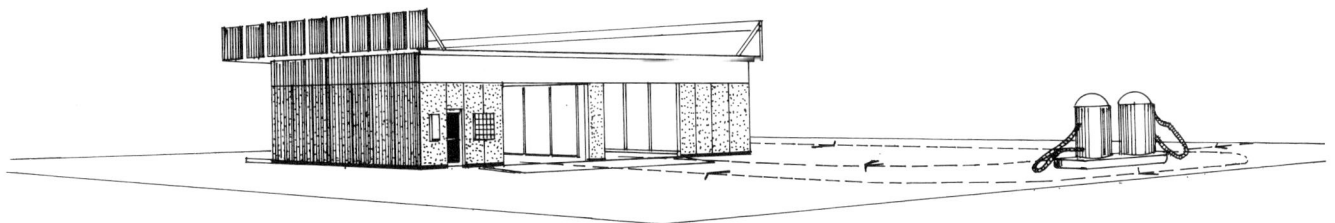

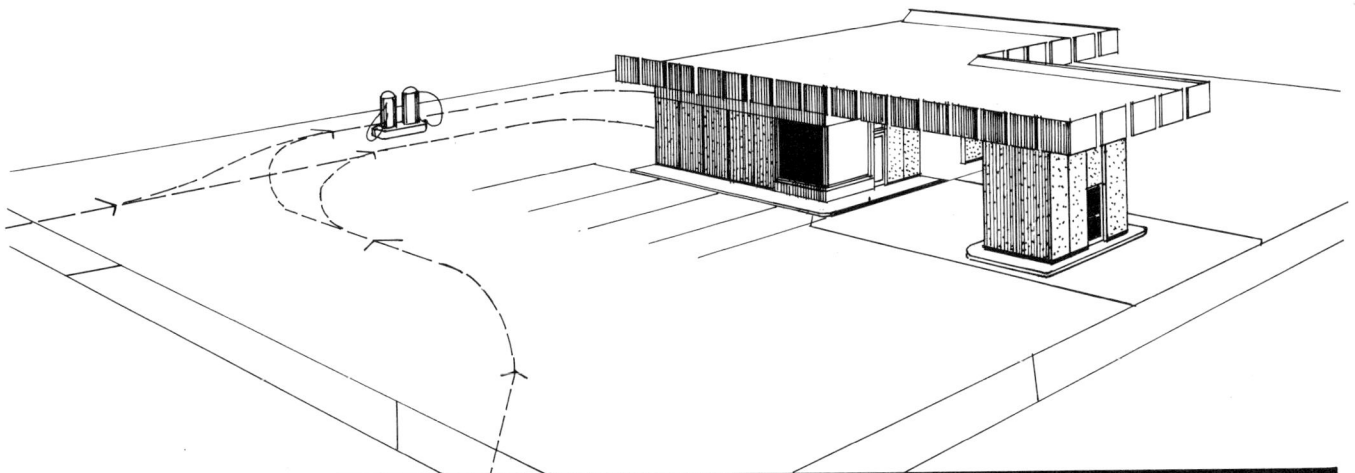

DRIVE-UP CONVENIENCE FOOD STORE

A drive-up service window is the most important feature in the success of a convenience food store. Three out of four customers prefer to stay in their cars rather than go into the store. Since most patrons buy only three or four basic items, the drive-up window is a distinct advantage. Dairy products and beer are the most sought after items; therefore, a separate walk-in cooler is adjacent to the window and cashier. One employee can operate this store based on the illustrated layout. A special large pass-through drawer is required for the larger products. The diagonal arrangement of gondolas provides the cashier unlimited visibility of the store and cooler and freezer section.

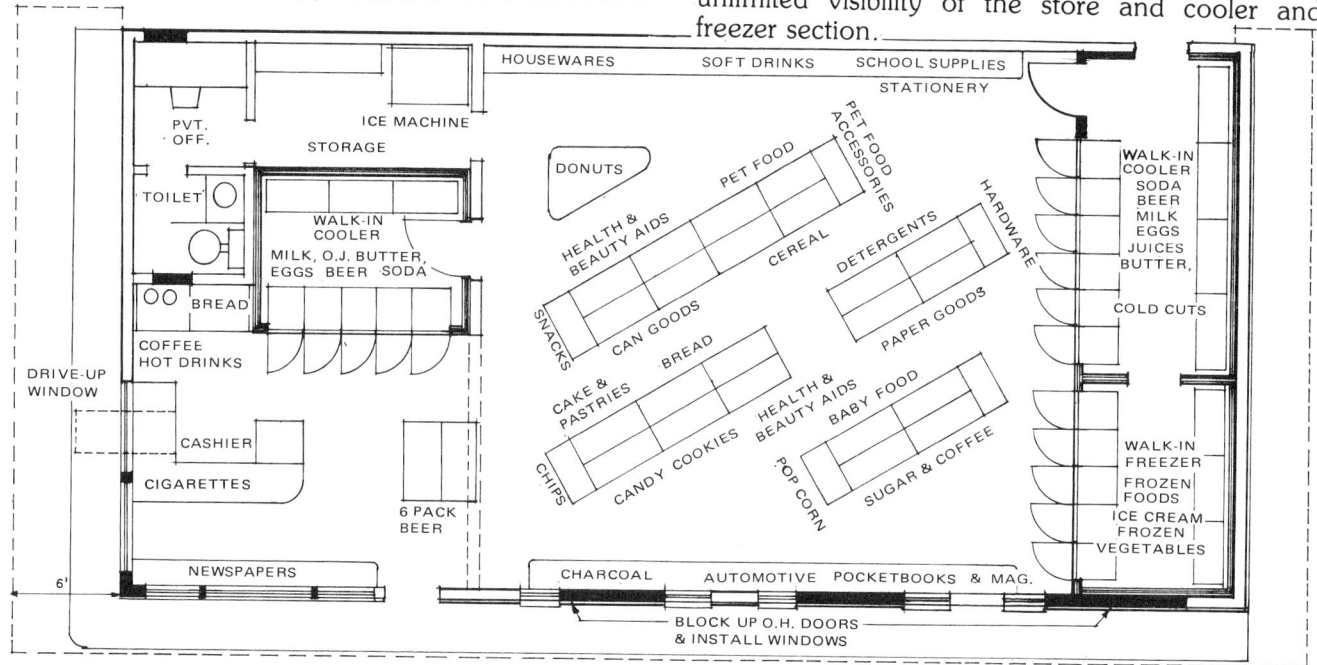

FLOOR PLAN SCALE 3/32" = 1'-0"

Floor plan labels: HOUSEWARES, SOFT DRINKS, SCHOOL SUPPLIES, STATIONERY, PVT. OFF., ICE MACHINE, STORAGE, DONUTS, PET FOOD, PET FOOD ACCESSORIES, HARDWARE, WALK-IN COOLER SODA BEER MILK EGGS JUICES BUTTER, TOILET, WALK-IN COOLER MILK, O.J. BUTTER, EGGS BEER SODA, HEALTH & BEAUTY AIDS, CEREAL, DETERGENTS, COLD CUTS, BREAD, SNACKS, CAN GOODS, PAPER GOODS, COFFEE HOT DRINKS, DRIVE-UP WINDOW, CAKE & PASTRIES, BREAD, HEALTH & BEAUTY AIDS, BABY FOOD, WALK-IN FREEZER FROZEN FOODS, CASHIER, CHIPS, CANDY COOKIES, POP CORN, SUGAR & COFFEE, ICE CREAM FROZEN VEGETABLES, CIGARETTES, 6 PACK BEER, NEWSPAPERS, CHARCOAL, AUTOMOTIVE, POCKETBOOKS & MAG., BLOCK UP O.H. DOORS & INSTALL WINDOWS, 6'

FRONT ELEVATION — QUICK FOOD MART, DIAGONAL WOOD SIDING, VERTICAL WOOD SIDING

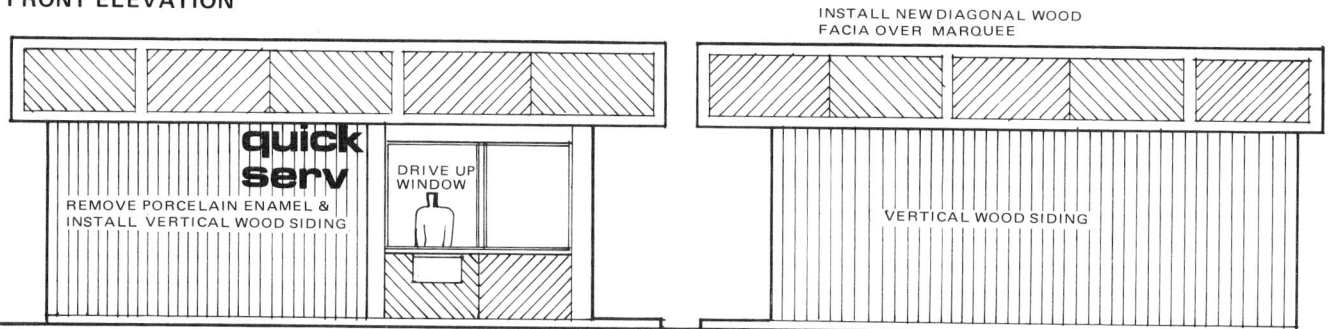

LEFT SIDE ELEVATION — quick serv, REMOVE PORCELAIN ENAMEL & INSTALL VERTICAL WOOD SIDING, DRIVE UP WINDOW

RIGHT SIDE ELEVATION — INSTALL NEW DIAGONAL WOOD FACIA OVER MARQUEE, VERTICAL WOOD SIDING

CONVENIENCE FOOD STORE, WITH DRIVE-UP WINDOW

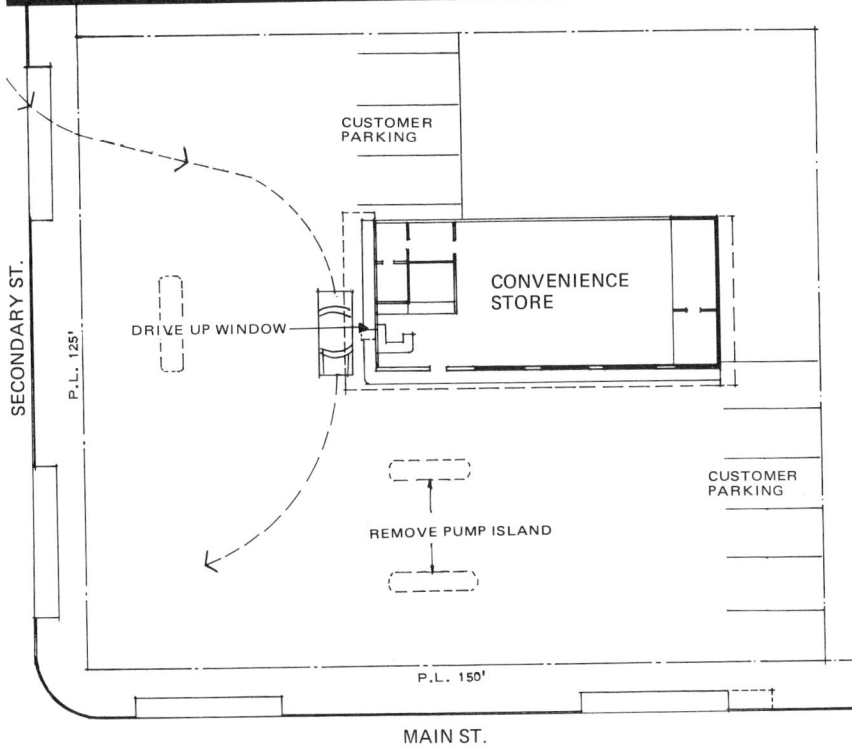

SECONDARY ST.

P.L. 125'

CUSTOMER PARKING

CONVENIENCE STORE

DRIVE UP WINDOW

CUSTOMER PARKING

REMOVE PUMP ISLAND

P.L. 150'

MAIN ST.

PLOT PLAN   SCALE 1'' = 40'

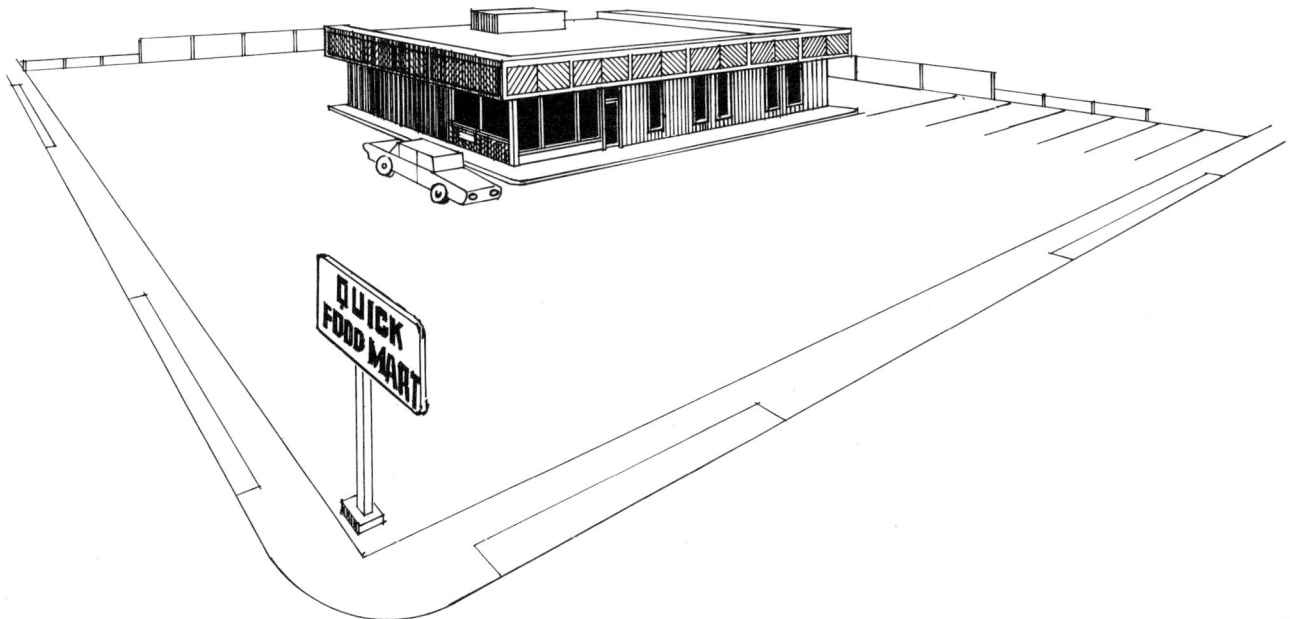

QUICK FOOD MART

CONVENIENCE FOOD STORE & QUICK PICK-UP

THE ABANDONED SERVICE STATION

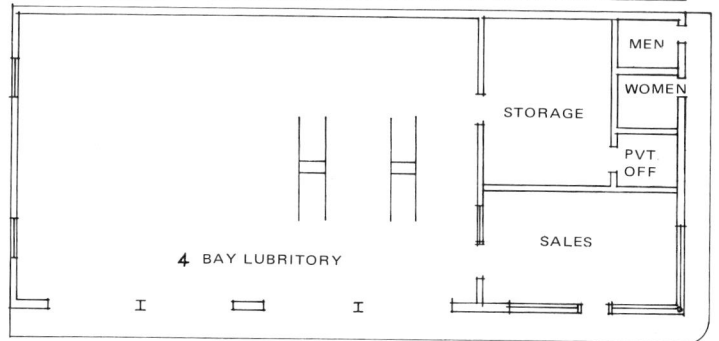

MEN
WOMEN
STORAGE
PVT OFF
SALES

4 BAY LUBRITORY

FLOOR PLAN
ABANDONED BUILDING

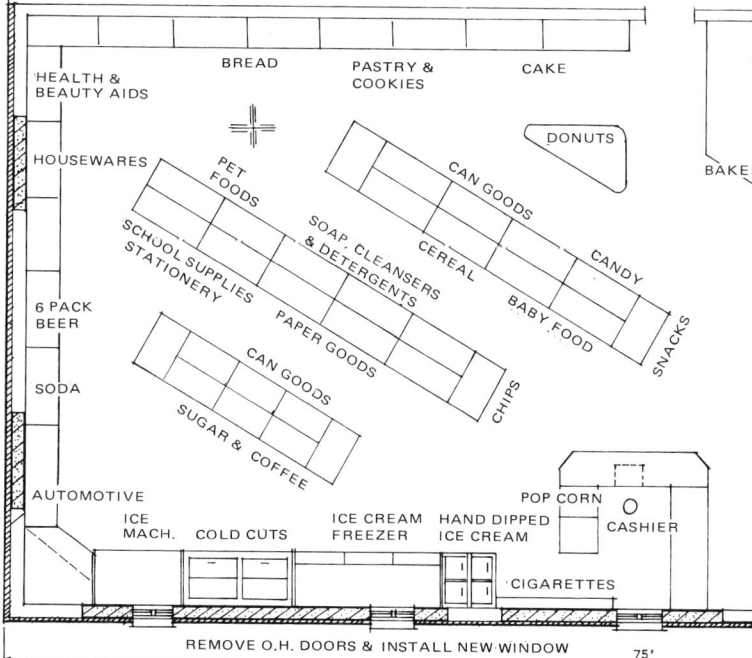

HEALTH & BEAUTY AIDS

BREAD   PASTRY & COOKIES   CAKE

HOUSEWARES

DONUTS

PET FOODS   CAN GOODS

SCHOOL SUPPLIES STATIONERY   SOAP, CLEANSERS & DETERGENTS   CANDY

CEREAL

6 PACK BEER   PAPER GOODS   BABY FOOD

SODA   CAN GOODS   SNACKS

SUGAR & COFFEE   CHIPS

AUTOMOTIVE

ICE MACH.   COLD CUTS   ICE CREAM FREEZER   HAND DIPPED ICE CREAM   POP CORN   CASHIER

CIGARETTES

CAKE & BREAD RACKS BAKERY SHOP   OVENS   COUNTER   KITCHEN   REFRIG.   TOILET

BAKERY & PASTRY ITEMS

PVT. OFF.

STORAGE

FAST FOODS
SANDWICHES
COFFEE HOT DRINKS
MICRO-WAVE OVEN

WALK-IN FREEZER   WALK-IN COOLER
MILK, O.J. BUTTER, EGGS BEER SODA

FROZEN FOODS

FROZEN CAKES & PIES   FROZEN VEGETABLES

NEWSPAPERS   FREEZERS

REMOVE O.H. DOORS & INSTALL NEW WINDOW   75'

FLOOR PLAN
SCALE   3/32' = 1'-0"

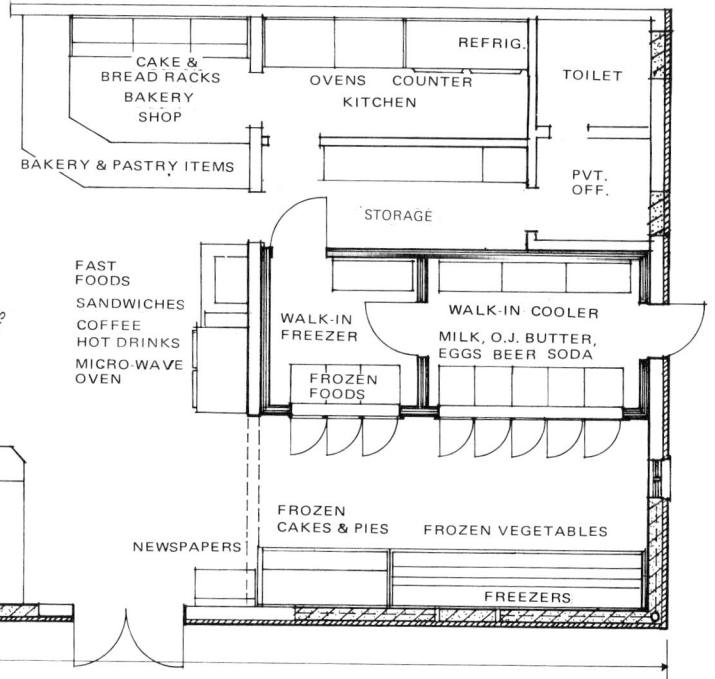

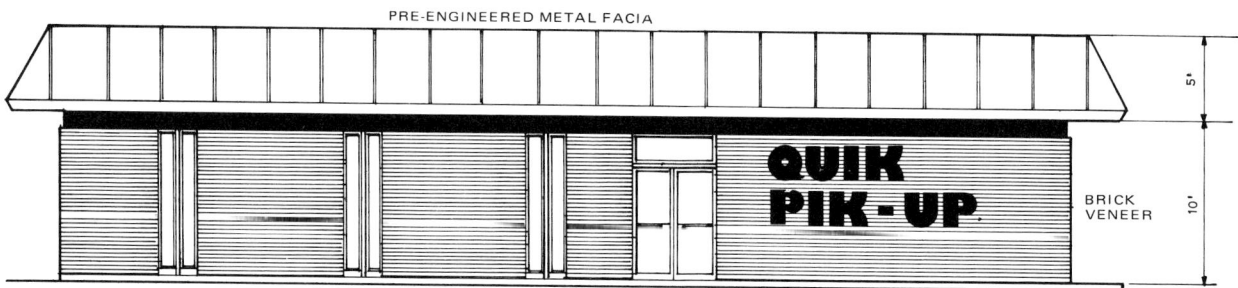

PRE-ENGINEERED METAL FACIA

QUIK PIK-UP

BRICK VENEER

5'

10'

FRONT ELEVATION

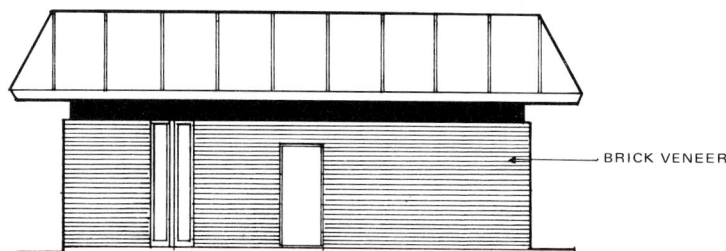

BRICK VENEER

RIGHT SIDE ELEVATION

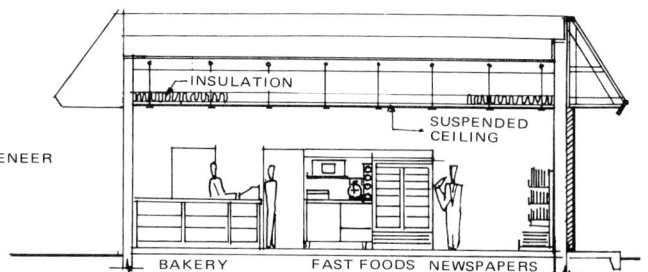

INSULATION
SUSPENDED CEILING

BAKERY   FAST FOODS   NEWSPAPERS

LONGITUDINAL SECTION

An abandoned service station, containing 2250 square feet of floor area, is an excellent convenience food store candidate. The building illustrated in the photograph is situated on a large triangular parcel with both major arteries being heavily travelled. A small but well equipped quick-service drive-up dairy building with one employee is an excellent profit center. The 300 square foot quick pick-up mini-mart caters to customers who wish to buy only two or three dairy items without leaving their car.

The layout of the interior provides maximum visibility of the store from the cashier's counter. A bakery shop features on premises baking. Fresh baked bread and pastries are important to the success of the store. Important departments that should be included in the store are fast foods and hand dipped ice cream cones. Convenience store featuring quality products and a quick pick-up mart will meet the need of the community.

Service stations were not designed with energy conservation in mind, therefore a conversion should include:

1. A suspended ceiling with a minimum of 8″ insulation.

2. A brick veneer exterior with narrow windows across the front of the building will reduce heat loss considerably.

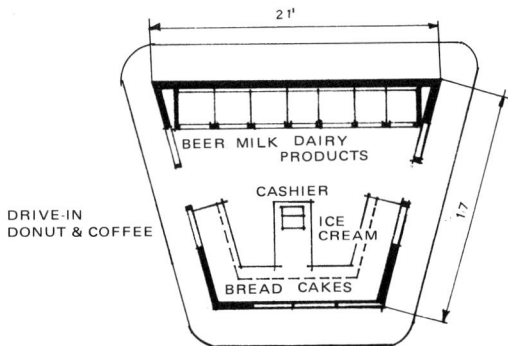

FLOOR PLAN

FRONT ELEVATION
QUICK-SERVICE DRIVE-IN

SIDE ELEVATION

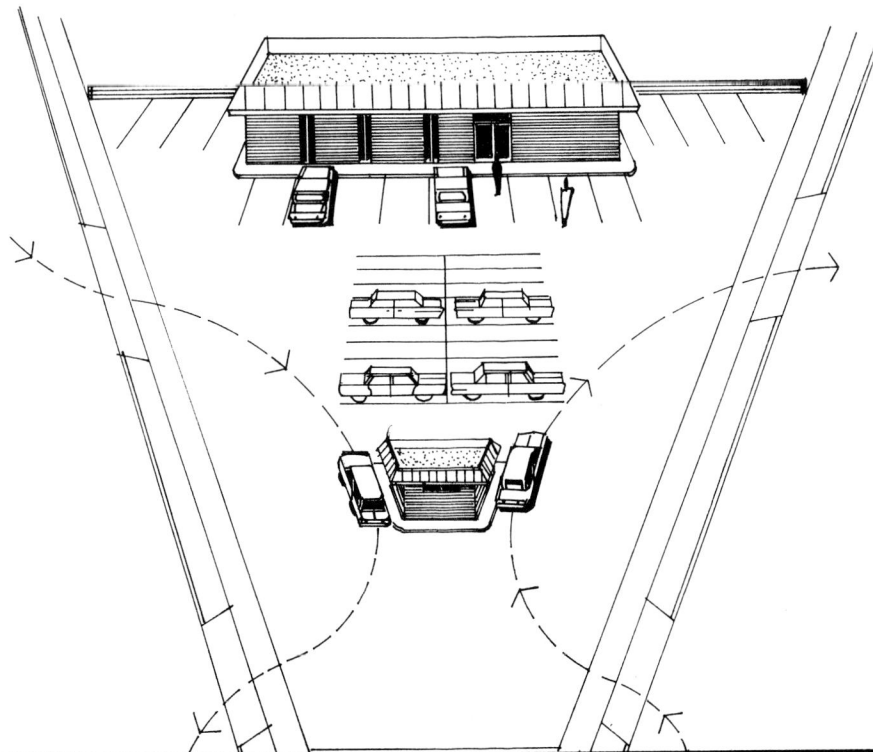

DRIVE-UP CONVENIENCE FOOD STORE

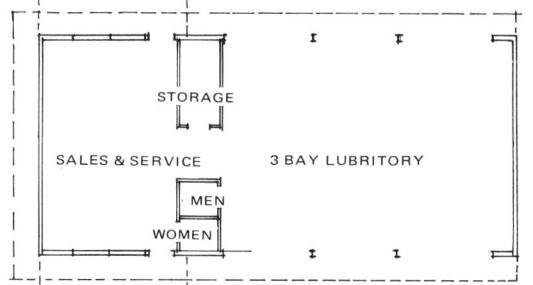

FLOOR PLAN ABANDONED BUILDING

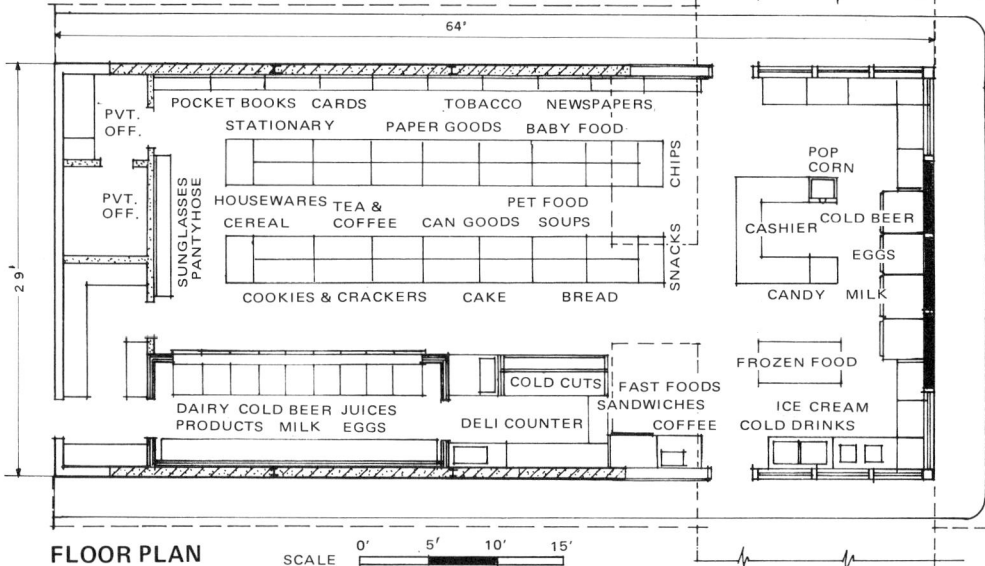

STORAGE

SALES & SERVICE   3 BAY LUBRITORY

MEN

WOMEN

FLOOR PLAN

64'

2 9'

PVT. OFF.

PVT. OFF.

POCKET BOOKS   CARDS   TOBACCO   NEWSPAPERS
STATIONARY   PAPER GOODS   BABY FOOD

CHIPS

SUNGLASSES PANTYHOSE

HOUSEWARES   TEA &   PET FOOD
CEREAL   COFFEE   CAN GOODS   SOUPS

SNACKS

COOKIES & CRACKERS   CAKE   BREAD

POP CORN

CASHIER   COLD BEER
EGGS
CANDY   MILK

FROZEN FOOD

DAIRY COLD BEER JUICES
PRODUCTS   MILK   EGGS

COLD CUTS

DELI COUNTER

FAST FOODS
SANDWICHES
COFFEE

ICE CREAM
COLD DRINKS

SCALE   0'   5'   10'   15'

FRONT ELEVATION

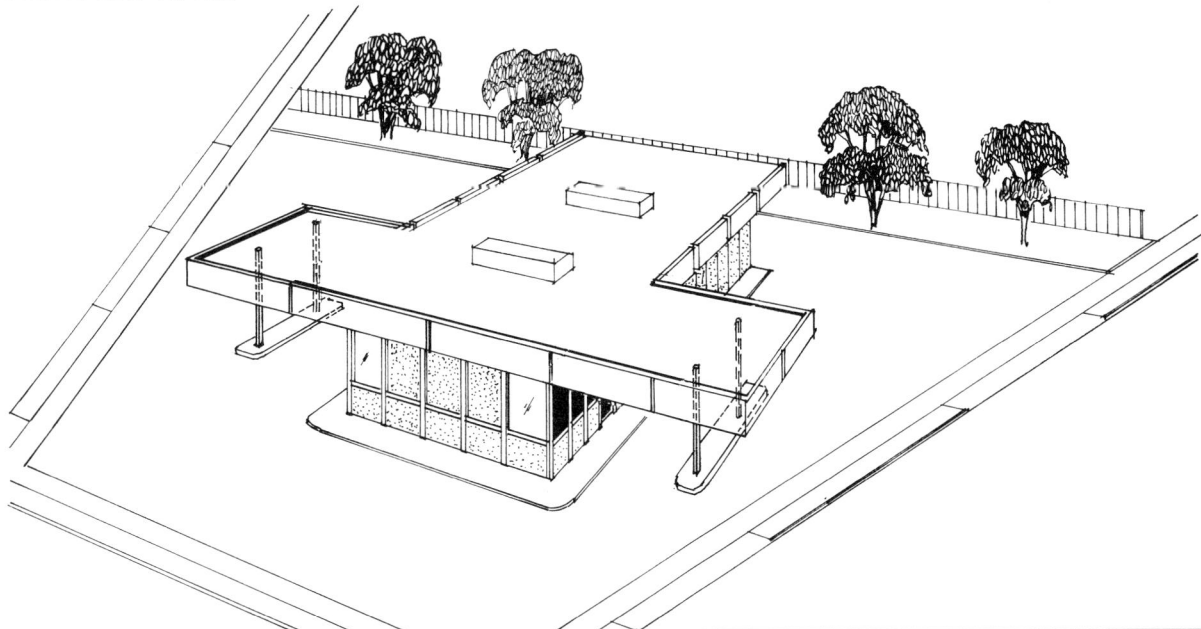

AGGREGATE PANELS   THE CAR HOP

A car-hop-type service is ideal for a small convenience store property where parking is limited to four or five cars. This abandoned three bay service station, converted to a convenience store, can justify two full-time employees; one on each side of the store. Cashier counters are positioned 6' from each doorway. The large glass area surrounding the front and two sides of the building affords the drive-up customer a full view of the merchandise within 25' of the register. The canopy area provides protective cover for the customer and clerk during inclement weather. Removal of the storage area and rest rooms and enlargement of the store front is required to provide customer visibility of the store merchandise.

COOKWARE STORE

THE ABANDONED SERVICE STATION

The space devoted to quality cookware in department stores is indicative of the demand for this product. As food prices escalate, eating out by families and couples will decline, resulting in a return to the popularity of home cooking. A closed two bay service station of 1200 square feet, situated in a retail district, with considerable foot traffic, parking, and visibility will make an ideal store. A flair for merchandising and developing a store image is essential for the business.

Attractive eye-catching display windows are extremely important, together with a warm atmosphere created by the interior layout. The floor area should be divided into a demonstration cooking center, areas for specialized foods, teas, and coffees, utensils, pots and pans, glassware, blenders and other small appliances. The removal of the walls as indicated will provide sufficient open area for the display of these items. The mechanical equipment must be mounted on the roof. Rustic paneling and beamed ceiling will help create the atmosphere desired for the store.

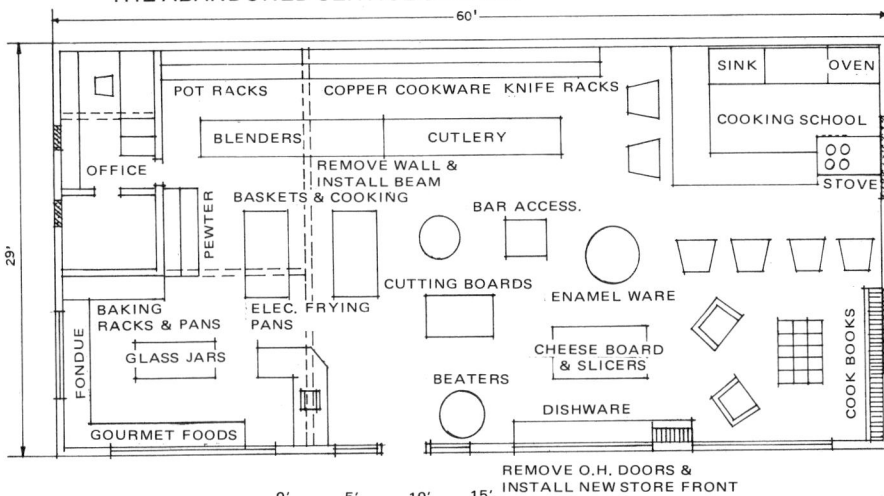

FLOOR PLAN SCALE 0' 5' 10' 15'

FRONT ELEVATION

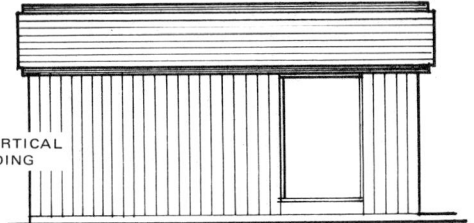

RIGHT SIDE ELEVATION

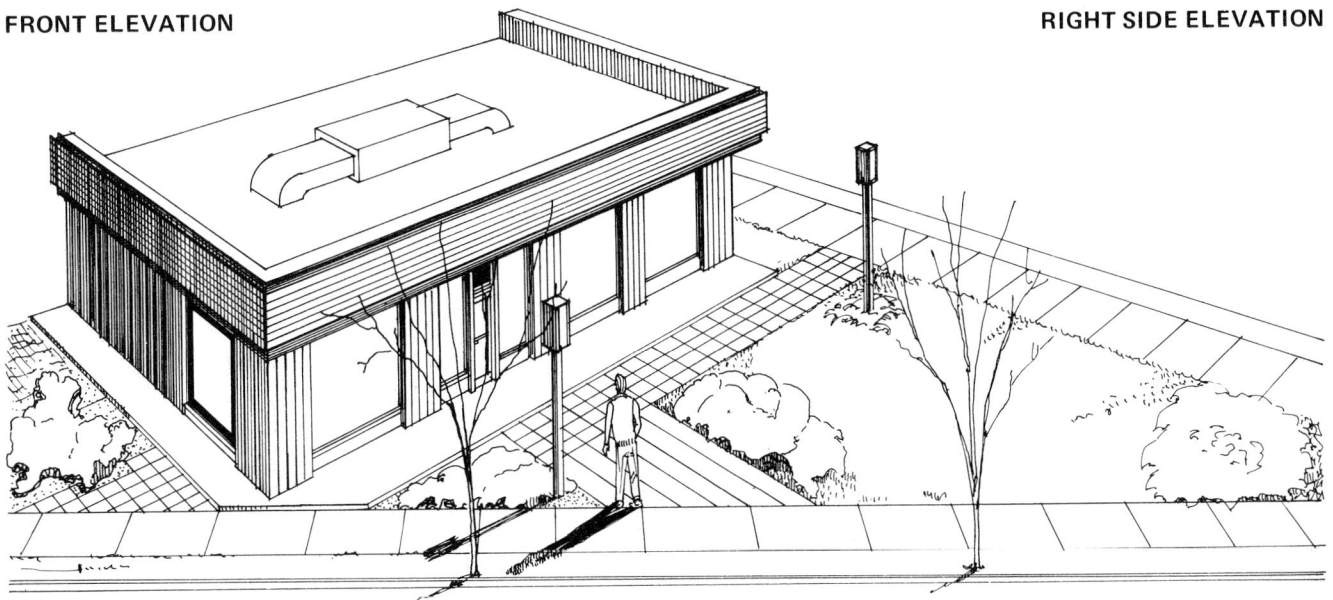

DRIVE-UP DAIRY MART

The fast service concept of limited dairy products is an excellent drive-up business use for an abandoned "self-service, gas-only" facility. After removal of the pump island and kiosk, an 18' x 30' dairy unit carrying perhaps a dozen products, is installed in the same area utilizing exisiting utilities. Room should be provided for a walk-in cooler for dairy products and soda; a freezer for ice cream; and a gondola for bread, cake and donuts. Appropriate remodeling is accomplished by installing a rough textured natural wood finish on the exterior, and on the canopy fascia, and enclosing the columns.

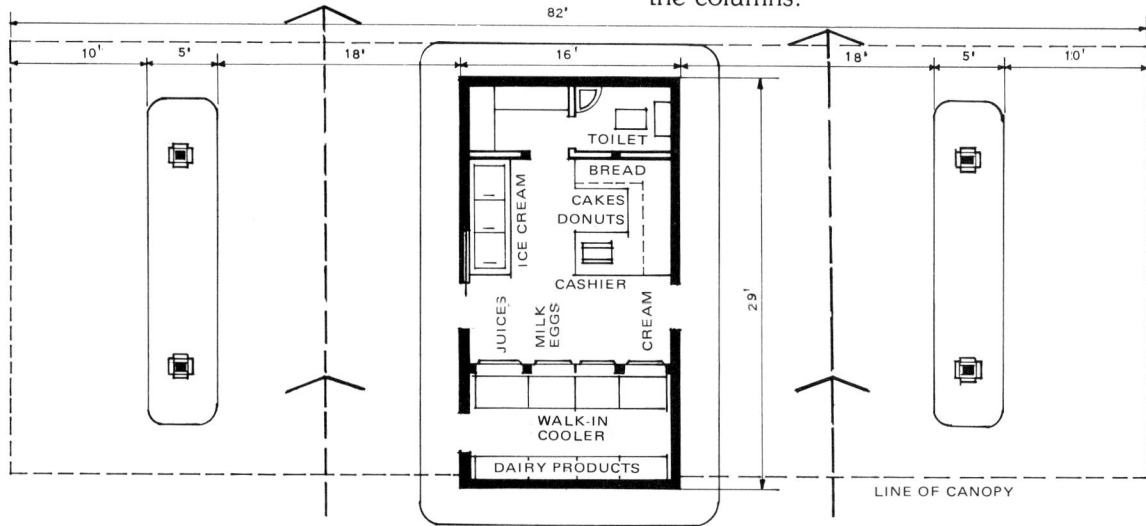

FLOOR PLAN   SCALE   0'   5'   10'   15'

82'

10'   5'   18'   16'   18'   5'   10'

TOILET
BREAD
CAKES
DONUTS
ICE CREAM
CASHIER
JUICES
MILK EGGS
CREAM
WALK-IN COOLER
DAIRY PRODUCTS
29'
LINE OF CANOPY

FRONT ELEVATION

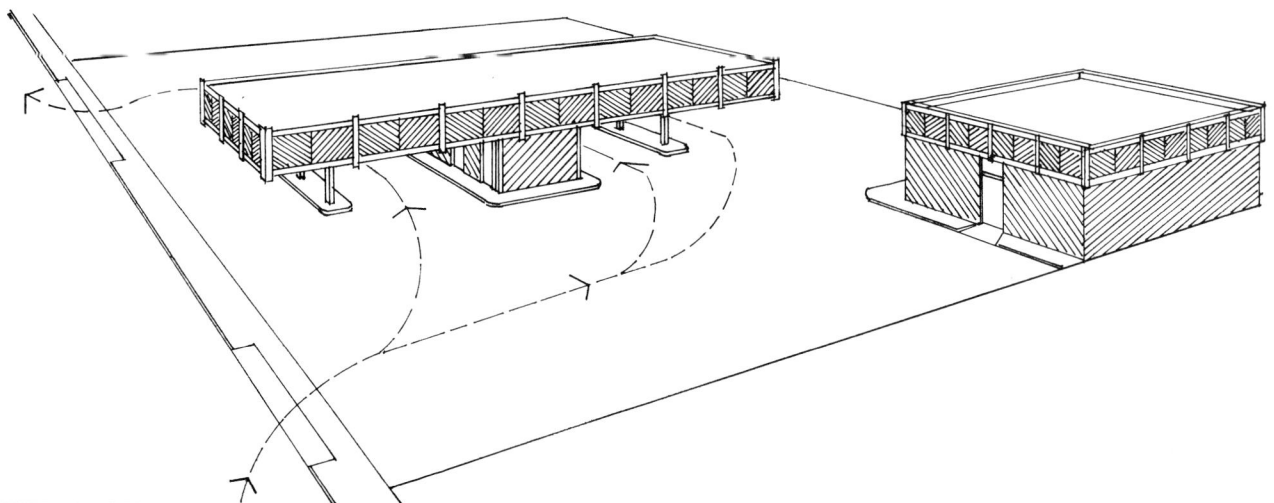

INSTALL NEW PRE-ENGINEERED FACIA OVER CANOPY

4'
12'

ENCLOSE COLUMNS WITH WOOD

DIAGONAL SIDING

DAIRY MART

HOME DECORATORS CENTER

Home decorating centers have become very popular in middle income residential areas. During periods of a depressed economy, the homeowner or condominium dweller will make many of his own improvements.

A canopy attached to a service station is an asset generally overlooked by owners when they convert the building to another business use. Enclosing the canopy area with walls or glass will enlarge the floor area of the building by one third. A glass enclosed canopy becomes an excellent showroom for the display of merchandise.

The interior of the service station should be converted exclusively with the materials and fixtures that are sold in the store. This building has been expanded from 1200 to 1800 square feet by enclosing the canopy. A 1000 square foot addition should be a prime consideration for the storage of wood paneling, floor and bathroom tile, electrical fixtures and paint and hardware.

The exterior has been remodeled with inexpensive brick veneer and epoxy sprayed sandstone panels. Lighting fixtures, sold by the store, are installed on the exterior of the building, and landscape fixtures are placed around the perimeter of the parking lot.

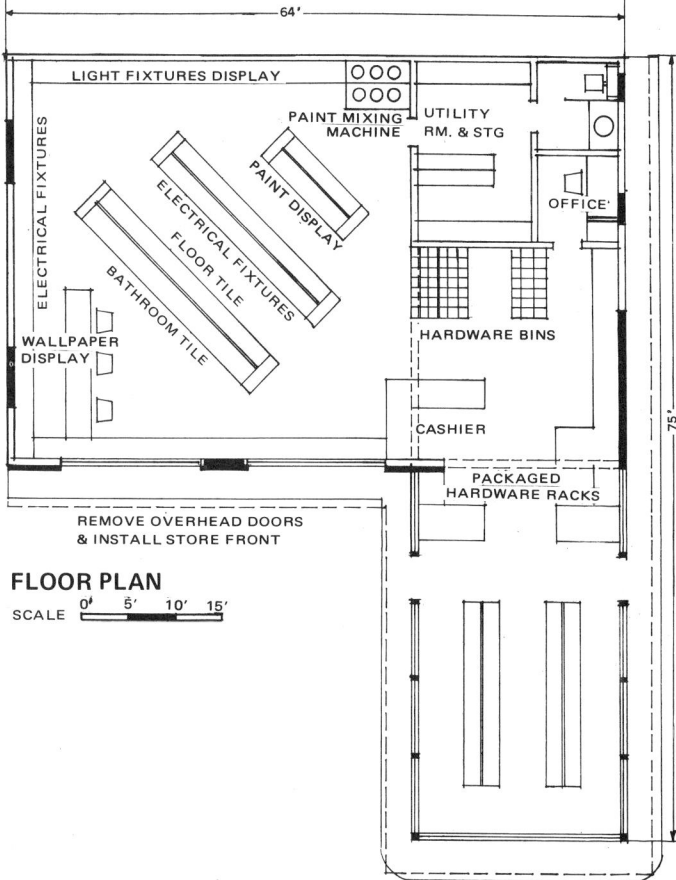

FLOOR PLAN

SCALE 0' 5' 10' 15'

LIGHT FIXTURES DISPLAY

PAINT MIXING MACHINE

UTILITY RM. & STG

OFFICE

ELECTRICAL FIXTURES

PAINT DISPLAY

ELECTRICAL FIXTURES

FLOOR TILE

BATHROOM TILE

WALLPAPER DISPLAY

HARDWARE BINS

CASHIER

PACKAGED HARDWARE RACKS

REMOVE OVERHEAD DOORS & INSTALL STORE FRONT

64'

75'

REMOVE PORCELAIN ENAMEL & INSTALL AGGREGATE PANELS

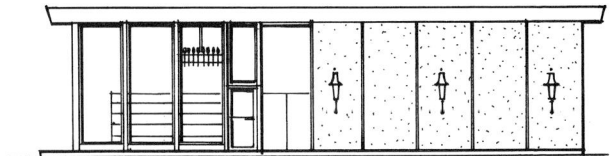

RIGHT SIDE ELEVATION

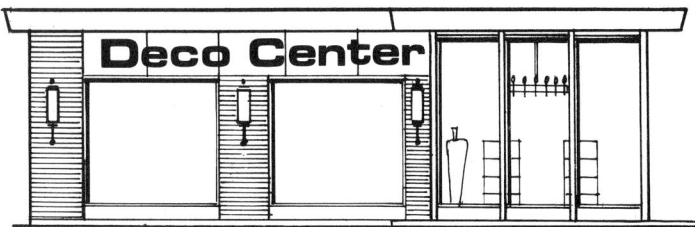

Deco Center

FRONT ELEVATION

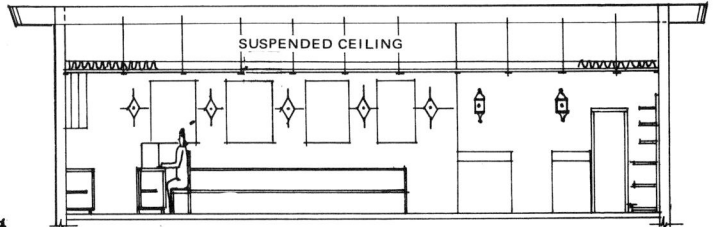

SUSPENDED CEILING

LONGITUDINAL SECTION

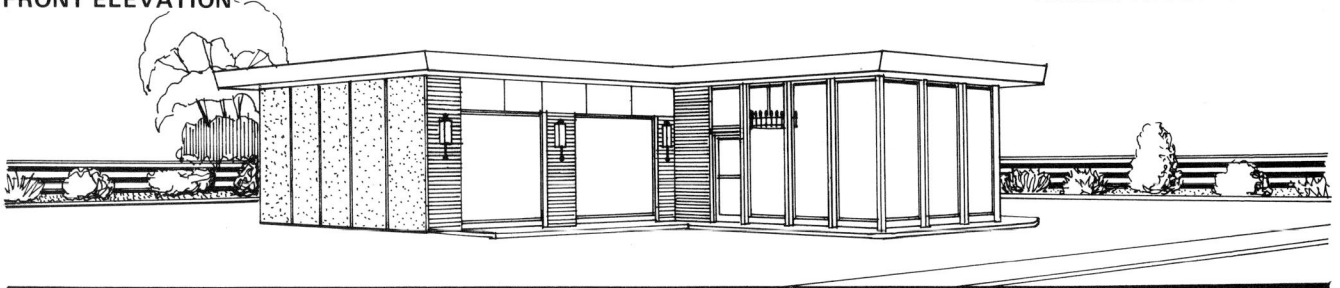

HOME DECORATORS CENTER

An enterprising business for an upper middle income area is the drapery outlet store. This concept is suitable for an abandoned three bay service station situated on a heavily travelled thoroughfare in the suburbs of an affluent community. The conversion of a typical three bay building should be accomplished in a traditional motif as illustrated. The interior of the building must be remodeled to display a variety of draperies, curtains, laminated valances, Austrian valances, padded cornice boards, shades, drapery and hardware. Beds and furniture arrangements are required to exhibit coordinated bedspreads, pillows and, if desired, carpeting. A small bathroom display for shower curtains, bathroom towels, guest towels, and accessories should also be included. A decorator service may be furnished by the store management.

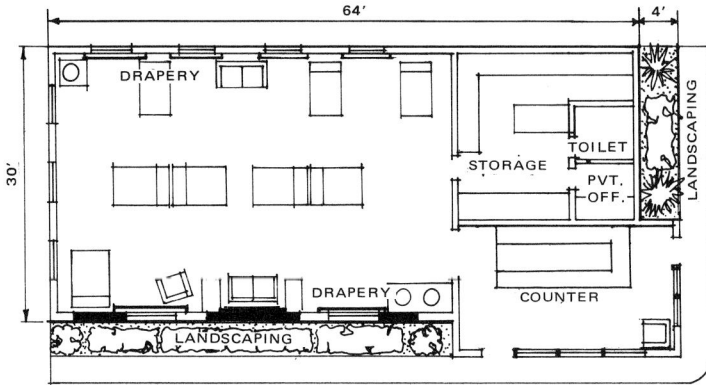

FLOOR PLAN

SCALE   0'   5'   10'   15'

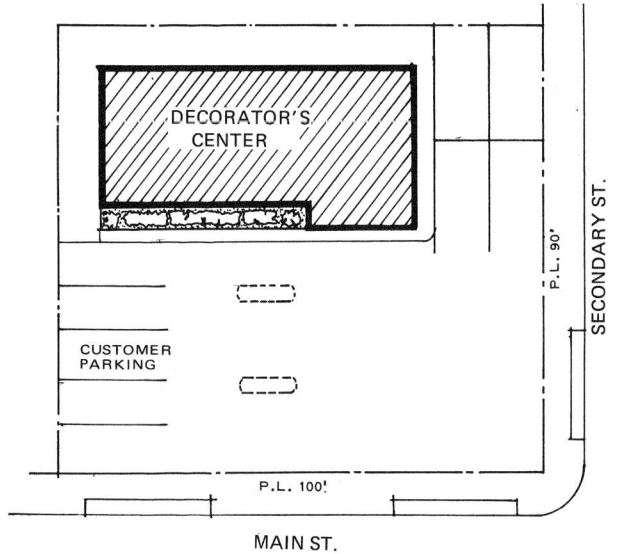

PLOT PLAN

SCALE 1" = 40'

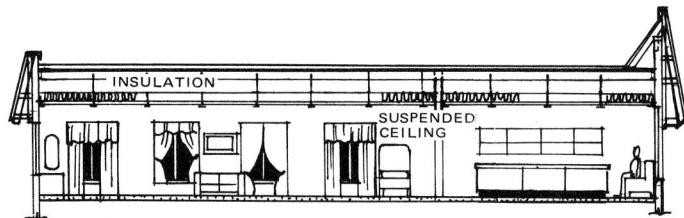

LONGITUDINAL SECTION

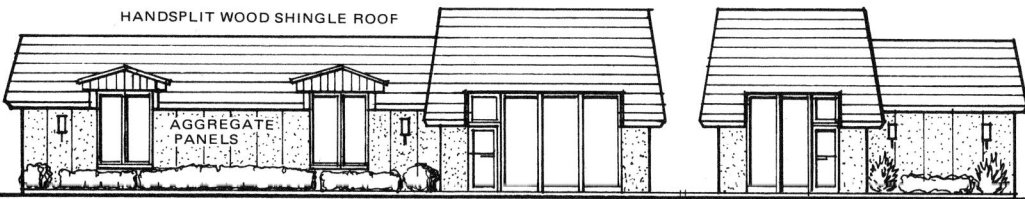

FRONT ELEVATION

RIGHT SIDE ELEVATION

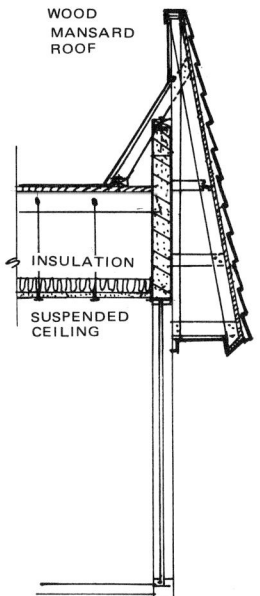

SECTION

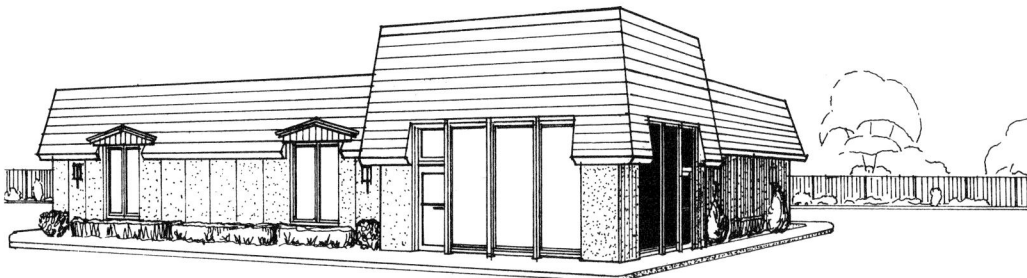

DONUT STORE

Interviews with numerous owners and/or franchisees revealed that a well located donut store is generally family operated and requires long working hours on the part of the principals. The successful store can yield a net income before taxes ranging from $35,000 to a high of $55,000 annually.

A unique profit center, for motorists in a hurry, is a drive-up kiosk providing coffee and hot fresh donuts.

It is estimated that a $40,000 to $60,000 savings in construction costs may be realized from a conversion of an abandoned service station as compared with a grass roots development. The construction time for a conventional store ranges from 120 to 180 days; however, a conversion can be completed in 45 to 60 days.

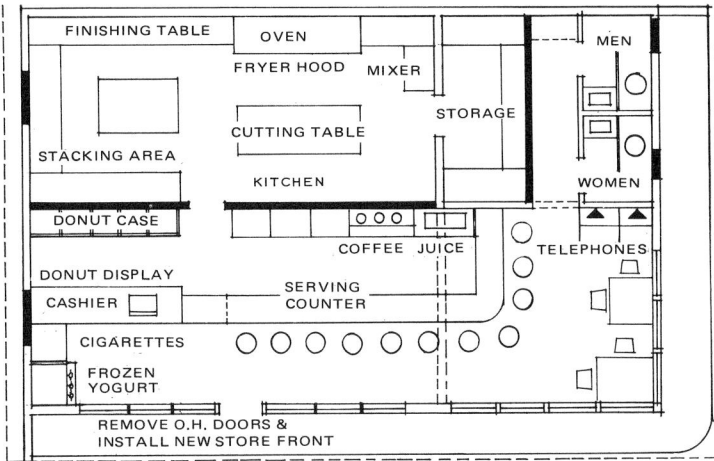

FLOOR PLAN

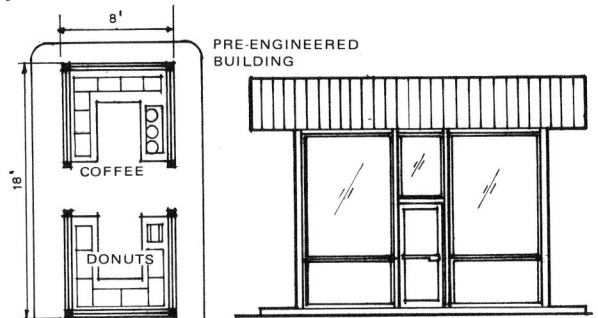

QUICK-PICK-UP   **FRONT ELEVATION**

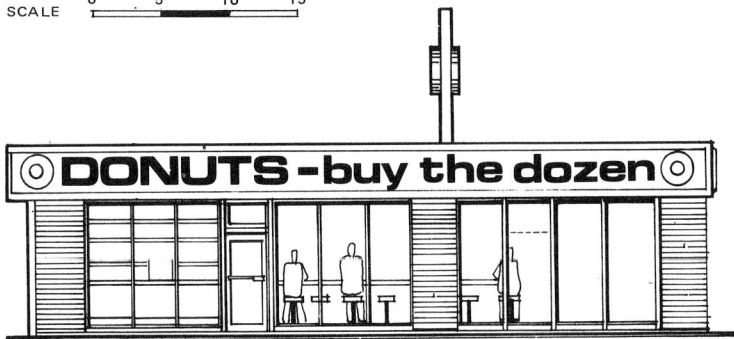

FRONT ELEVATION

RIGHT SIDE ELEVATION

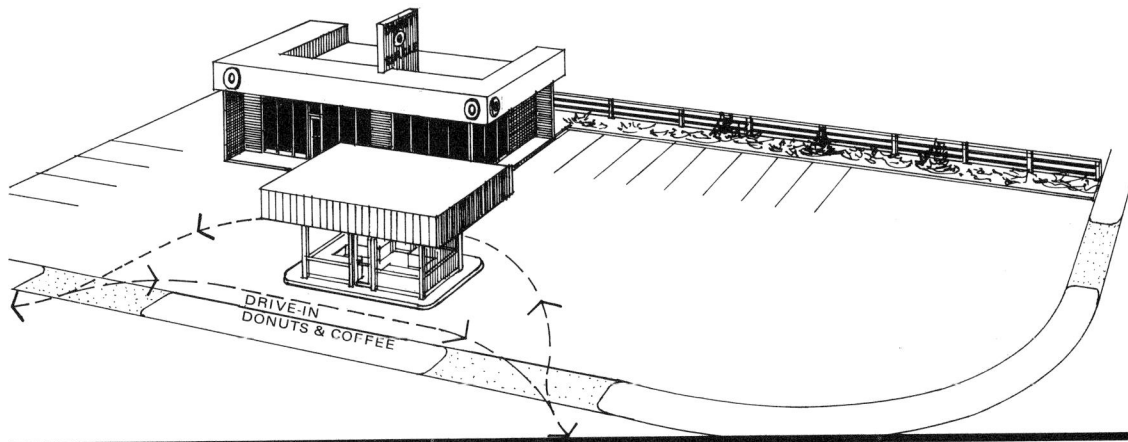

DONUT STORE

THE PYLON IS USED FOR LOGO PURPOSES

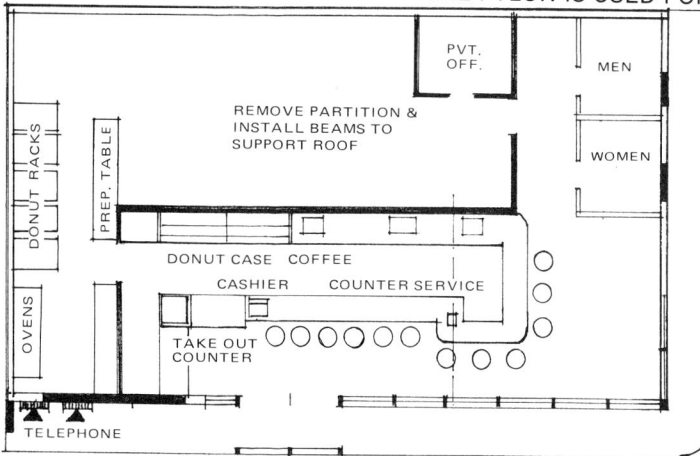

FLOOR PLAN

Inside floor plan labels:
DONUT RACKS
PREP. TABLE
OVENS
TELEPHONE
PVT. OFF.
MEN
WOMEN
REMOVE PARTITION & INSTALL BEAMS TO SUPPORT ROOF
DONUT CASE   COFFEE
CASHIER   COUNTER SERVICE
TAKE OUT COUNTER

THE ABANDONED SERVICE STATION

THE EARLY REMODELING STAGES OF A TWO BAY SERVICE STATION

Courtesy Donut Queen, Westbury, N.Y.

DONUT STORE

In 1979 the owner of an abandoned service station, situated on an excellent 150′ x 100′ plot, negotiated a lease to convert the building into a donut store. A survey revealed an excellent traffic count on the major artery that would support the new business use. The tenant, a knowledgeable operator, realized that she would save from 30% to 35% by remodeling the service station rather than demolishing the building and erecting a grass roots facility. The tenant wisely retained an architect to prepare plans for the extensive alteration work. The photos taken during the remodeling illustrate the nature of the conversion.

The proprietors of the donut shop are conscientious and experienced in the management of the business. The husband tends to the baking while his wife takes charge of the store sales. The thriving business reflects the quality of the product and the excellent service. The adjacent shopping center and high traffic count on both streets assure the success of this new business conversion.

THE COUNTER AREA

THE DONUT DISPLAY CASE AT THE CASHIER AREA

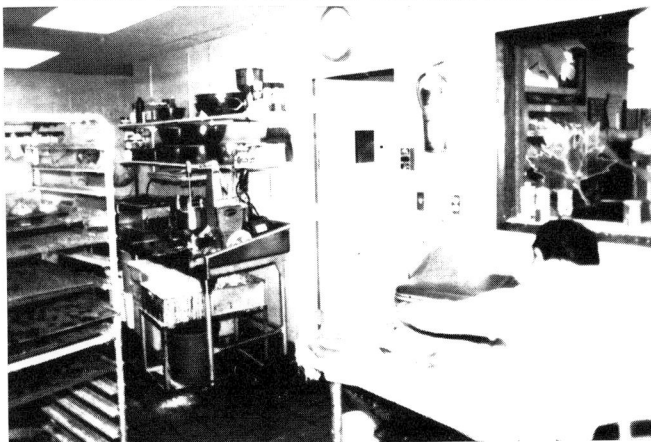

THE DONUT RACK AND KITCHEN

Courtesy Donut Queen, Westbury, N. Y.

DONUT STORE

THE NEWLY COMPLETED DONUT STORE

THE LATEST ARCHITECTURAL DESIGN OF A MAJOR FRANCHISE

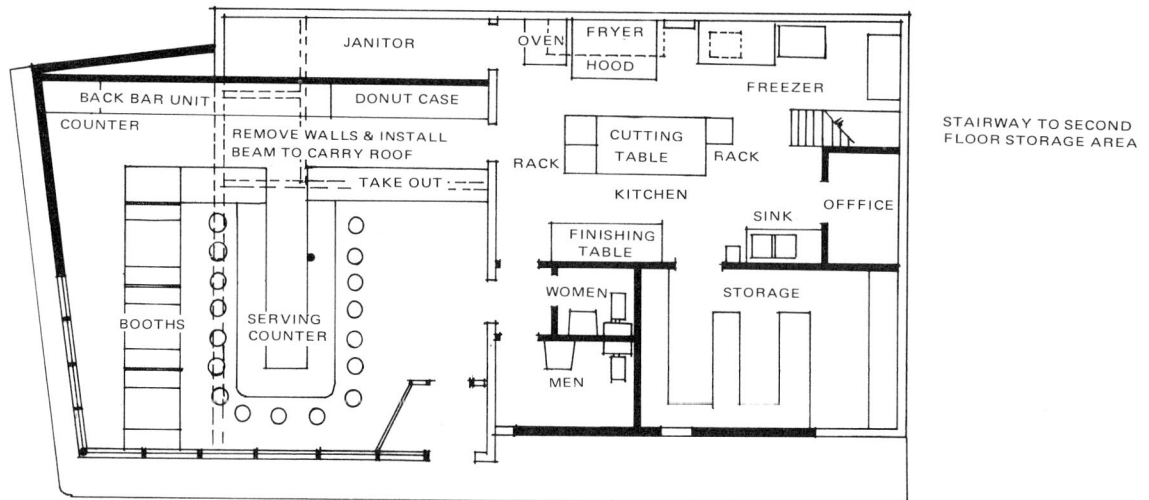

FLOOR PLAN

Several years ago, in an historic New England community, a two bay service station was converted to a formal wear shop. The business subsequently moved to another location, and the building was eventually leased at a rental of $900 per month for conversion to a donut shop. In 1980 an extensive interior and exterior remodeling was accomplished for approximately $104,000. The cost to furnish and install the equipment package was an additional $40,000. The conversion was completed in the new image of Dunkin' Donuts. The franchisor advised that his firm has converted approximately fifty service stations to donut shops.

As the customer enters the store he faces a full line display of a variety of handmade bakery products. Adjacent to the takeout area is the counter and booth seating area for in-shop patrons.

The architect for the franchisee stated that savings were estimated at $20,000 when compared with the cost of constructing a new building. Most of the savings were in the site work.

Courtesy Dunkin' Donuts, Randolph, Mass.
Photographs Courtesy of Gerard J. Bryce.

DONUT STORE

THE ATTRACTIVE NEW INTERIOR OF THE FORMER SALES WING OF THE SERVICE STATION

SEATING FOR 29 CUSTOMERS

THE CLEAN COUNTER AREA

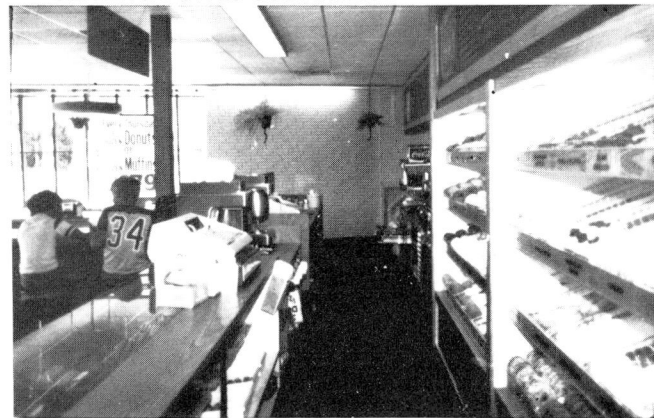

THE DONUT DISPLAY CASE

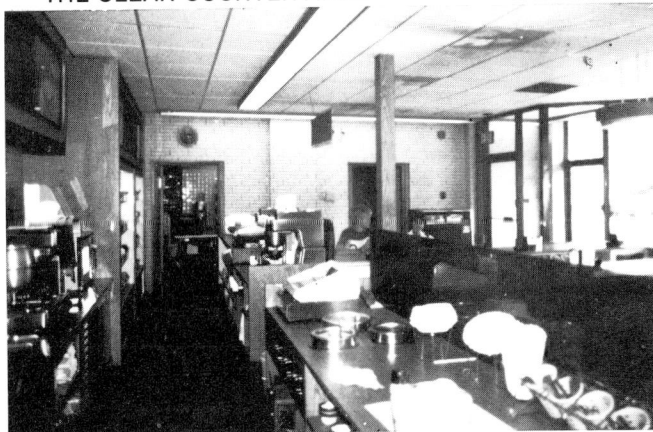

CERAMIC TILE COVERS THE OLD LUBRITORY WALLS

A SPACIOUS WELL ORGANIZED KITCHEN IS ESSENTIAL FOR SUCCESS

A VIEW OF THE OVEN, FRYER, PROOFER RACKS, FREEZER AND CUTTING TABLE

Courtesy Dunkin' Donuts, Randolph, Mass.
Photographs Courtesy of Gerard J. Bryce.

ENERGY CONSERVATION CENTER

An excellent new business concept for an abandoned service station is a store that sells insulation, storm windows and doors, weatherstripping, fireplaces and a variety of energy conservation hardware and materials for the "do-it-yourself" home-owner. Because the average three bay service station does not have sufficient floor space for this business, an 850 square foot addition across the rear of the building will provide area for material storage.

Certain interior walls must be removed to create an open floor plan for the display of fireplace designs,

wood burning stoves, fans, accessories, storm doors and windows, insulation, filters, weatherstripping, duct tape, siding and shingles. It is important that the businessman or manager of the store be qualified to make energy audits. This capability provides a customer service that will increase product sales. The outdoor sale of firewood is recommended as an accommodation for the one-stop consumer. Interior and exterior wall resurfacing, with energy related materials, is an excellent way to display products.

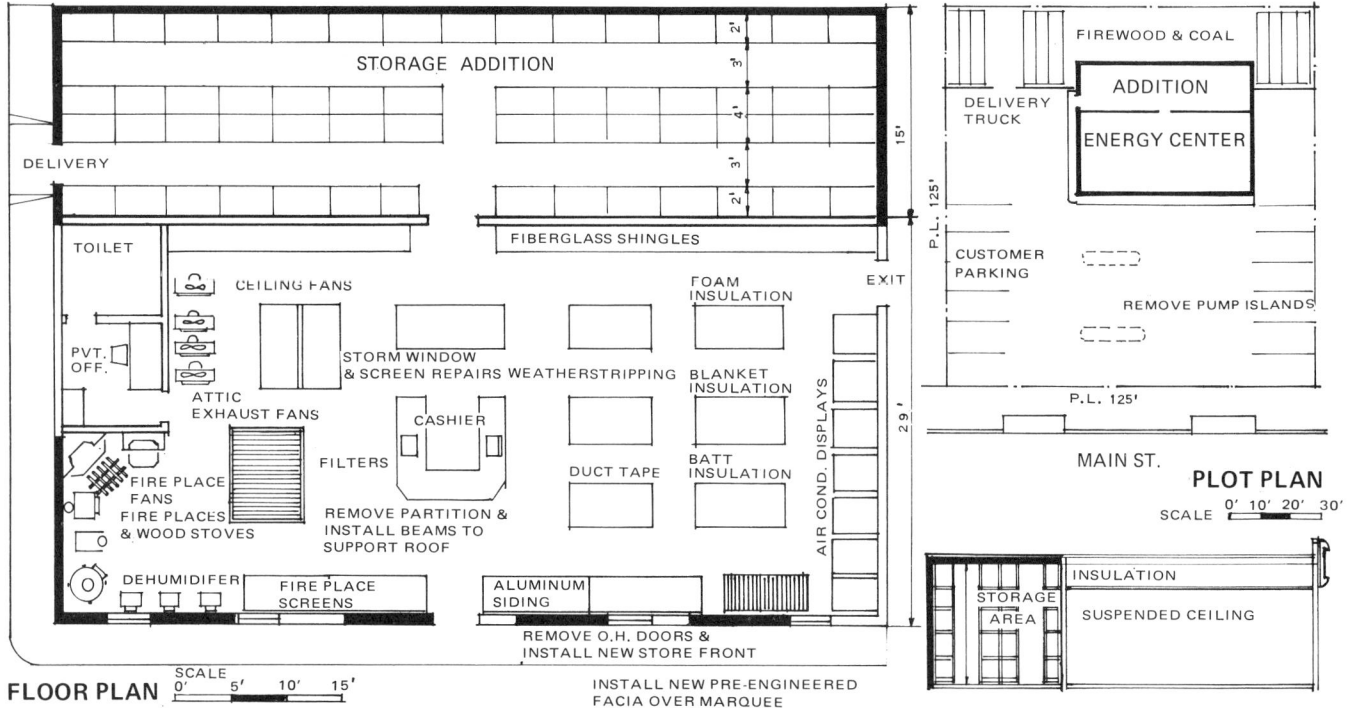

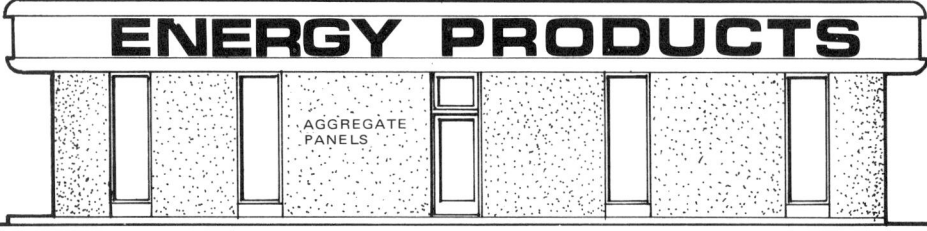

FLOOR PLAN

SCALE 0' 5' 10' 15'

PLOT PLAN

SCALE 0' 10' 20' 30'

CROSS SECTION

FRONT ELEVATION

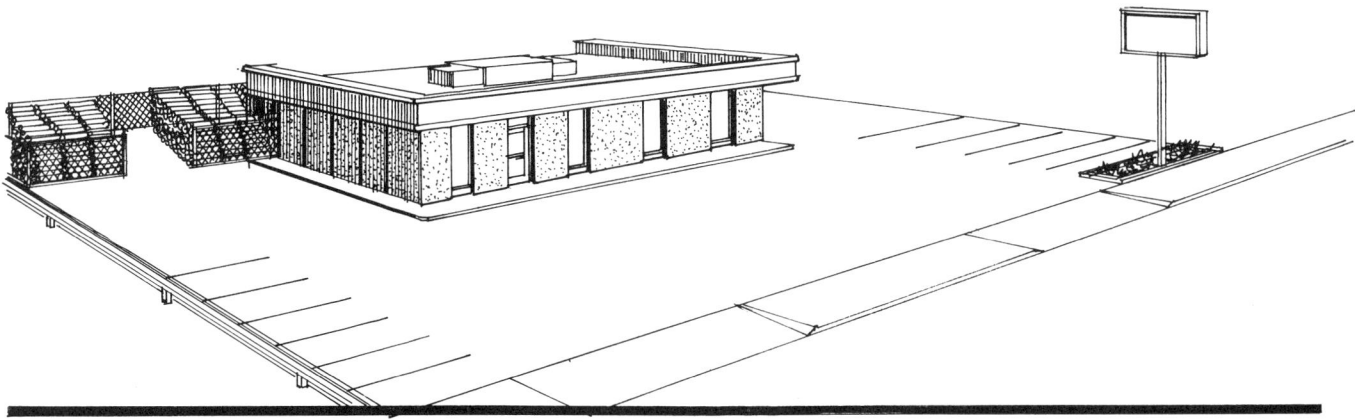

ENERGY PRODUCTS

ENTERTAINMENT CENTER

FLOOR PLAN

PVT. OFF.
TOILET
CASHIER
BUMPER POOL
BILLIARD TABLES
REMOVE WALLS
ELECTRONIC GAMES
SOFT DRINKS
BILLIARD TABLES
ELECTRONIC GAMES
ELECTRONIC GAMES
ENCLOSE CANOPY AREA WITH NEW MASONRY WALLS

SCALE 0' 5 10' 15'

The days of the old pinball arcade are vanishing, and are being replaced by electronic games that are popular and yield high returns. The abandoned service station, with a canopy, can be converted into an entertainment center, containing 2700 square feet of floor area. The center should be situated in a heavily populated neighborhood. In addition to the electronic games, hockey, bumper pool and billiards should be provided.

The Wall Street Analyst has predicted that the electronic games market will not feel the recession pinch at all, and ·has soared in 1980 to $400 million, up from $150 million in 1978. According to one of the largest manufacturers of electronic games, there are already two million coin operated amusement games in the marketplace. These games are placed in almost every type of location imaginable. The average game located in a store can earn as much as $2600 in one year.

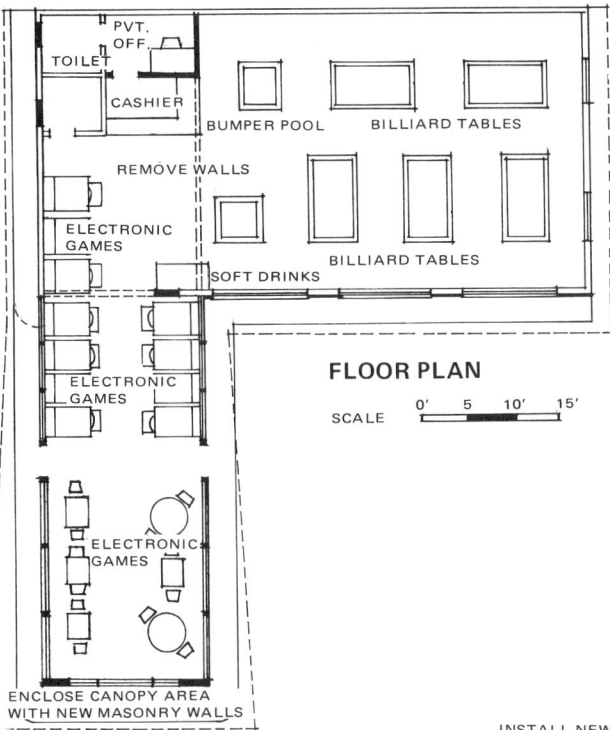

INSTALL NEW PRE-ENGINEERED FACIA OVER CANOPY & MARQUEE

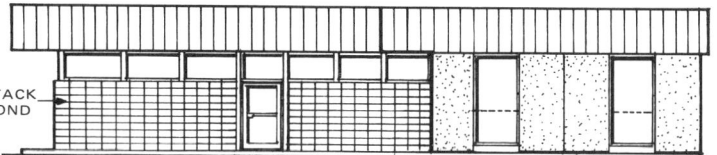

Game Room

STACK BOND

FRONT ELEVATION

RIGHT SIDE ELEVATION

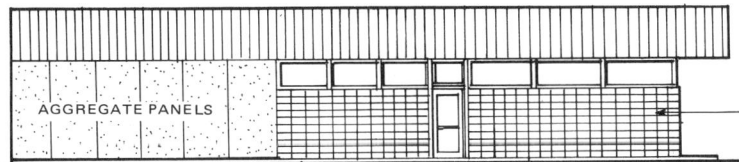

AGGREGATE PANELS

STACK-BOND MASONRY WALL

LEFT SIDE ELEVATION

BILLIARD TABLE

CROSS SECTION

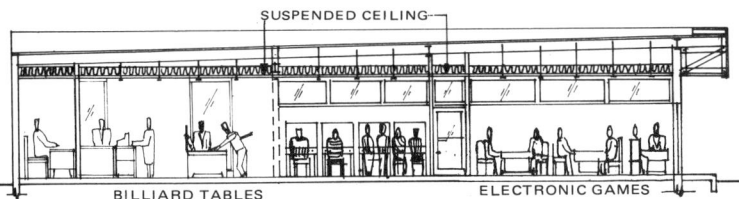

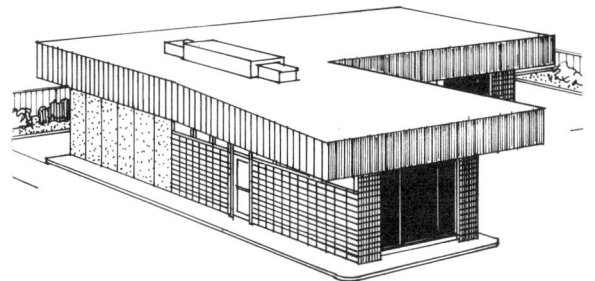

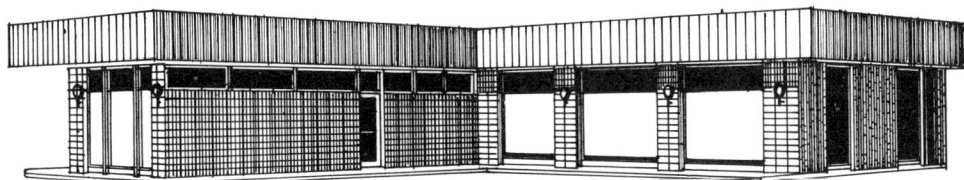

SUSPENDED CEILING

BILLIARD TABLES   ELECTRONIC GAMES

CROSS SECTION

EQUIPMENT RENTAL

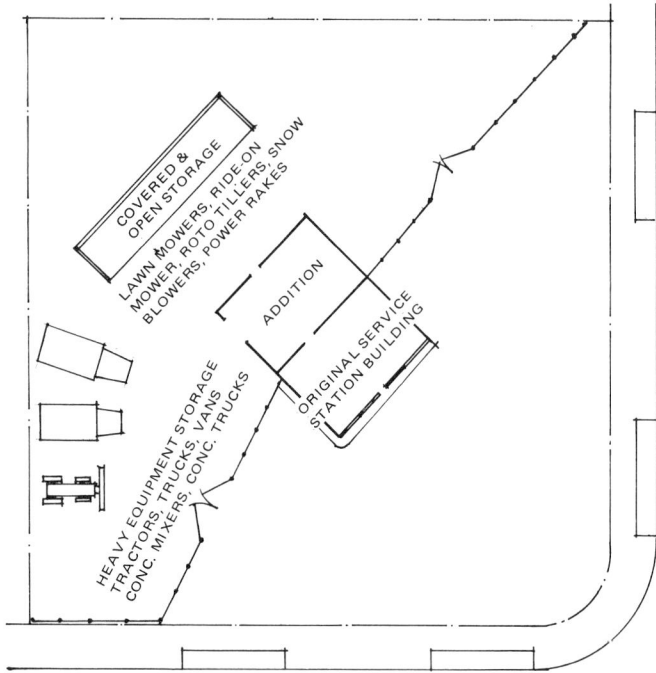

PLOT PLAN MAIN ST.

Equipment rental franchises offer excellent business opportunities for the entrepreneur. This is a homeowner type of business that will thrive in a growing community. A tremendous variety of equipment is leased to the do-it-yourself homeowner. The drawings and photographs illustrate the conversion of an abandoned two bay stucco service station, situated on a 42,000 square foot plot. The major highway carries extremely heavy traffic from the business section of a large city and major shopping centers to the rapidly growing suburbs.

In 1976 the owner added a 1200 square foot storage wing, doubling the area of the original building. The cost of this addition, resurfacing the yard area with 2″ of asphalt paving, installing a new sewage disposal system, and replacing the roof of the existing building was $25,000. The size of the building is still inadequate for the business and enlargement will ultimately be necessary.

THE STORAGE YARD FOR HEAVY EQUIPMENT AND TRACTORS

THE RENTAL CENTER AND STORAGE ADDITION

THE FENCED IN OUTSIDE STORAGE AREA AND RENTAL CENTER

YARD AREA FOR PARKING AND RENTAL TRUCK DISPLAY

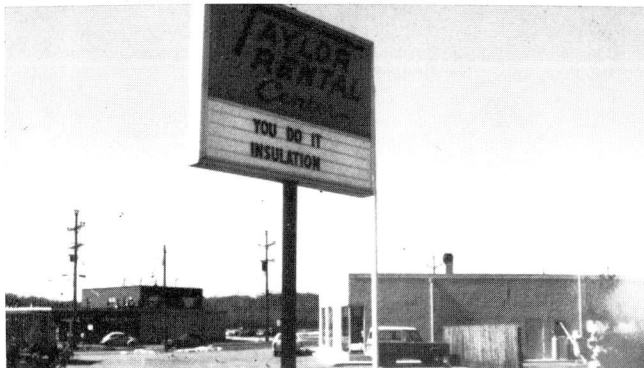
THE RENTAL CENTER SIGN AND READER BOARD

VIEW OF THE BUILDING FROM THE MAIN HIGHWAY

Courtesy Taylor Rental Center, Mount Laurel, N. J.

EQUIPMENT RENTAL

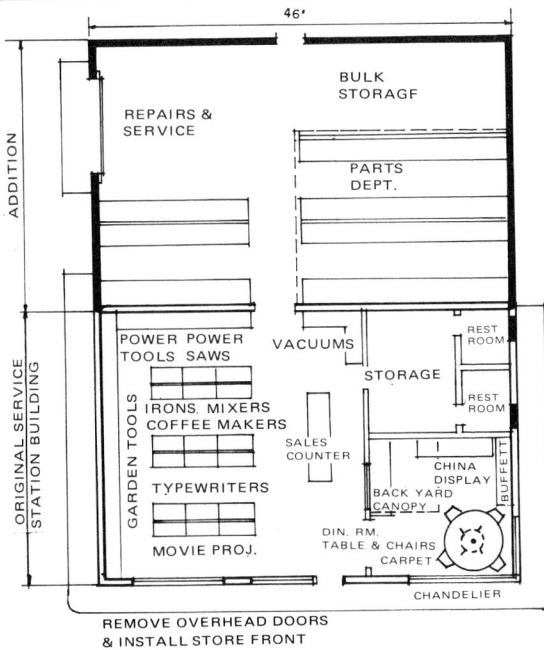

FLOOR PLAN

A limited amount of work was performed to convert the interior of the former service station building to a display area for equipment. An acoustical suspended ceiling was installed and masonry walls were painted. Display shelving and gondolas permit segregation of rental items, such as luggage, movie projectors, typewriters, a variety of appliances, power tools and saws, vacuums and garden tools. The old sales room was remodeled into a dining room, complete with table, chairs, buffet, china closet, carpet and chandelier, together with utensils and a set of china. For outdoor dining, a canopy is on display. The addition to the rear of the building is used for bulk storage, parts and repairs.

The rear portion of the property has been fenced in to include an area for truck rentals; small sheds provide cover for equipment rental, ranging from wheel barrows to bulldozers. There is no limit to equipment and material rentals.

In the interest of economy, the owner and his associates performed the majority of the construction of the new addition and sublet the remainder of the work to various mechanical and other trades.

EQUIPMENT RENTAL DISPLAY

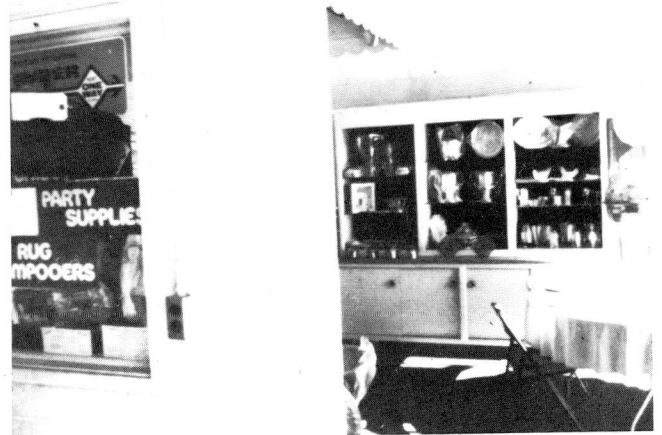

RENTALS FOR PARTIES AND DINING AREAS

CAMERA EQUIPMENT AND SALES COUNTER

GARDEN EQUIPMENT DISPLAY

Courtesy Taylor Rental Center, Mount Laurel, N. J.

EQUIPMENT SALES

A RUSTIC MOTIF FOR A FORMER SERVICE STATION

THE SECOND FLOOR USED FOR STORAGE PURPOSES

AN EXCELLENT DISPLAY AREA FOR TRACTORS AND LAWN MOWERS

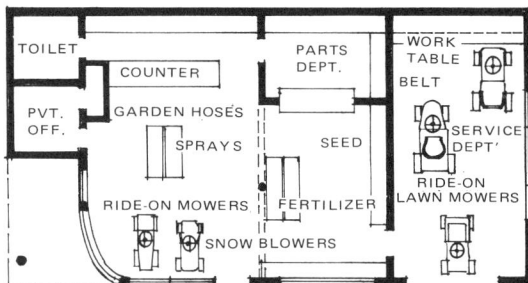

FLOOR PLAN

THE END LUBRITORY BAY SERVES AS THE SERVICE DEPARTMENT

This thirty year old service station building has been remodeled into a garden and farm equipment store. The large property is ideal for the display of snow plows, snowblowers, tractors, tillers, ride-on mowers, and wood barn buildings. The potential growth of the surrounding area supports the owner's selection of this property. The renovation of the two bay building has been performed with a unique gambrel roof, distressed wood siding at the gable ends, roof shingles and stone veneer. Only a trained eye, familiar with service station designs during the past thirty years, would detect that the curved glass store front was a former standard for the largest of the major oil companies. The old sales and lubritory areas now display garden center equipment, such as garden hoses, seed, fertilizer and water sprinklers. The end bay has been utilized for the repair of the farm and garden equipment.

Courtesy Miller Equipment Co., Robbinsville, N. J.

EQUIPMENT SALES

SALES COUNTER

SEED AND FERTILIZER

SALES DISPLAYS

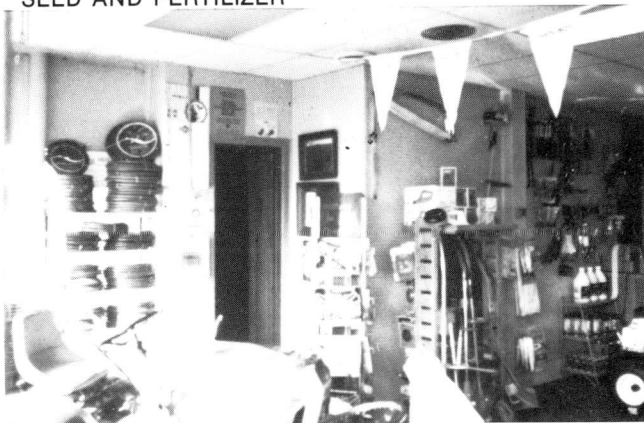

TRACTORS AND LAWN MOWER DISPLAY

THE PARTS DEPARTMENT AND DISPLAYS

Courtesy Miller Equipment Co., Robbinsville, N. J.

FLORIST STORE

FLOWER BARN

AN INEXPENSIVE REMODELING

Several years ago, a real estate speculator bought an abandoned two bay service station situated on a 150′ x 100′ parcel on a prime corner for $21,000. Thirty days later, he sold the property to the present owner for $40,000. The new owner invested $5,000 to install the mansard roof and remodel the interior with shelves and counters. The pavement in front of the store was resurfaced at a cost of $3,000. The Flower Barn has been a success because of the efforts of the owner and the location of the property. In the author's opinion, the value of the property in 1980 was in excess of $125,000.

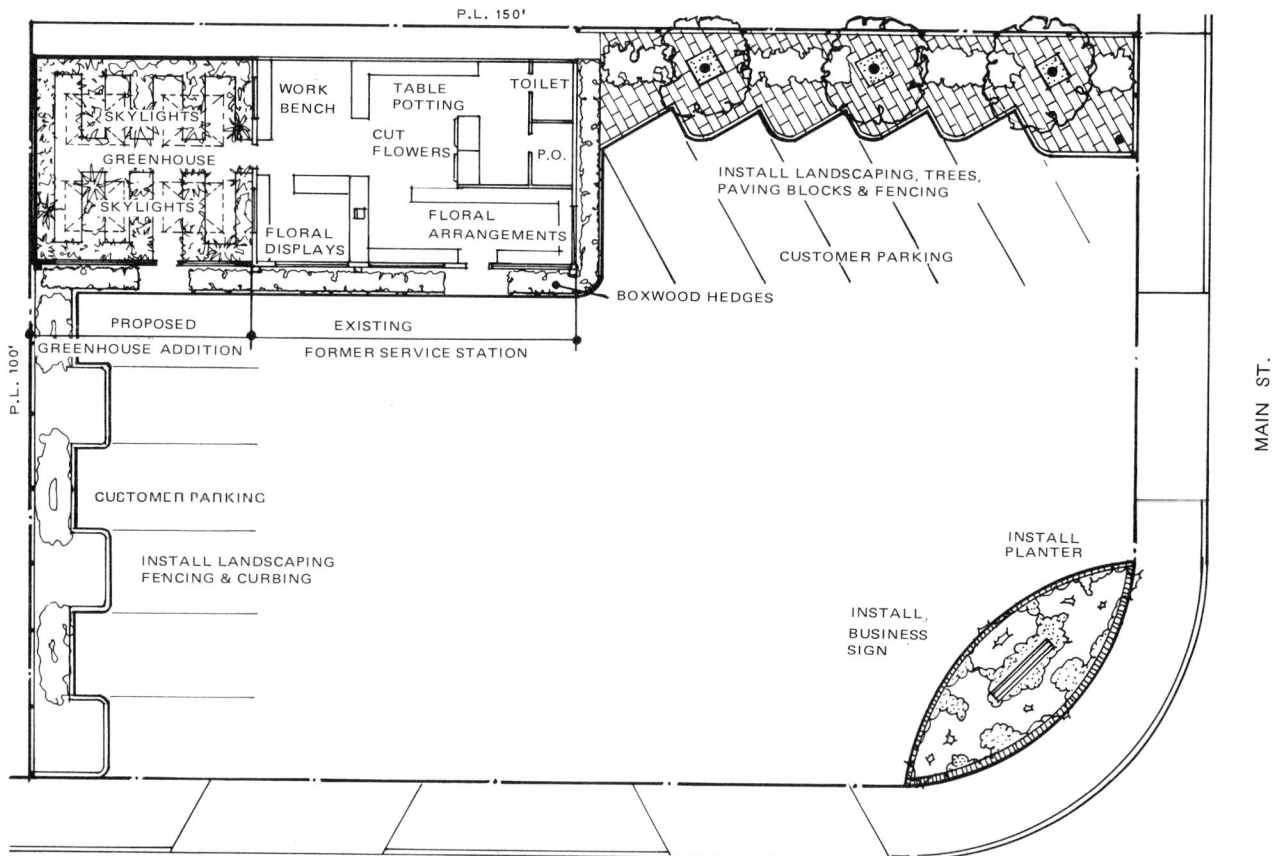

PLOT PLAN & FLOOR PLAN OF PROPOSED IMPROVEMENTS   MAJOR ST.

Courtesy Bonnie's Azud Jr., Florist, Raritan, N.J.

FLORIST STORE

The author has prepared drawings of a remodeling that illustrates the potential of the present building and property. This business may be enlarged to include a greenhouse for house plants. The architectural motif may be extended across the new addition. Ample sunlight for the plants is provided, which is a major requirement for a greenhouse. This is accomplished by installing skylights. The outdoor space should be landscaped along the side and rear property lines and at the intersecting corner. Curbing along the landscaped areas can be used to create interesting patterns for customer parking area. It is important that parking be limited along the rear and side of the premises so that the view of the store and greenhouse is not blocked. The planting of dwarf fruit trees, as illustrated on the drawings, will provide a colorful background and attract the attention of the motoring public. The planting area at the corner should be limited to low growth shrubs, such as creeping juniper.

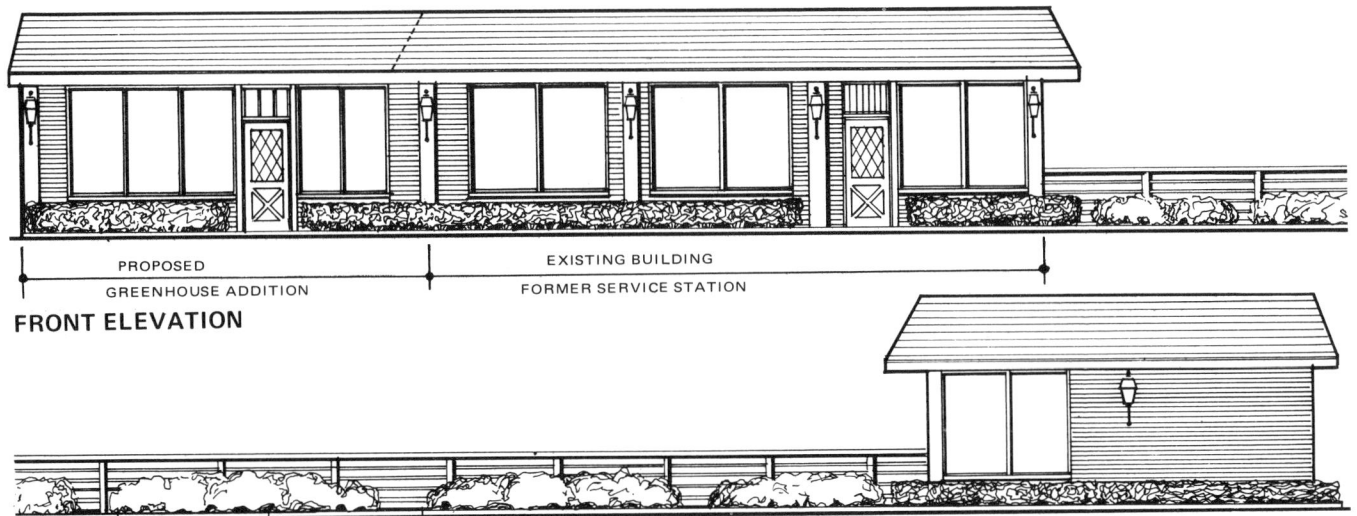

PROPOSED
GREENHOUSE ADDITION

EXISTING BUILDING
FORMER SERVICE STATION

FRONT ELEVATION

RIGHT SIDE ELEVATION

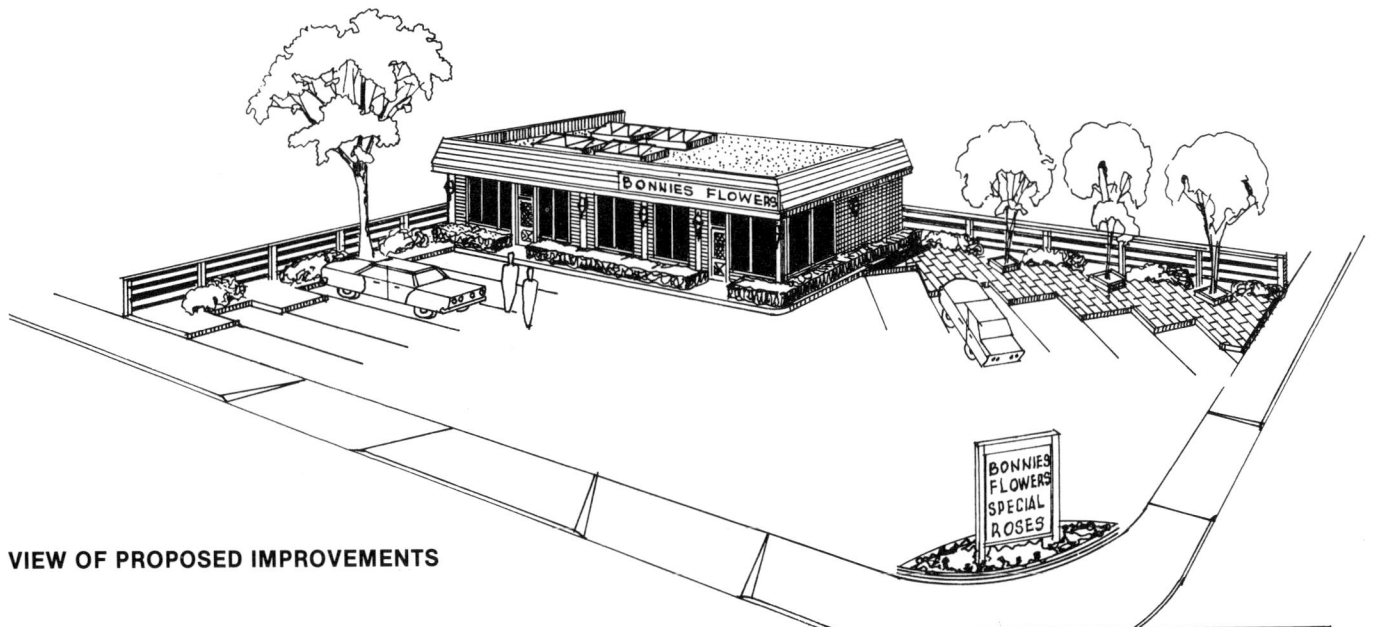

VIEW OF PROPOSED IMPROVEMENTS

Courtesy Bonnie's Azud Jr., Florist, Raritan, N.J.

FLORIST STORE

A COTTAGE STATION CONVERTED TO A FLORIST SHOP

ATTRACTIVE WINDOWS AND PLANTERS HAVE REPLACED THE OVERHEAD DOORS AND OLD STOREFRONT

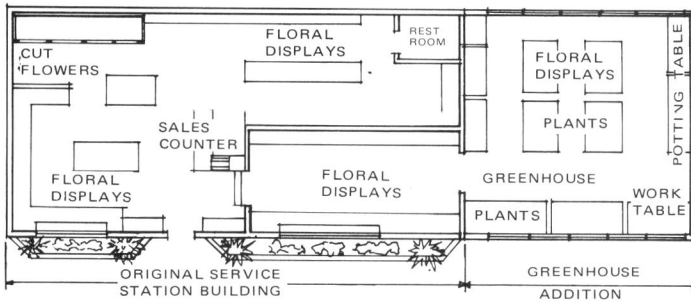

FLOOR PLAN

In 1930, the "cottage" was the most popular service station design in the petroleum industry. It can also be converted into one of the most attractive business buildings, such as the florist shop. This 90' x 125' interior parcel is situated on a main artery in a populated area with an excellent growth potential. The greenhouse addition was erected in 1976 to accommodate increased sales. This is a highly visible store where the owner has creative window displays, particularly during holiday and Easter seasons.

SALES COUNTER AND DISPLAY

GIFT DISPLAYS

FLORAL DISPLAY AND REFRIGERATOR CASE FOR CUT FLOWERS

GREENHOUSE PLANTS

Courtesy Rich Mar, Allentown, Pa.

FACTORY CLOTHING OUTLET STORE

Factory outlet stores are very popular due to their discount practices. A men's clothing store requires at least 2900 square feet of floor area to provide a full line of clothing for customers. The 40' x 30' addition can be integrated into the overall design without adversely affecting the layout. A distinctive conversion of a colonial building was developed by specifying diagonal redwood siding for the exterior, completely changing the architectural style of the former three bay service station. The removal of the gable roof over the old sales area and installation of a wood shingle roof completes the transformation and contributes strongly to the store's contemporary image.

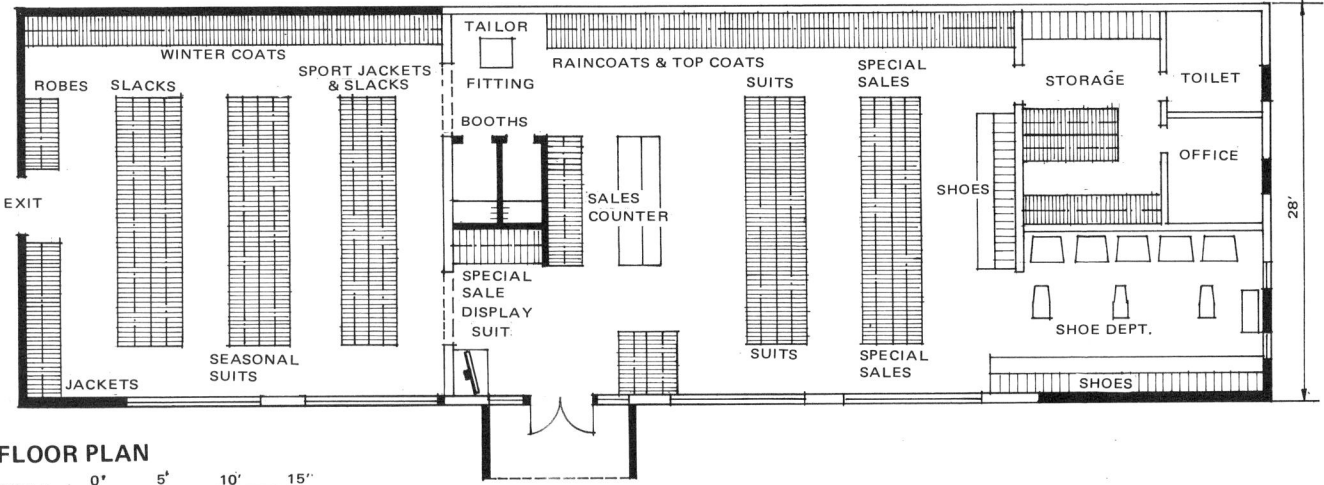

FLOOR PLAN

SCALE 0' 5' 10' 15'

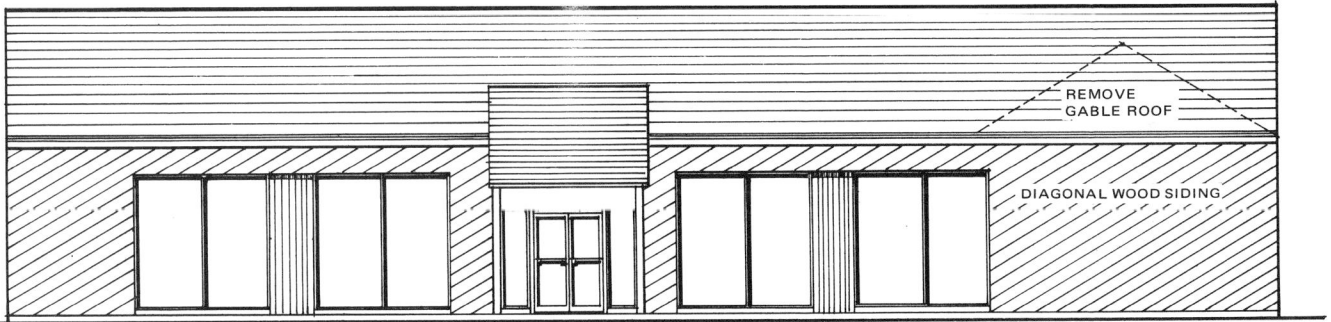

FRONT ELEVATION

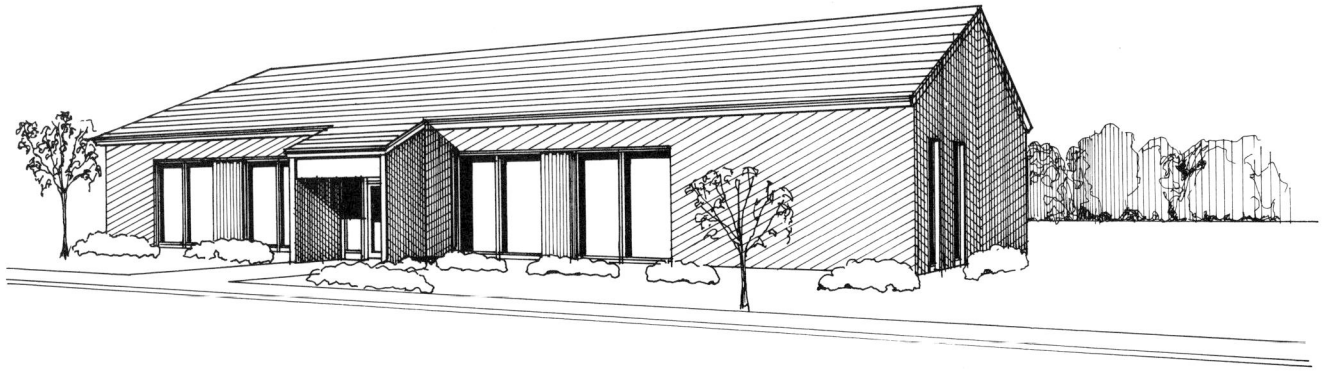

ONE STOP STORE

The energy crunch and inflation force conservation, and a one-stop store concept will become an increasingly popular and profitable business use for abandoned service stations. The two bay station, with a canopy extended over two pump islands, is ideal for this new use. This is not a true convenience store. It is more like a general store which carries not only diary products and newspapers, but a wide variety of products as indicated on the floor plan.

A 8′ x 16′ quick-serve drive-up marketing building replaces the old pump island, and the building is enlarged 480 square feet to provide areas for the fast food center and newspapers, paperbacks and magazines. The drive-up building provides quick services for the consumer on his way to work in the morning with a newspaper, coffee and donut. In the summer when the worker is driving home, he can buy the evening paper, ice cream, soda, beer, milk and bread without leaving his car.

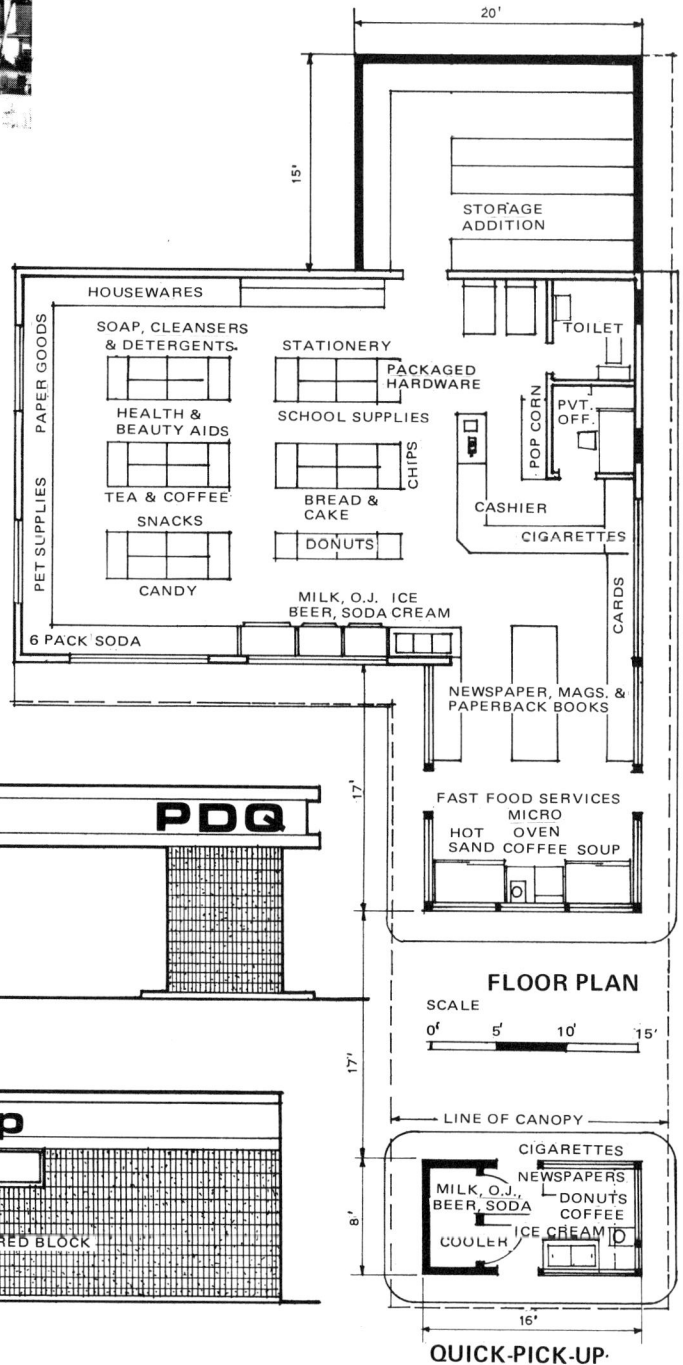

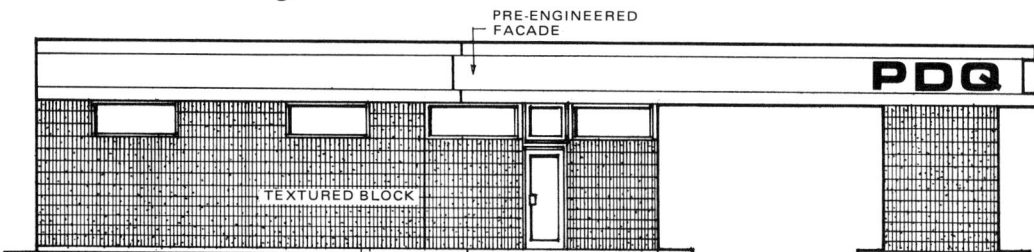

FLOOR PLAN

SCALE

LEFT SIDE ELEVATION

RIGHT SIDE ELEVATION

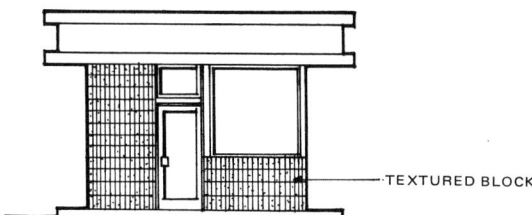

QUICK-PICK-UP

FRONT ELEVATION   QUICK-PICK-UP

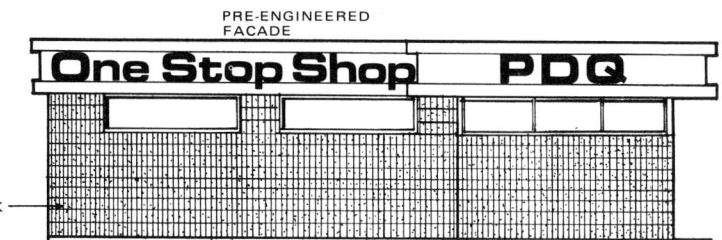

FRONT ELEVATION

ONE STOP STORE

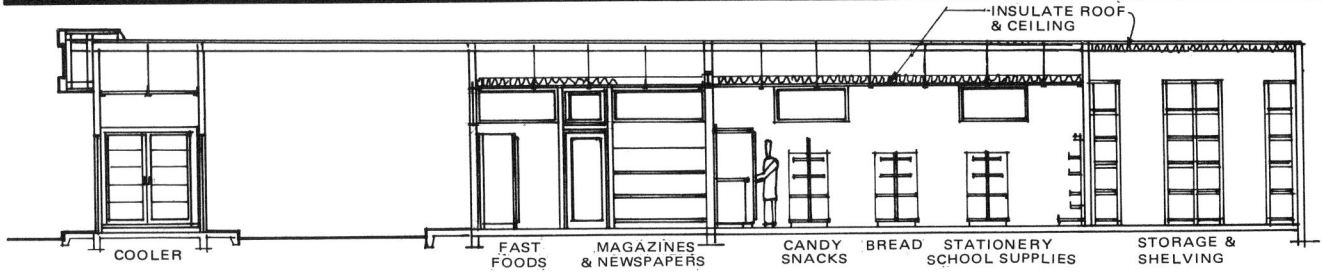

COOLER   INSULATE ROOF & CEILING

FAST FOODS   MAGAZINES & NEWSPAPERS   CANDY SNACKS   BREAD   STATIONERY SCHOOL SUPPLIES   STORAGE & SHELVING

CROSS SECTION

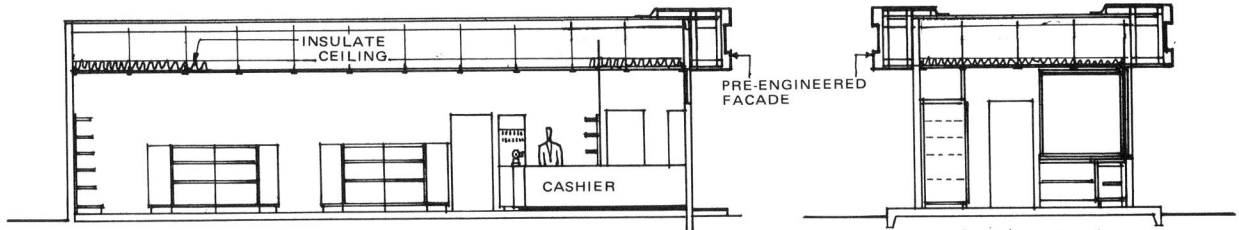

INSULATE CEILING

CASHIER

PRE-ENGINEERED FACADE

LONGITUDINAL SECTION

QUICK-PICK-UP SECTION

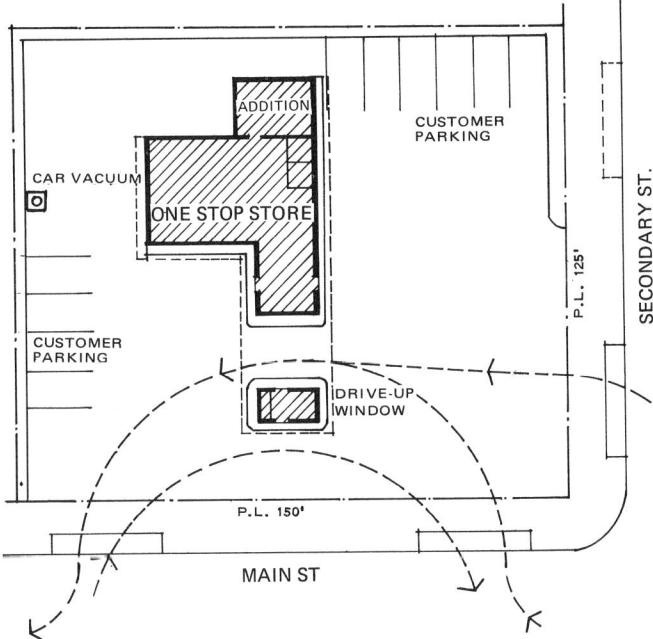

ADDITION

CUSTOMER PARKING

CAR VACUUM

ONE STOP STORE

P.L. 125'

SECONDARY ST.

CUSTOMER PARKING

DRIVE-UP WINDOW

P.L. 150'

MAIN ST

PLOT PLAN

SCALE   0'   5'   10'   15'

An exterior enclosed area can be utilized for seasonal sales, such as firewood, Christmas trees, plants, grass seed and other popular impulse items. Space for two electronic video games, popcorn, a key center and a film and photo processing center is also available.

A 600 square foot storage room must be added to provide the necessary inventory backup. A split concrete block veneer covers the front and sides, and a prefinished formed metal fascia is attached to the canopy.

A large plot that provides a parking area for at least ten cars is also important.

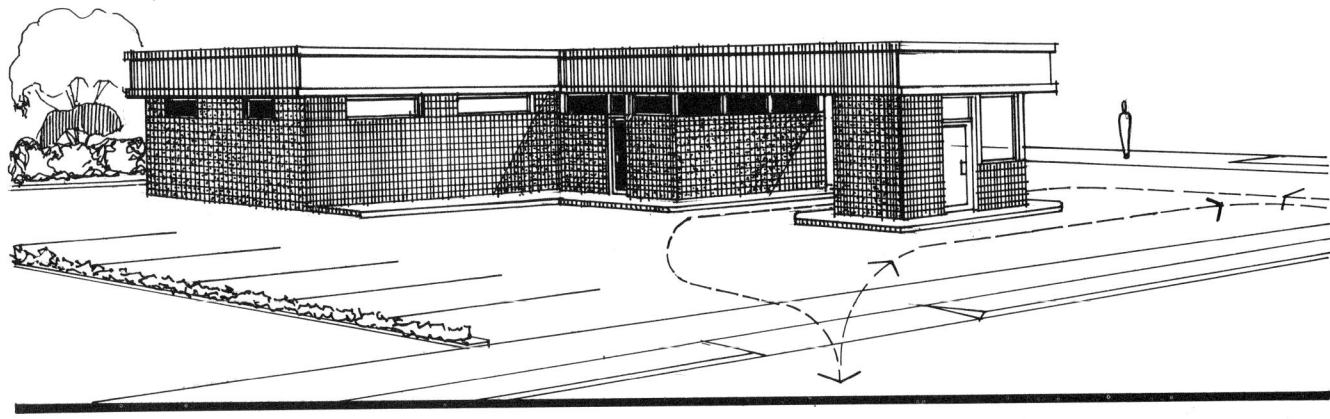

FOOD STORE AND CRAFT SHOP

Schandler's Pickle Barrel

THE CONVERSION OF AN OLD COTTAGE SERVICE STATION

THE DINING TERRACE IN FRONT OF THE STORE

AN IMPORTANT INTERSECTION IN TOWN

Courtesy Schandler's Pickle Barrel & A. & L.'s Hobbicraft, Asheville, N. C.

FOOD STORE AND CRAFT SHOP

A VARIETY OF DELICACIES ARE AVAILABLE

In 1958, the Redevelopment Commission notified a local businessman and his wife that their food store was to be purchased and then demolished by the city authority. Searching for a new location in the business district led the owners to an old cottage service station situated on a 100' x 100' plot in the business center of the city. They purchased this station for $35,000 and remodeled it for their two businesses--the food store and a craft shop--at an additional cost of $15,000.

The stucco exterior was removed and walls sand-blasted to expose a rustic brick structure that would have been difficult to duplicate. The interior was gutted and new partitions erected for the two business uses. To quote Fine Foods and Beverages International, "The charming red brick cottage-type edifice on Broadway in Asheville, North Carolina, looks like the inspiration for a post card bought in an English village, and not at all like what it is--a converted gas station." The remodeling was designed and executed by the owners. They maintain two separate business enterprises in the building. The food store occupies the old sales wing and a portion of the lubritory. The craft shop is situated in the former end bay and storage room.

The food store is a gourmet shop and delicatessen with fine wines, gift fruit baskets, candy, cheese and pate. Take-out sandwiches and beverages, as well as a continental breakfast, are also available. In good weather, patrons can eat at tables on the terrace in front of the store. The interior is attractively arranged with display gondolas, wine racks, and a deli counter. The interior brick walls have been restored and create a warm friendly atmosphere.

The Hobbicraft Shop is operated by the owner's talented wife who conducts classes in tole painting, oil painting and other crafts. It is located in two of the former repair bays of the old service station. The shop houses a complete line of arts and crafts materials, from paints of all types to needlework supplies and special tools and accessories.

Courtesy Schandler's Pickle Barrel & A. & L.'s Hobbicraft, Asheville, N. C.

FOOD STORE AND CRAFT SHOP

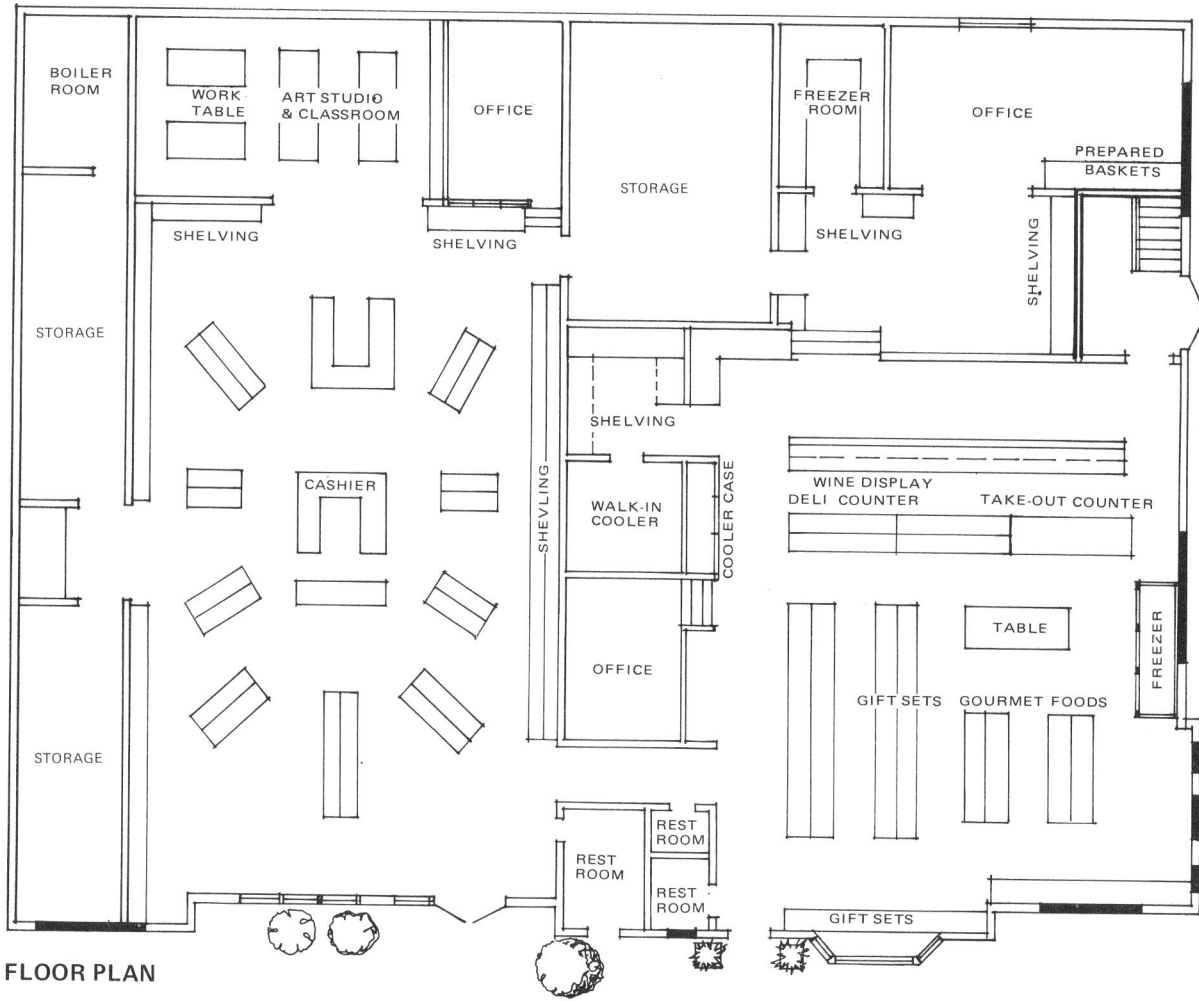

FLOOR PLAN

Labels within the floor plan:

BOILER ROOM — WORK TABLE — ART STUDIO & CLASSROOM — OFFICE — FREEZER ROOM — OFFICE — PREPARED BASKETS — SHELVING — STORAGE — SHELVING — SHELVING — SHELVING — STORAGE — SHELVING — SHELVING — CASHIER — WALK-IN COOLER — COOLER CASE — WINE DISPLAY — DELI COUNTER — TAKE-OUT COUNTER — TABLE — FREEZER — STORAGE — OFFICE — GIFT SETS — GOURMET FOODS — REST ROOM — REST ROOM — REST ROOM — GIFT SETS

THE GOURMET STORE

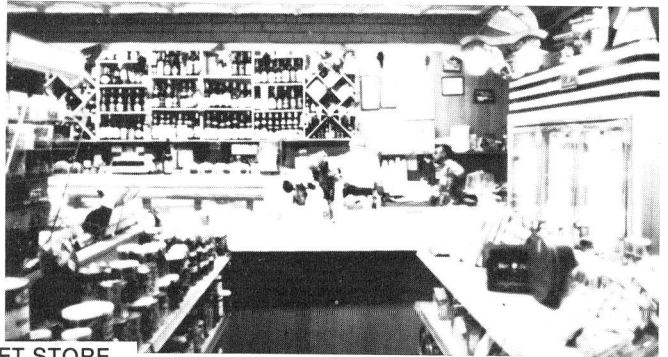

THE WINE DISPLAY AND DELI CASE

NEAT WELL STACKED SHELVES

Courtesy Schandler's Pickle Barrel & A & L's Hobbicraft, Asheville, N.C.

FOOD STORE AND CRAFT SHOP

THE ART STUDIO AND CLASS ROOM

THE INTERIOR OF THE CRAFT SHOP

AN ORGANIZED CRAFT SHOP

Courtesy Schandler's Pickle Barrel & A. & L.'s Hobbicraft, Asheville, N. C.

SEAFOOD STORES

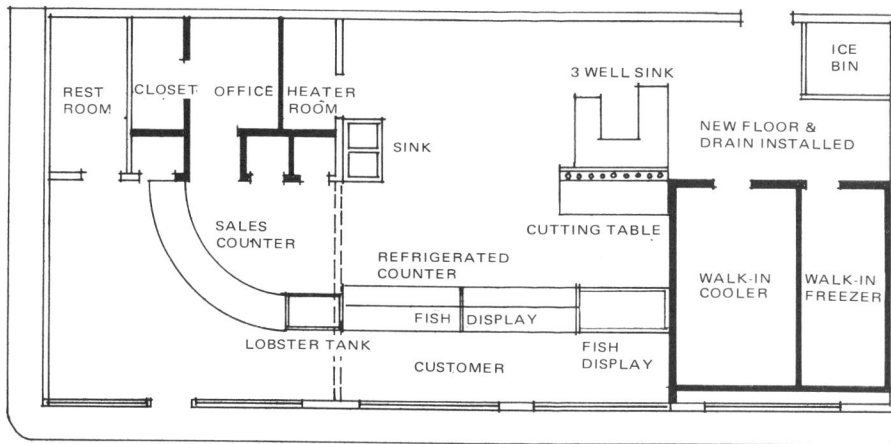

FLOOR PLAN

PARKING FOR MORE THAN 20 CARS HAS BEEN PROVIDED

AN ATTRACTIVE WINDOW DISPLAY

THE COUNTER FOR FRESH CAUGHT FISH

SEAFOOD STORES

The owners of the Seafood Markets have concentrated their efforts in increasing their number of businesses by acquiring abandoned service stations. They are very selective in buying new facilities within a geographic delivery area of their warehouse. Two of their four stores are converted service stations. The stores are situated in residential areas and/or on heavily travelled roads. The owners estimated a savings of 35% to 40% by converting an existing structure to their needs and they designed and implemented the plans for remodeling themselves. The store illustrated on these two pages is situated on a heavily travelled major street leading to both business and residential areas. The property is large and provides parking for more than twenty cars.

The property was acquired and remodeled in 1981. The interior of the building was recycled into a seafood store specializing in the sale of fresh and frozen fish. Forty-five feet of counters and freezer cases stretch across the former three lubritory bays. A cutting and cleaning table, surrounded with ceramic tile, (for ease of cleaning), is situated adjacent to the daily fresh catch display. Two large freezers and coolers, for fish storage, were installed in the end lubritory bay. A lobster tank is located adjacent to the cashier's counter. To the rear, a freezer, containing chowder and other prepared seafood delicacies, is positioned near the checkout counter. Glazed storefronts have replaced overhead doors. A large sailfish is positioned in one of the display areas to attract motorists. The remodeling and site plan work was completed in two months at a cost of $54,000.

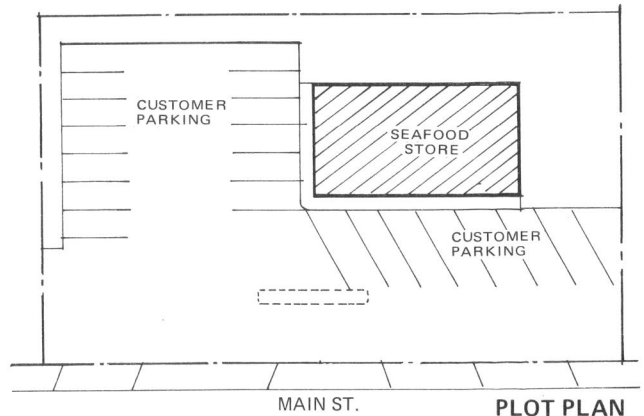

PLOT PLAN

Courtesy Bob's Seafood Markets, Haddonfield, N.J.

Courtesy Bob's Seafood Markets, Linwood, N.J.

SEAFOOD STORES

The seafood store illustrated on these two pages was formerly a two bay service station. The site is situated at the intersection of a heavily travelled major highway and a busy secondary artery. Although the service station was not successful this location is excellent for a retail business use. In 1979 the property was acquired and the building was converted into a seafood store.

A mansard roof with wood shingles was added in the standard seafood market image. Lobsters and lobster trap mockups were mounted on the roof of the mansard to accent the seafood image. The former service station sales room now contains a lobster tank, a refrigerator case for fresh seafood delicacies, and chowder and the cashier's counter. A 35 foot counter containing a large variety of fresh seafood products now occupies the two lubritory bays. Behind the counter are cutting tables and sinks for cleaning freshly caught fish. The building conversion was completed in 45 days at a cost of $35,000.

The parking area was one of the main reasons this site was selected. On Fridays and weekends the parking area is filled to capacity. The stores are open seven days a week--so important to the success of the business.

AN EXCELLENT IDENTITY SIGN

THE ABANDONED SERVICE STATION

THE OWNER'S STANDARD MANSARD CONVERSION

THE LOBSTER TANK IS PLACED IN A
CONSPICIOUS PLACE NEXT TO THE CASHIER

THE PARKING AREA IS ALWAYS FILLED ON FRIDAY

Courtesy Bob's Seafood Markets, Williamstown, N. J.
Courtesy Bob's Seafood Markets, Linwood, N. J.

SEAFOOD STORES

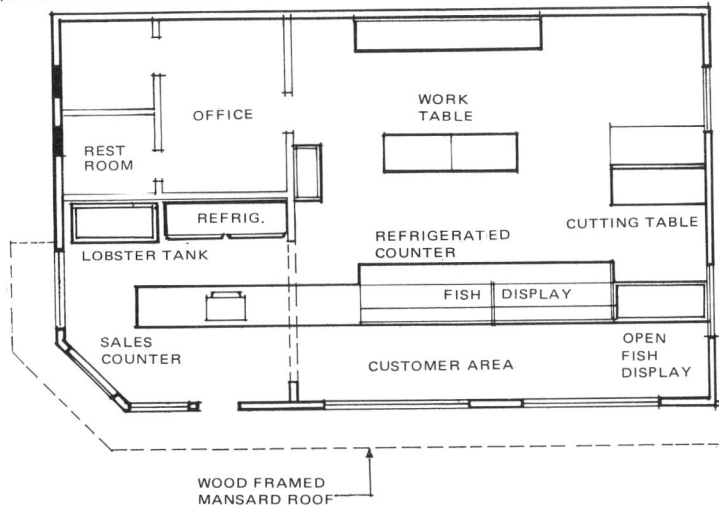

FLOOR PLAN

35 LINEAL FEET OF COUNTERS DISPLAYING SEAFOOD

SIX EMPLOYEES WAIT ON CUSTOMERS

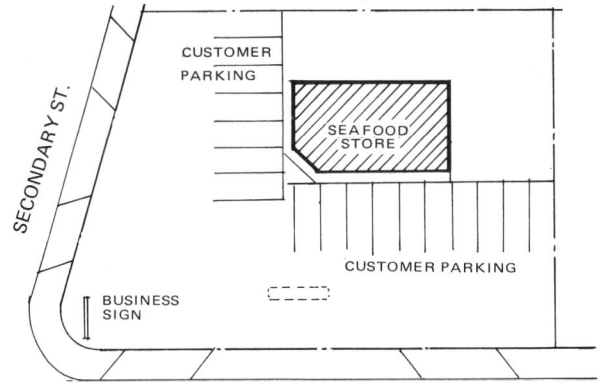

PLOT PLAN

Courtesy Bob's Seafood Markets, Williamstown, N.J.
Courtesy Bob's Seafood Markets, Linwood, N. J.

MEAT MARKET

THE ABANDONED SERVICE STATION

A NEW LIFE FOR THE BUILDING

A MATCHING RED BRICK WAS USED
TO CLOSE-UP THE OVERHEAD DOOR OPENINGS

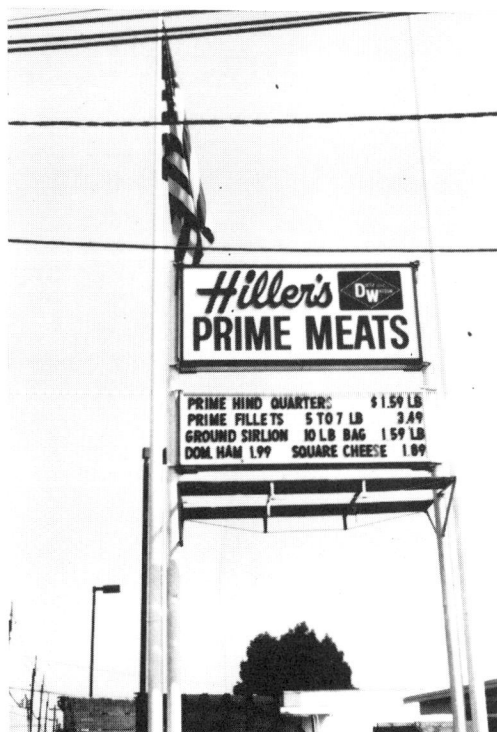

A HIGH-RISE SIGN AND READER BOARD IS VERY
IMPORTANT IN ATTRACTING CUSTOMERS

THE MEAT CUTTING AND STORAGE AREA WAS
ADDED TO THE BUILDING

Courtesy Hillers Meat Market, Cinnaminson, New Jersey

MEAT MARKET

In 1979 a businessman, with a background in managing a meat store, purchased this abandoned three bay service station. In 1981 the owner constructed an addition to the existing building and converted the service station portion into a meat store with other food items. The addition has facilities for hanging meats, a complete cutting department and a refrigeration unit. This permits the owner to purchase top grade beef from the midwest and provide his customers with prime cuts of meat at reasonable prices.

The owner retained the services of an architect to execute his conceptual design of the building remodeling. The contemporary style of the former service station building was not changed. The following work was performed:

1. A two story addition to the rear of the existing building for meat cutting, storage and refrigeration.

2. A suspended acoustical ceiling.

TWO ROWS OF MEAT CASES PROVIDE THE CONSUMER WITH A WIDE RANGE OF SELECTED MEAT

THE MEAT CASES ARE NEATLY ARRANGED

COLD DRINKS, CHIPS AND NUTS ARE DISPLAYED

THE SIDE WALL DISPLAYS ARE DEVOTED TO THE POULTRY DEPARTMENT

Courtesy Hillers Meat Market, Cinnaminson, New Jersey

MEAT MARKET

3. Ceramic tile walls and quarry tile floors.

4. Store front and electric entry and exit doors.

5. Two check-out counters with electronic cash registers.

6. Refrigerated display counters for meat, poultry, cold cuts and dairy products; freezers for ice cream and frozen packaged products; a meat counter for custom cuts, slicing and cutting machines.

7. Asphalt paving, curbing and landscaping.

8. A high rise advertising sign.

The total cost to purchase the property, construct a two story brick addition, remodel the existing building, purchase and install all equipment and complete yard paving, landscaping and sign installation exceeded $500,000.

The family operation of the store is one of the reasons for the success of the business.

Courtesy Hillers Meat Market, Cinnaminson, New Jersey

MEAT MARKET

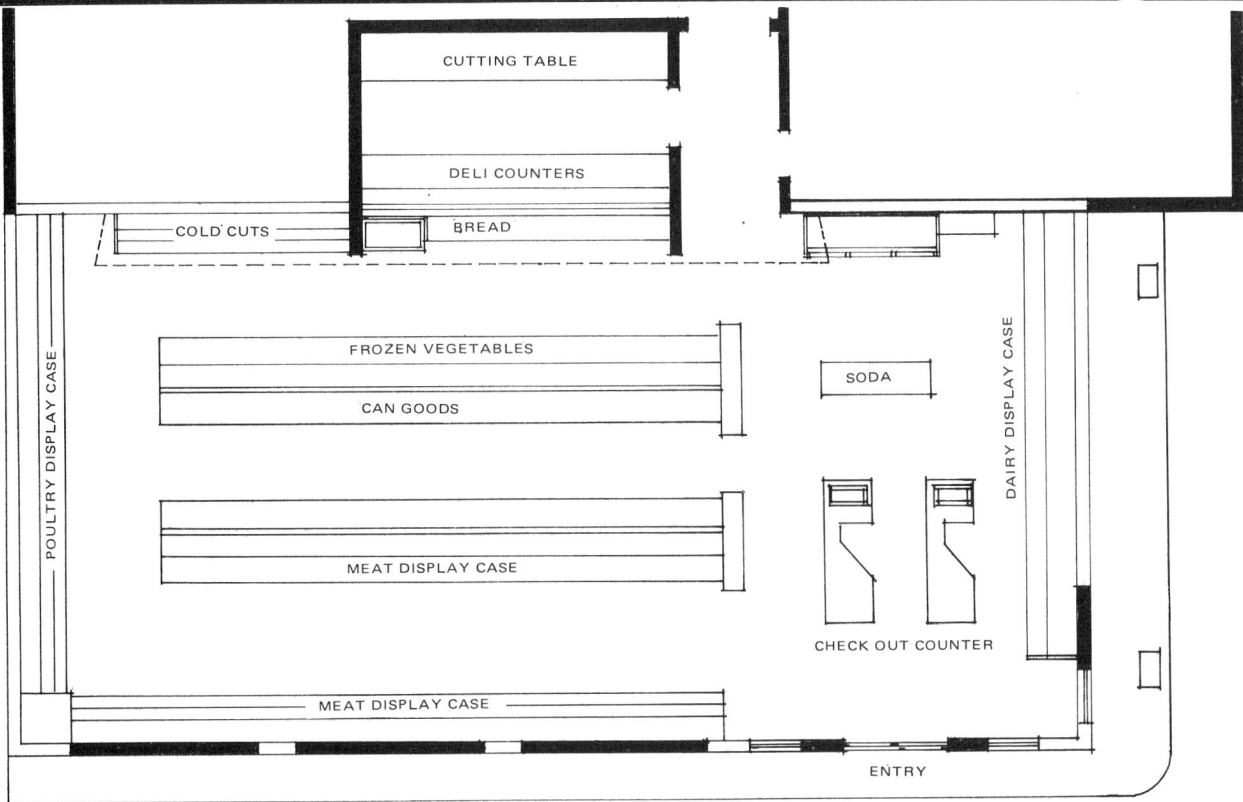

FLOOR PLAN

Courtesy Hillers Meat Market, Cinnaminson, New Jersey

FURNITURE RESTORATION STORE

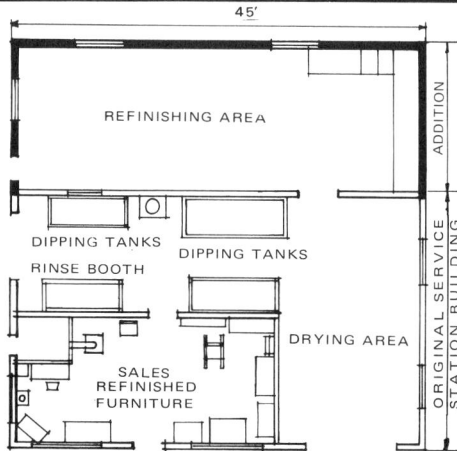

FLOOR PLAN

A PRACTICAL CONVERSION OF AN OLD SERVICE STATION

CHRISTMAS TREE AND FIREWOOD DISPLAYS

THE RINSE BOOTH

THE DIPPING TANKS

In 1977, a residential contractor was searching for a location where he could establish a furniture refinishing business. He wisely purchased an abandoned two bay station situated on a Main Street. Although the property has limited frontage and is situated on an interior parcel, it has excellent exposure to pedestrian and motorist traffic. The property is deep, with sufficient room for building expansion and customer parking.

The cost of the property and the two bay service station was $42,000. Approximately $20,000 was spent to remodel the interior and exterior and to erect a 25' x 45' addition, increasing the building size by 50%.

This is a recession-proof business that improves during periods of uncertain economy and inflation. High furniture prices result in the restoring and refinishing of old furniture. The rear of the original two bay building contains dip tanks and rinse areas, and the end bay is used for drying. The new addition is used for the refinishing of furniture and an office for the owner. The sales room has been paneled and converted into an excellent display area for the sale of refinished furniture.

There are eight to ten antique shops located within four miles of this site, which is another reason it was selected by the owner.

Courtesy The Strip Joint, Dunellen, N.J.

FURNITURE RESTORATION STORE

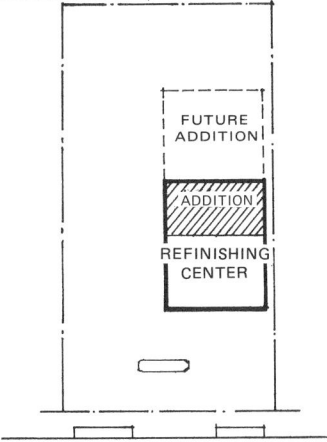

PLOT PLAN

THE OUTSIDE AREA

THE DRYING AREA

DISPLAY OF REFINISHED FURNITURE

Courtesy The Strip Joint, Dunellen, N.J.

HAIR SALON

A unisex hair center, which includes hair cutting, hair styling and beauty salon operation, is one of the growth businesses of the future. It is an excellent use for the abandoned three bay service station, located in or adjacent to a community shopping area where there is high pedestrian and motorist traffic. This layout is based on the remodeling of a 1800 square foot three bay building into a spacious hair styling salon with four shampoo stations and eleven cutting and styling chairs. The interior decor should be colorful with contrasting upholstered furniture. An attractive exterior can be achieved with the versatility of the many patterns of redwood.

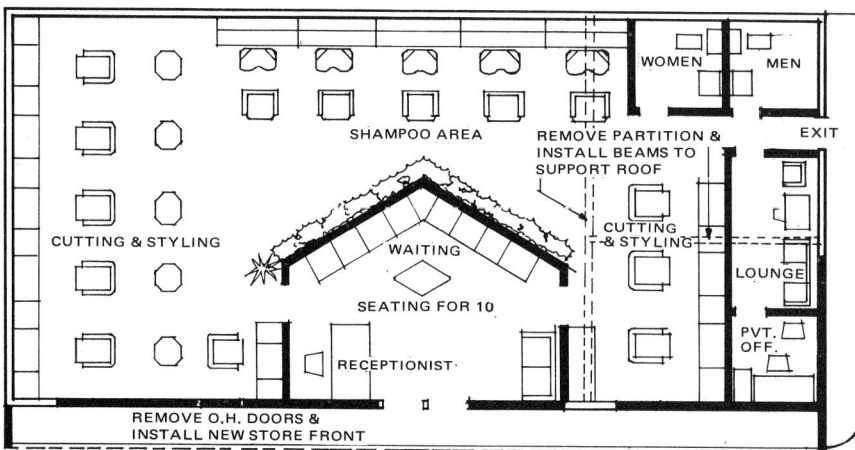

FLOOR PLAN SCALE 0' 5' 10' 15'

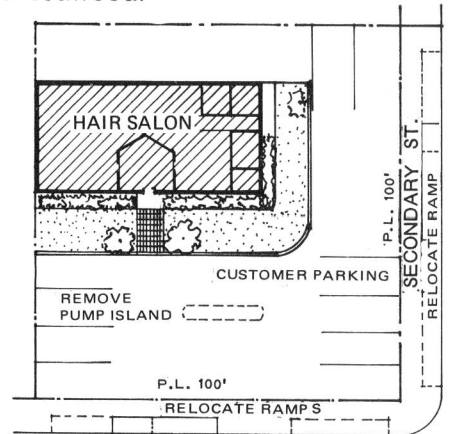

MAIN ST.

PLOT PLAN SCALE 1" = 50'

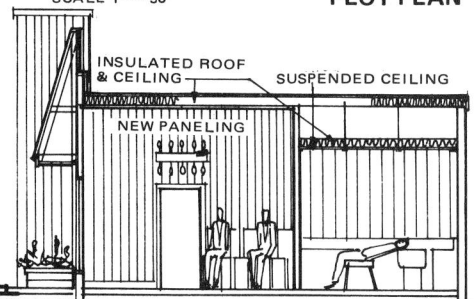

FRONT ELEVATION

CROSS SECTION

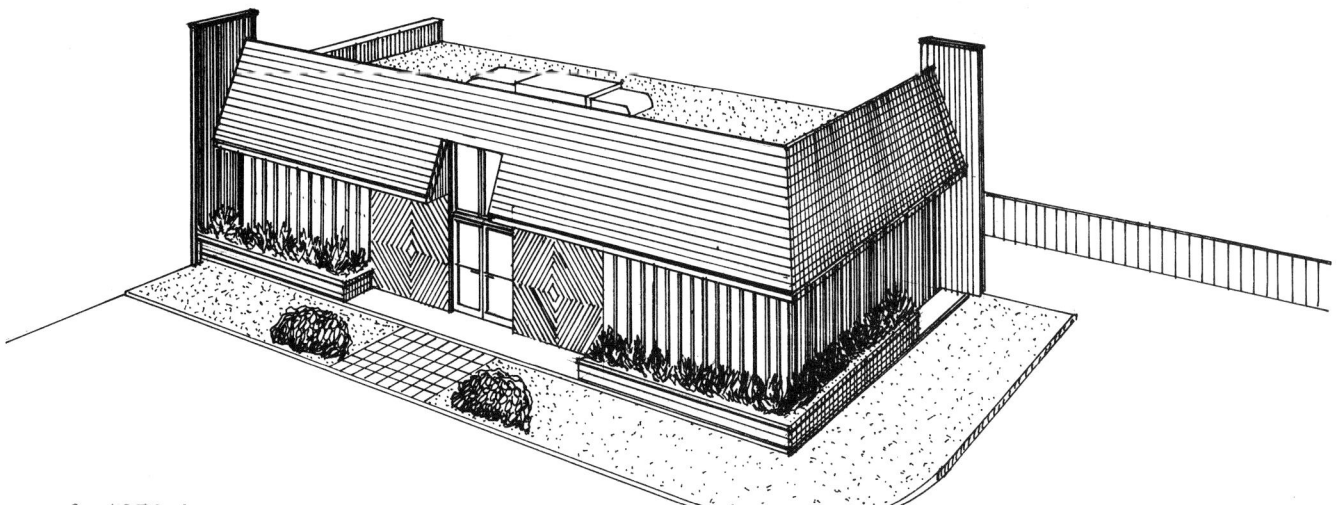

In 1979 the Department of Commerce stated there were 1264 hair salons with sales of $199 million.
It is estimated that the number of hair salons will increase to 1862 in 1981 with sales of $328 million.

HARDWARE STORE

During periods of recessions or sluggish economy, the "do-it-yourself" market prospers. Many repairs, alterations, decorating and other home improvements are performed by the home owner. This multi billion dollar market creates opportunities for the growth of many small businesses, such as a hardware center. The center must have a large storage room to allow the business to feature items such as a variety of insulation products, screens, storm windows and weatherstripping. Important profit centers are key and lock area featuring burglar and smoke alarms, an equipment rental center for power tools, waxing, sanding and spray painting machines, carpet steam cleaners, a paint department with a mechanical mixture machines, wood paneling, a large hardware department, and screen and storm window repair.

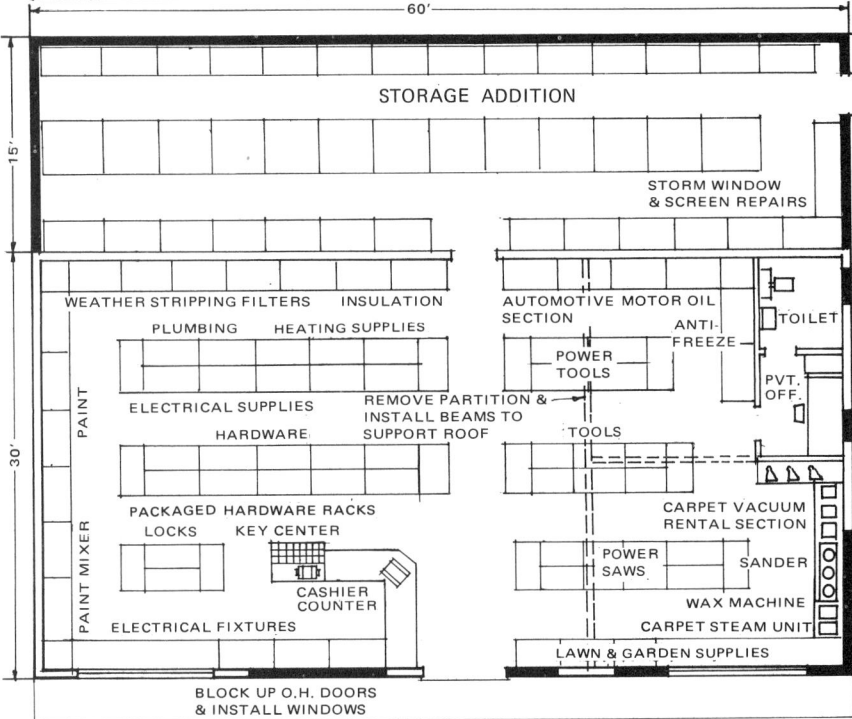

FLOOR PLAN

SCALE 0' 5' 10' 15'

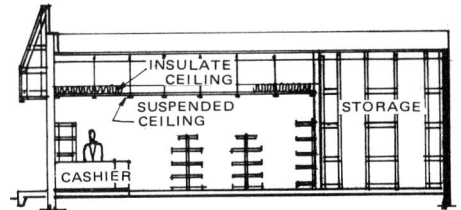

CROSS SECTION

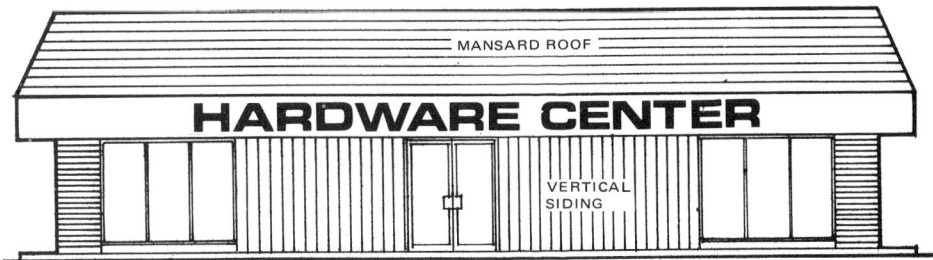

FRONT ELEVATION

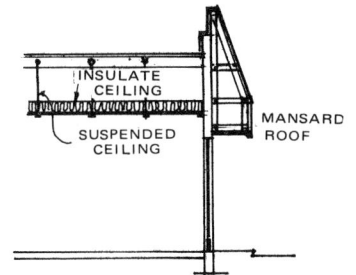

SECTION

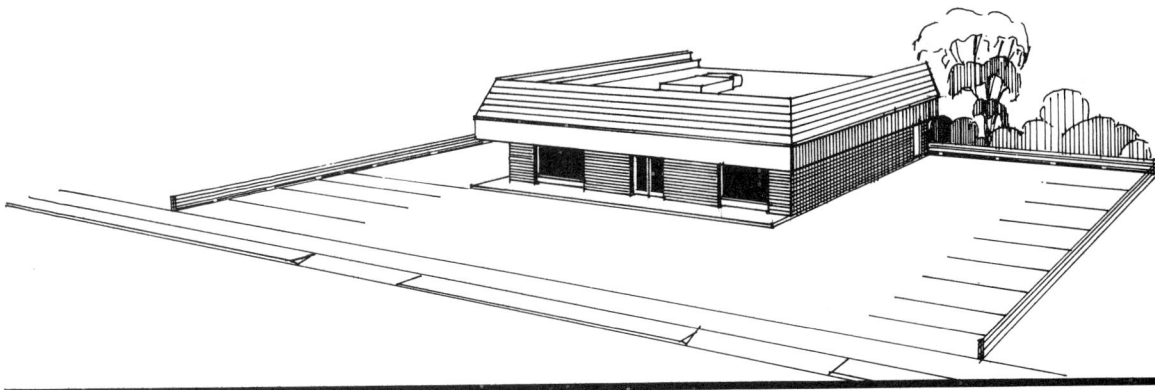

HEALTH FOOD AND ATHLETIC SHOE STORE

THE ABANDONED SERVICE STATION

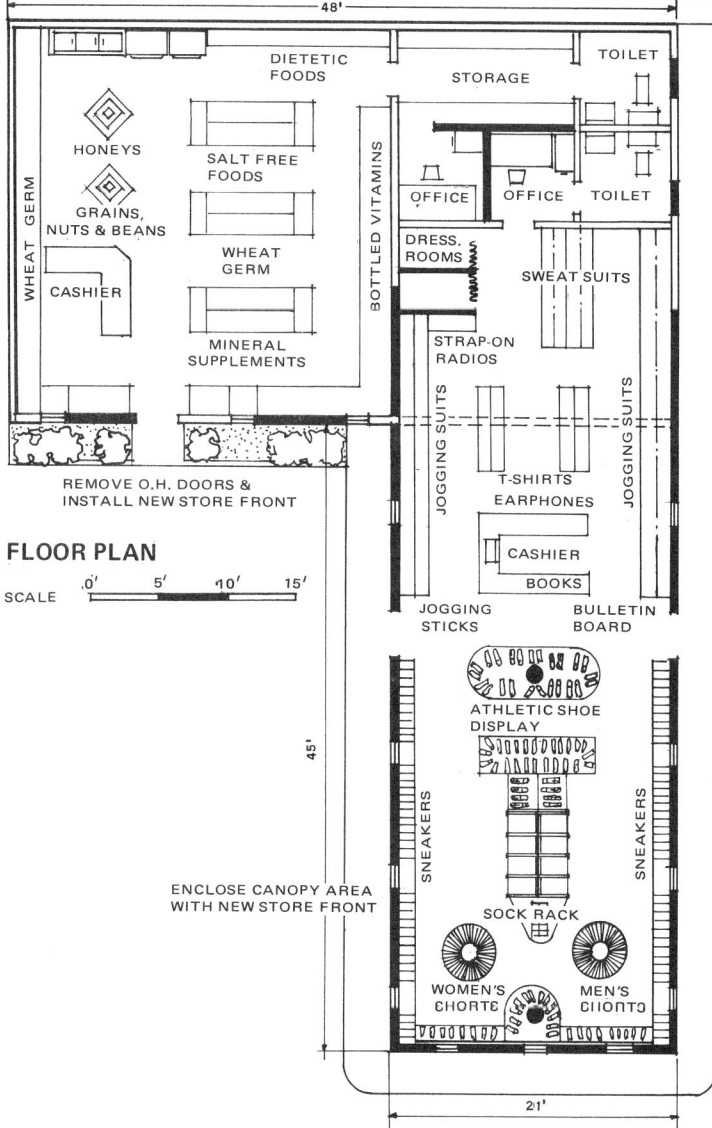

48'

DIETETIC FOODS
STORAGE
TOILET

HONEYS

SALT FREE FOODS

GRAINS, NUTS & BEANS

WHEAT GERM

CASHIER

MINERAL SUPPLEMENTS

WHEAT GERM

BOTTLED VITAMINS

OFFICE
OFFICE
TOILET

DRESS. ROOMS

SWEAT SUITS

STRAP-ON RADIOS

JOGGING SUITS

JOGGING SUITS

T-SHIRTS
EARPHONES

CASHIER
BOOKS

JOGGING STICKS
BULLETIN BOARD

REMOVE O.H. DOORS & INSTALL NEW STORE FRONT

FLOOR PLAN

SCALE   0'   5'   10'   15'

45'

ENCLOSE CANOPY AREA WITH NEW STORE FRONT

ATHLETIC SHOE DISPLAY

SNEAKERS

SNEAKERS

SOCK RACK

WOMEN'S SHORTS

MEN'S SHORTS

21'

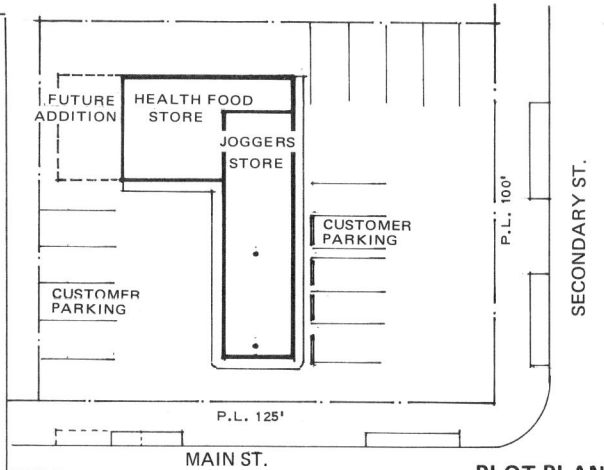

FUTURE ADDITION

HEALTH FOOD STORE

JOGGERS STORE

CUSTOMER PARKING

P.L. 100'

SECONDARY ST.

CUSTOMER PARKING

P.L. 125'

SCALE 1" = 50'

MAIN ST.

PLOT PLAN

Two compatible business ventures, worthy of consideration by the conservative entrepreneur, are a shoe store for joggers and a health food store. Of course a suitable location is essential for the success of any business. The business uses complement each other and are excellent during periods of sluggish economy. More than $150 million worth of jogging shoes are sold annually. Physical fitness is very popular and more than 10% of the population is running to improve their health. In addition to athletic shoes, sales of sweatsuits, shorts, socks, pants and T-shirts supplement the income.

A small health food store offers personal service for the patrons. An inventory of vitamins, herbs, nuts, beans, and grains can be displayed on 12" pine wall shelving and gondolas. Wood siding installed over the old masonry walls provides a simple and inexpensive facade. Ample area should be available on both sides or rear of the building for future expansion.

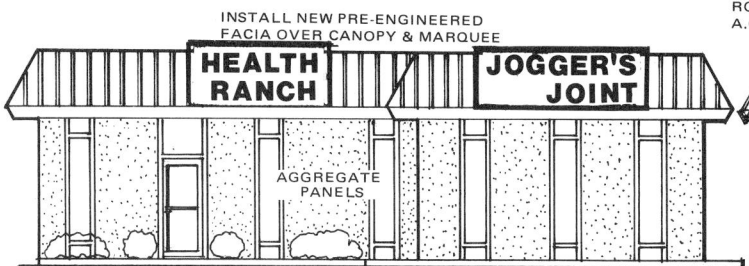

INSTALL NEW PRE-ENGINEERED FACIA OVER CANOPY & MARQUEE

HEALTH RANCH

JOGGER'S JOINT

AGGREGATE PANELS

FRONT ELEVATION

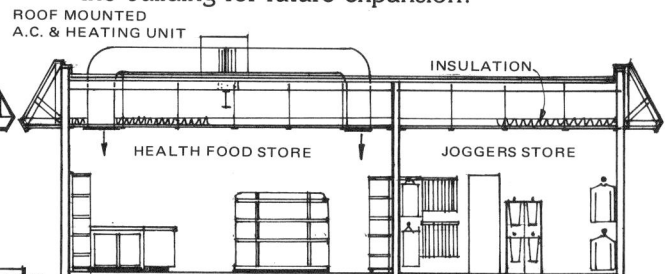

ROOF MOUNTED A.C. & HEATING UNIT

INSULATION

HEALTH FOOD STORE

JOGGERS STORE

CROSS SECTION

HEALTH FOOD AND ATHLETIC SHOE STORE

LEFT SIDE ELEVATION

AGGREGATE PANELS

INSTALL NEW PRE-ENGINEERED
FACIA OVER CANOPY & MARQUEE

RIGHT SIDE ELEVATION

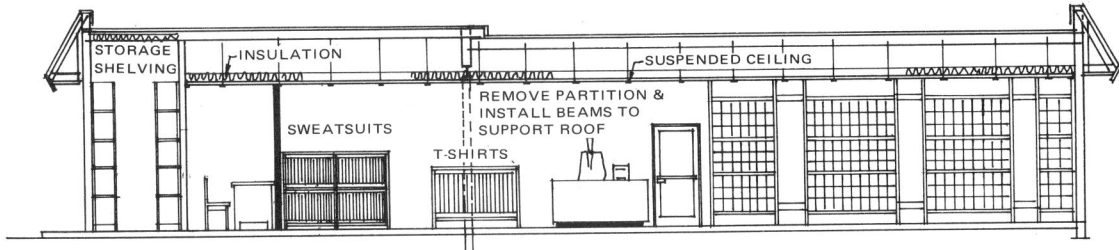

AGGREGATE PANELS

LONGITUDINAL SECTION

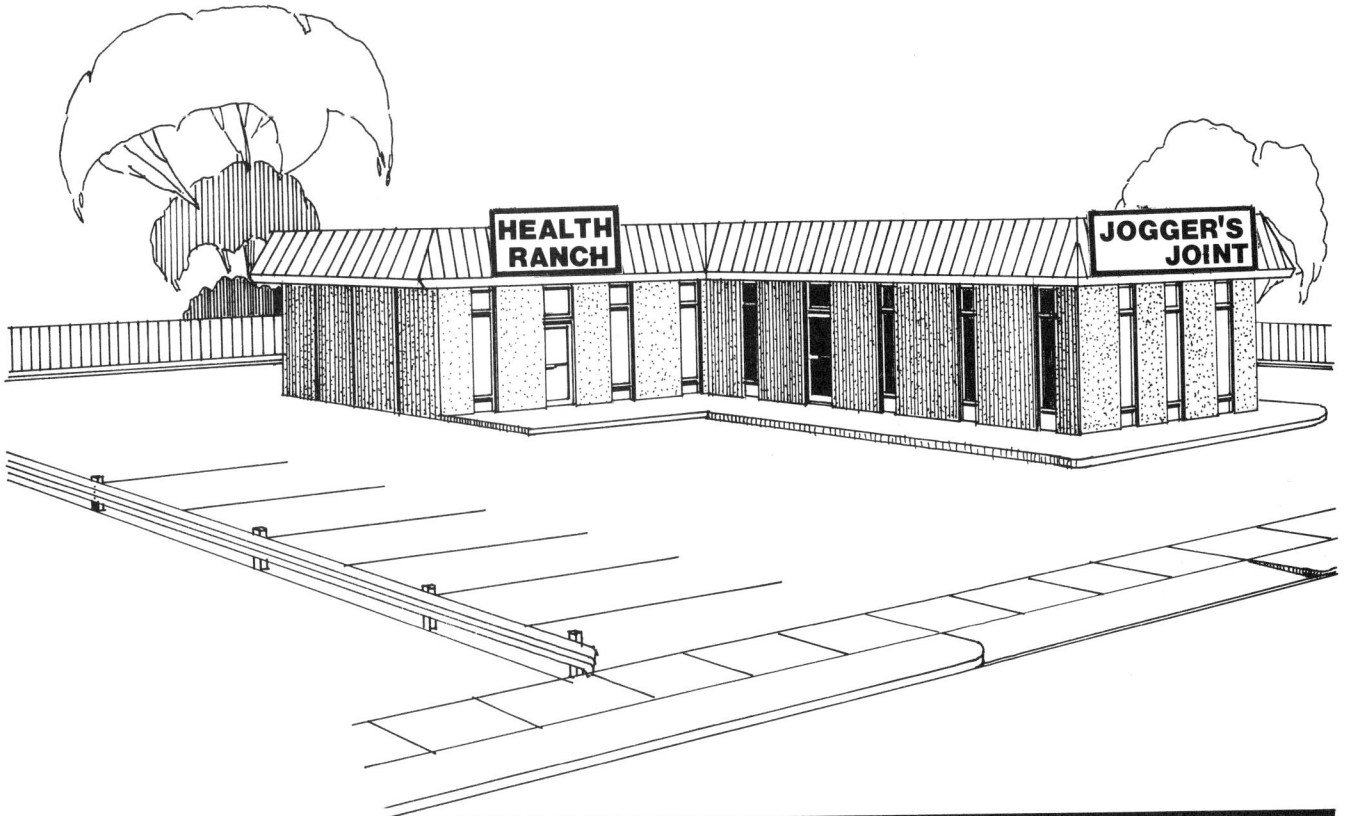

STORAGE SHELVING

INSULATION

SUSPENDED CEILING

SWEATSUITS

REMOVE PARTITION &
INSTALL BEAMS TO
SUPPORT ROOF

T-SHIRTS

HEALTH RANCH

JOGGER'S JOINT

LAUNDRY AND DRYCLEANER

Laundry and dry cleaner services are one of the fastest growing franchises according to the Department of Commerce. The number of franchised stores has increased from 2,692 in 1979 to 2,872 in 1981. The estimated 1981 sales volume of $317 million was up from $268 million in 1979. The average annual sales per store is estimated at $93,660 for 1981. Many franchised laundry and dry cleaning stores have broadened their services to include drapery processing, rental units to clean rugs at home, and alteration work.

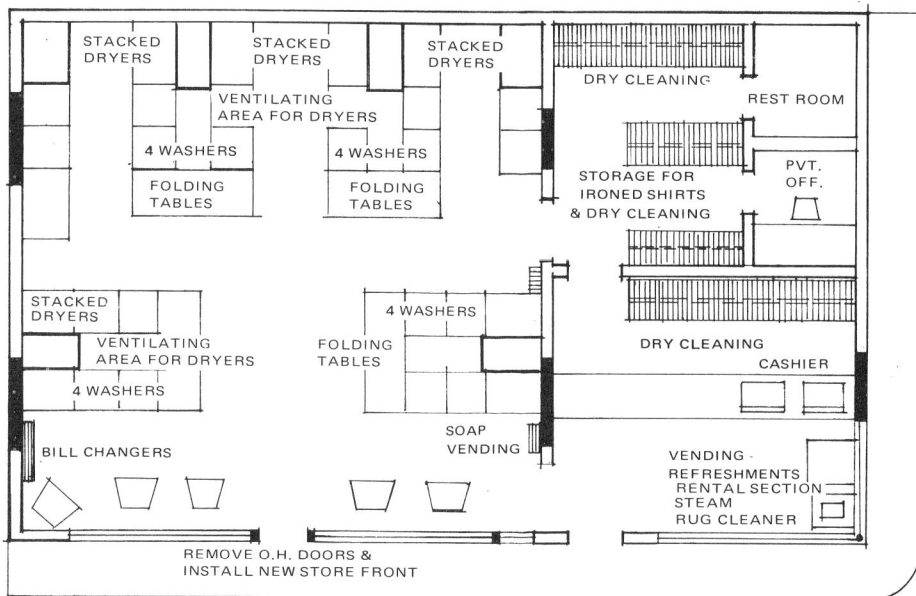

FLOOR PLAN

SCALE   3/32"= 1'-0"

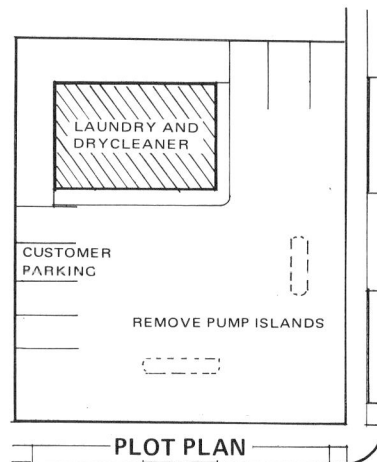

PLOT PLAN

Permits for laundries are difficult to obtain in many regions of the country. An abandoned service station should not be overlooked because it offers the businessman an opportunity to acquire a good location with sewer and water utilities and a potential permit. The design of a new "homestyle laundry system" and the concept of stacked dryers, adjacent to washers, is advocated by a major manufacturer of laundry equipment. This concept is illustrated in the conversion of a conventional two bay 1200 square foot abandoned service station.

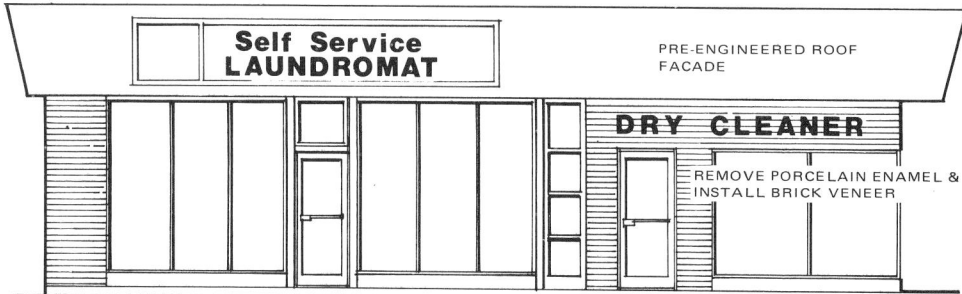

FRONT ELEVATION

ICE CREAM STORES

The entrepreneur shopping around for a recession proof business should not overlook the ice cream store. The following facts indicate that sales continue to increase during periods of sluggish economy. In 1979, according to the Department of Commerce there were 5355 ice cream stores with sales of $598 million, and in 1981 it was estimated that the number of outlets had increased to 6399 with sales of $763 million. The average investment per unit ranges from $60,000 to $105,000, considerably lower than other franchises. The sales per unit were expected to average $119,230 in 1980.

The profitability of an ice cream store is dependent on the following factors: the type of product sold; i.e., soft or "hard" ice cream; the hours and days of operation; the location of the store; and the managerial skills. Location is the single most important factor. A good traffic flow during an extended period, and a property with ample parking, are essential for success. Try to situate adjacent to a densely populated residential area with young families of moderate income. It is important to determine the age factor of the marketplace to insure a potential high volume of sales. Zero in on the locations where the consumer age ranges from the teens to under forty. This site criteria is one of the reasons many abandoned service stations offer the entrepreneur excellent business opportunities.

ICE CREAM STORE

A three bay service station contains approximately 1800 square feet of floor area, two rest rooms, adequate electric service, and water and sewerage facilities. Approximately 30% of the floor area should be devoted to the manufacture of ice cream. Another 30% should be allotted to the serving counter and back wall unit. Booths and round tables for two and four are a must. Also consider an area that can be closed off with folding partitions for small parties. Fifty percent of the business will be of the take out variety.

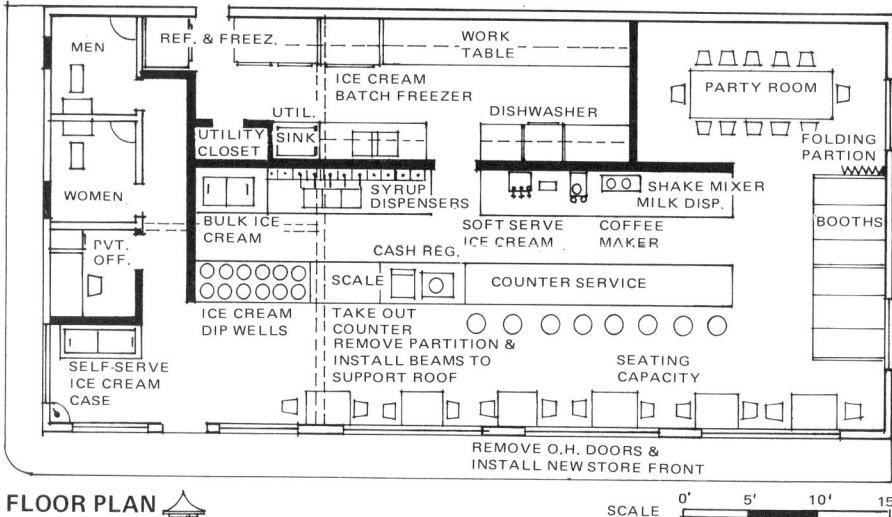

FLOOR PLAN

SCALE 0' 5' 10' 15'

RIGHT SIDE ELEVATION

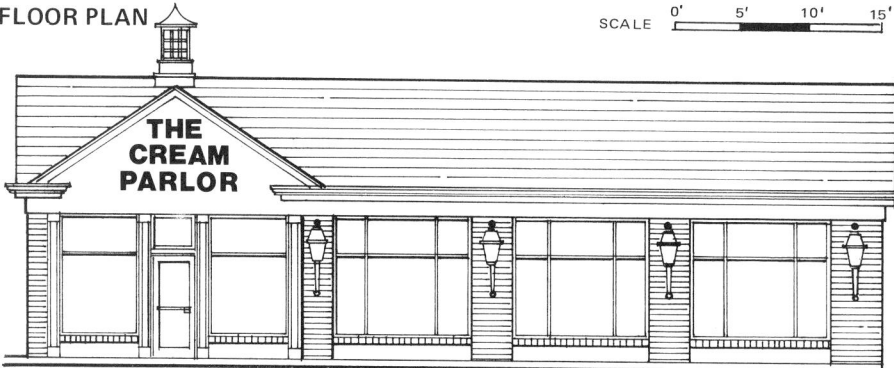

FRONT ELEVATION

REMOVE O.H. DOORS &
INSTALL NEW STORE FRONT

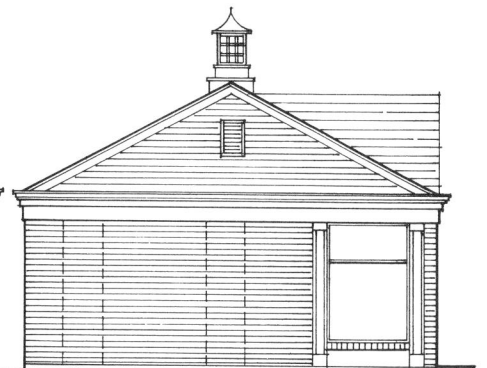

LEFT SIDE ELEVATION

The decor of the interior and exterior should reflect the turn of the century ice cream parlor atmosphere. Walls should be covered with wood paneling and brightly colored washable vinyl wallpaper. The floor should be easy to clean quarry tile.

In addition to the traditional ice cream parlor, a self-serve concept that permits customers to make their own sundaes has become popular. The experimental system appeals to many "do-it-yourself" consumers. The choice of selecting ice cream flavors and toppings gives the customers an opportunity to prepare creative and unusual combinations.

An unusually large service station site provides the businessman with an option to add an outside patio area for customer convenience and relaxation. In certain communities, this feature can be an asset; however, it can also lead to loitering. Therefore, the entrepreneur must consider the surrounding neighborhood before he includes the patio.

Attractive landscaping and colonial area and low level lighting will complement the "old fashioned" look. The application of an impervious asphalt sealer will preserve the service station paving. It is also necessary to paint stripes to control the parking lot. Entrance and exit signs are important to assist the circulation of traffic.

ICE CREAM STORE

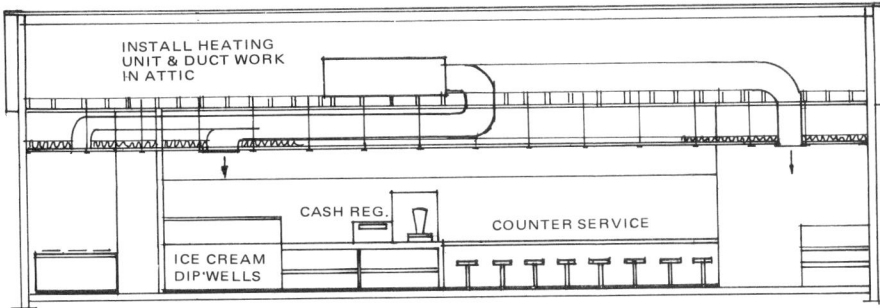

INSTALL HEATING
UNIT & DUCT WORK
IN ATTIC

CASH REG.

COUNTER SERVICE

ICE CREAM
DIP'WELLS

LONGITUDINAL SECTION

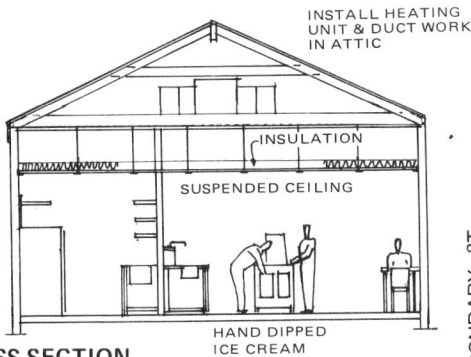

INSTALL HEATING
UNIT & DUCT WORK
IN ATTIC

INSULATION

SUSPENDED CEILING

CROSS SECTION

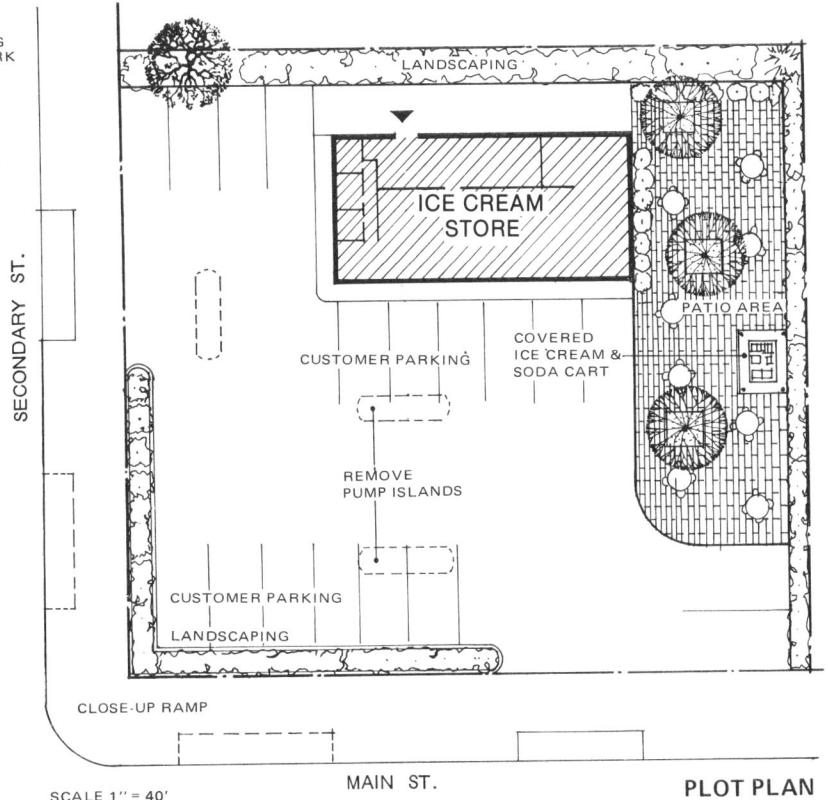

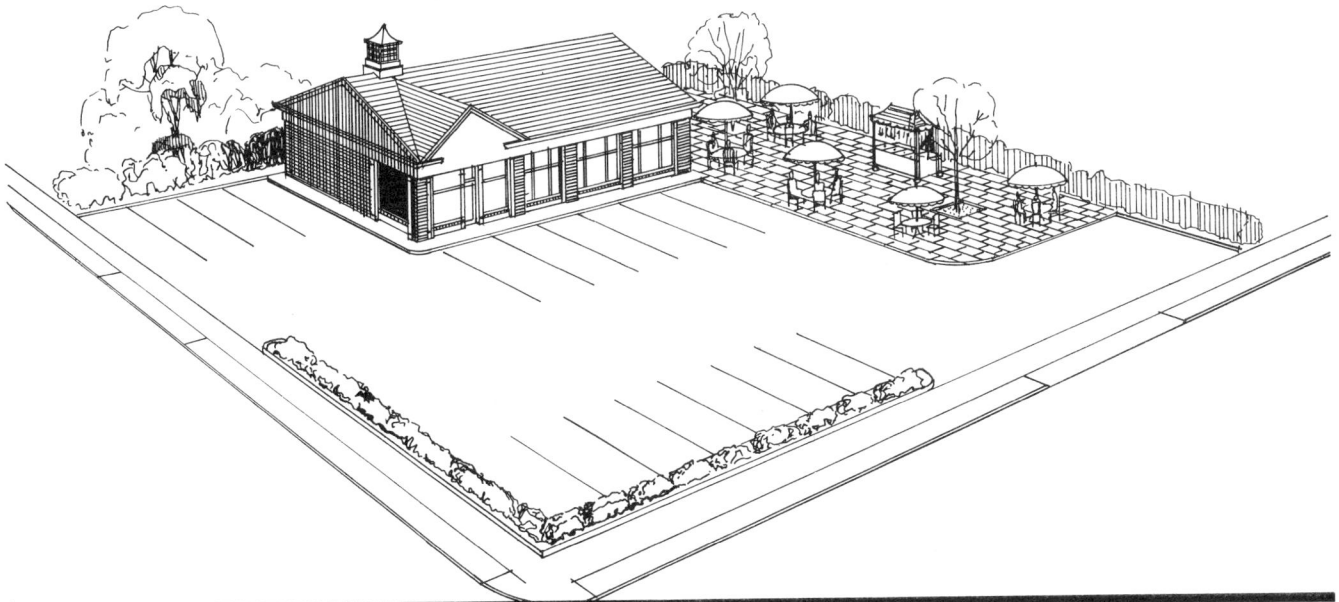

HAND DIPPED
ICE CREAM

LANDSCAPING

ICE CREAM
STORE

PATIO AREA

SECONDARY ST.

CUSTOMER PARKING

COVERED
ICE CREAM &
SODA CART

REMOVE
PUMP ISLANDS

CUSTOMER PARKING

LANDSCAPING

CLOSE-UP RAMP

SCALE 1" = 40'

MAIN ST.

PLOT PLAN

ICE CREAM STORE

Two businessmen with vision and an idea conducted their own survey for the right location for an ice cream store. They acquired this 75' x 55' irregular property and vintage service station, circa 1938. The property has a limited parking area and is bounded by three streets. The heavily travelled intersection is conveniently located for this conversion to an ice cream store. The end lubritory bay has been converted to a covered drive-thru with two service windows. Two cars may be served simulateneously. This helps to relieve the problem of limited parking for customers. A three to four car stack-up lane is tastefully landscaped with dividers that channel the vehicle to the service window. The entrepreneurs wisely utilized the services of a professional to prepare a space plan. No zoning problems were encountered in the change of business use. The interior of the building was remodeled; new fixtures were installed; walls and ceilings were restored; flooring was raised; new water lines installed; and plumbing and electrical work was completed. The yard area was completely repaved, and curbing, brick walkways and landscaping were installed. The cost of this work in 1978 was $30,000. This was $10,000 above the original budget.

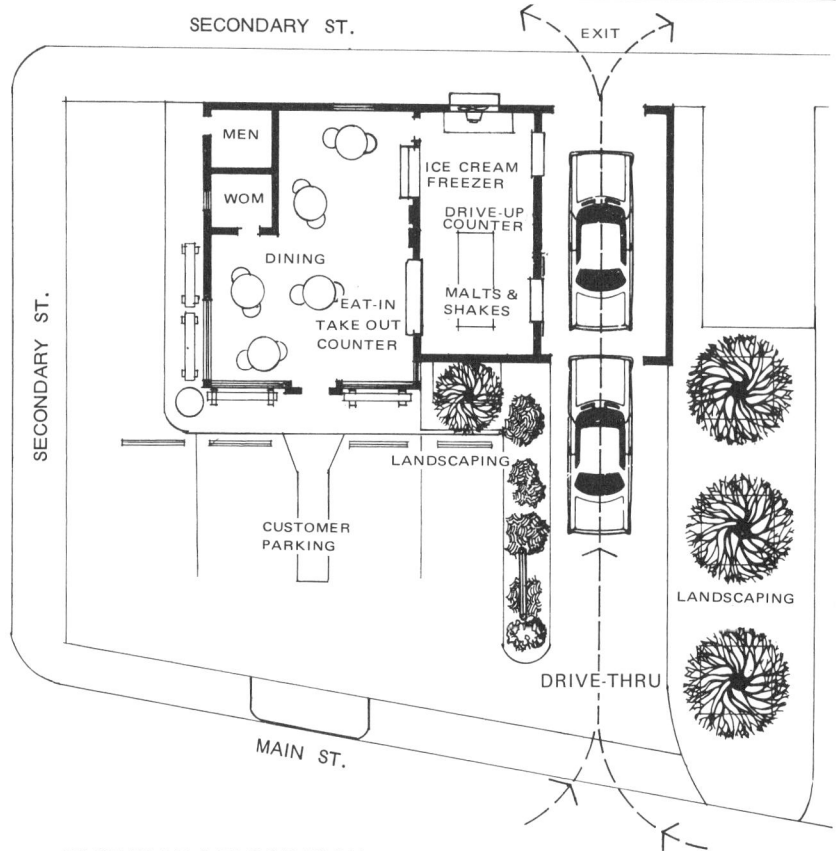

PLOT PLAN & FLOOR PLAN

The owners sell all natural ice cream. They make soft ice cream and ice cream cakes; the latter are displayed in a freezer case located in the parlor area. At least one-half of the business is conducted at the drive-up windows. During busy periods as many as six employees will work during one shift.

THE HOP

THE SUCCESSFUL CONVERSION OF A SUBSTANDARD SITE AND A 45 YEAR OLD BUILDING

Courtesy The Hop, Asheville, N. C.

ICE CREAM STORE

A LIMITED PARKING AREA

THE DRIVE-THRU BAY AND STACK-UP AREA

CUSTOMER EXIT TO THE SIDE STREET

THE TWO SERVICE WINDOWS

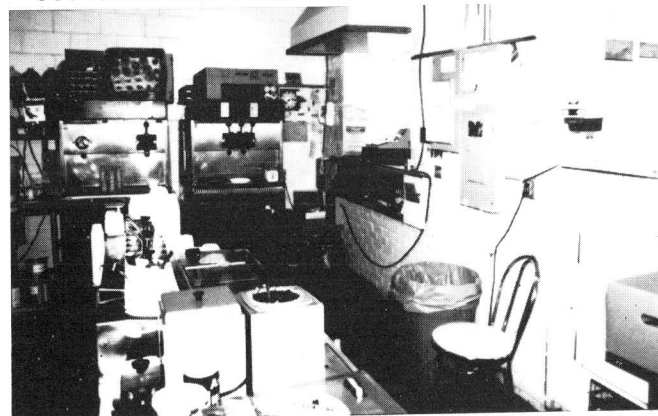

A FUNCTIONAL KITCHEN

THE TAKE-OUT COUNTER

THE ICE CREAM PARLOR

Courtesy The Hop, Asheville, N. C.

ICE CREAM STORE

AN EARLY PHOTO OF THE ABANDONED COTTAGE SERVICE STATION

THE CREAMERY

During the 1950's a cottage service station was abandoned and later remodeled into a new business use. For the next twenty years this building was converted to a number of business uses. In the summer of 1979 a businessman bought the property for $65,000 and remodeled the building into an old fashioned creamery. The property is situated in a historic restoration district and was remodeled in an early period style. Handsplit cedar shakes now cover the roof and the stucco and wood siding was repaired or replaced. The remodeling and landscaping was accomplished for $45,000 and equipment furnished and installed for $28,000.

THE OUTDOOR TABLES AND BENCHES ARE VERY POPULAR

AN ATTRACTIVE REMODELING AND LANDSCAPING TRANSFORM THE FACILITY INTO A COMMUNITY ASSET

A VIEW OF THE ICE CREAM PARLOR AND STORAGE AREA

Courtesy of Colorado City Creamery, Old Colorado City, Colorado Springs, Colorado.

ICE CREAM STORE

A BRICK WALK AND LANDSCAPING CREATE A
PICTURESQUE SETTING

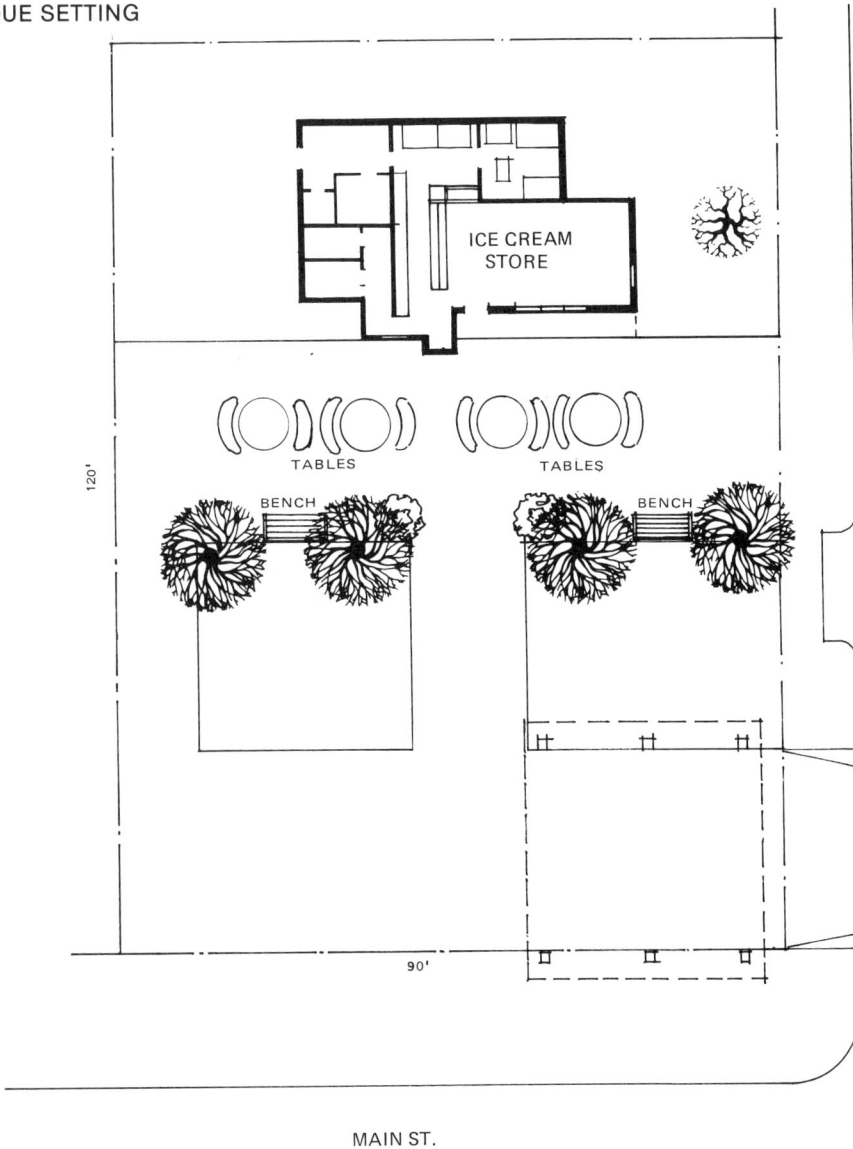

ICE CREAM
STORE

TABLES

TABLES

BENCH

BENCH

120'

90'

SECONDARY ST.

MAIN ST.

PLOT PLAN

Courtesy of Colorado City Creamery, Old Colorado City, Colorado Springs, Colorado.

ICE CREAM STORE

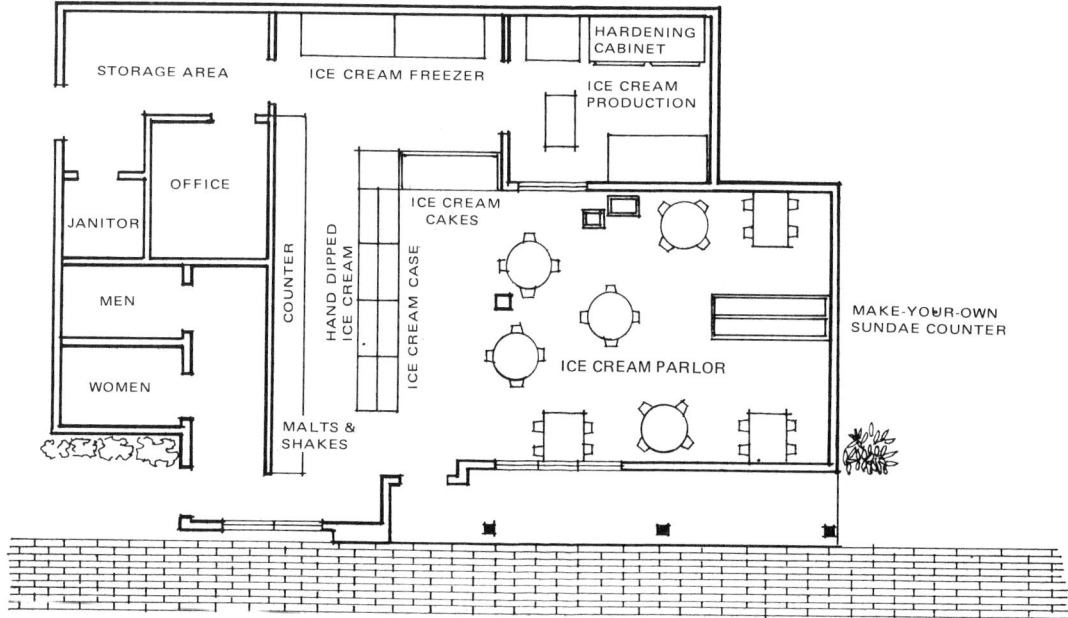

FLOOR PLAN

This business was a success from the start and for two years it has improved steadily. During June and July, sales reach their peak when "the door to the store rarely closes". According to the owner sales currently average 125 gallons of ice cream per day and the owner is now planning additional freezer capacity to store ice cream. The success of "The Creamery" is due to a number of factors:

1. Convenience of location in a restored historic area.

2. Quality ice cream made by the Creamery, specialty cones, ice cream cakes and a make-your-own-sundae bar.

3. Courteous service.

4. Attractive indoor dining area and an outdoor patio with tables and benches.

The owner stated that when they build another store it will be completed in the same architectural style as the current store.

THE ICE CREAM COUNTER

Courtesy of Colorado City Creamery, Old Colorado City, Colorado Springs, Colorado.

ICE CREAM STORE

ICE CREAM MANUFACTURING

THE "MAKE YOUR OWN SUNDAE BAR"

THE ICE CREAM PARLOR

A BUSY DAY BEHIND THE COUNTER

Courtesy of Colorado City Creamery, Old Colorado City, Colorado Springs, Colorado.

LAUNDROMAT

A LARGE NEW LAUNDROMAT

THE DRY CLEANING SERVICE, WAITING AREA AND VENDING

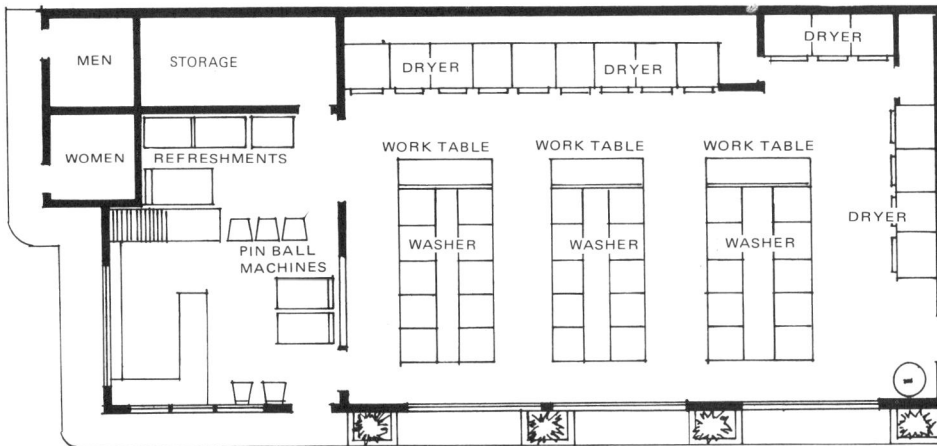

FLOOR PLAN

An enterprising petroleum marketer converted this former three-bay service station into a laundromat with wash and fold service. The owner felt that he saved approximately $80,000 remodeling the service station, rather than building a grass roots facility. The conversion was accomplished in December, 1978, at a cost of $30,000 plus equipment. The owner claims that the only disadvantage is that the building is too small; however, this is the reason he had a low investment base. The entrepreneur conducted his own survey to assure the success of his business.

A CLEAN LAUNDROMAT IS IMPORTANT

SALES COUNTER

THREE ROWS OF WASHING MACHINES AND DRYERS

FOLDING TABLE AT THE DRYER AREA

Courtesy The Wash House, Statesville, N. C.

LAUNDROMAT & DRY CLEANER

PARKING IS IMPORTANT FOR ALL DRIVE-IN BUSINESS USES

A businessman acquired a two bay porcelain service station in a middle income residential neighborhood and remodeled the building into a profitable laundromat and dry cleaning store at a nominal cost. There are twenty-four washing machines and nine dryers. The owner has one attendant to maintain the laundromat and operate the dry cleaning business. Vending machines dispensing laundry aids, a dollar bill changer and soap dispenser have been provided for the customers.

The interior is attractively decorated and is neat and clean, as a laundromat should be. Both interior and exterior have excellent illumination--an important feature for a community laundromat.

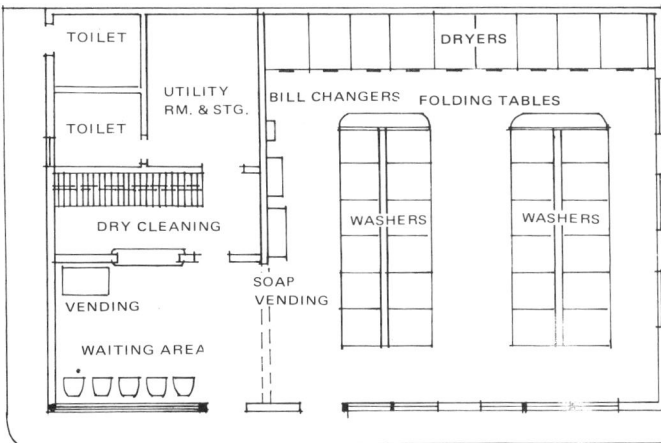

FLOOR PLAN

WASHERS AND DRYERS

DRY CLEANING COUNTER, WAITING AREA AND SODA MACHINE

Courtesy The Orange Blossom Laundromat, Allentown, Pa.

LIQUOR STORE, WITH DRIVE-UP WINDOW

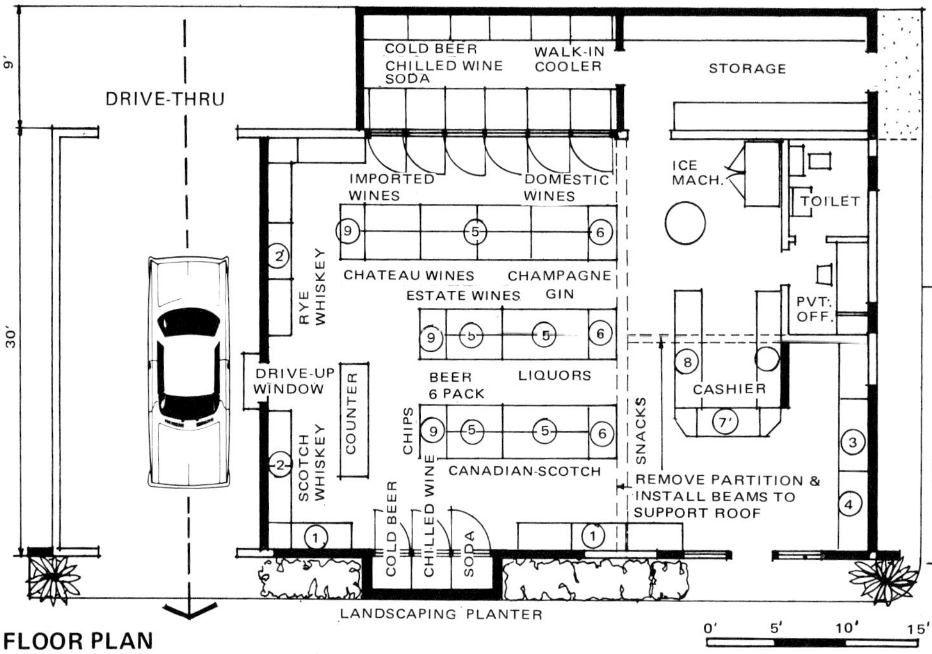
DRIVE-THRU
9'
30'
COLD BEER
CHILLED WINE
SODA
WALK-IN
COOLER
STORAGE
IMPORTED
WINES
DOMESTIC
WINES
ICE
MACH.
TOILET
CHATEAU WINES
CHAMPAGNE
ESTATE WINES
GIN
PVT.
OFF.
RYE
WHISKEY
DRIVE-UP
WINDOW
COUNTER
BEER
6 PACK
LIQUORS
CASHIER
CHIPS
SNACKS
SCOTCH
WHISKEY
CANADIAN-SCOTCH
REMOVE PARTITION &
INSTALL BEAMS TO
SUPPORT ROOF
COLD BEER
CHILLED WINE
SODA
LANDSCAPING PLANTER
0'
5'
10'
15'

FLOOR PLAN

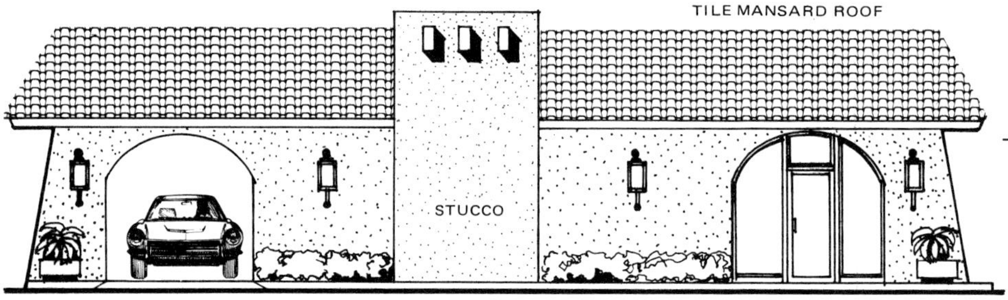
TILE MANSARD ROOF
STUCCO

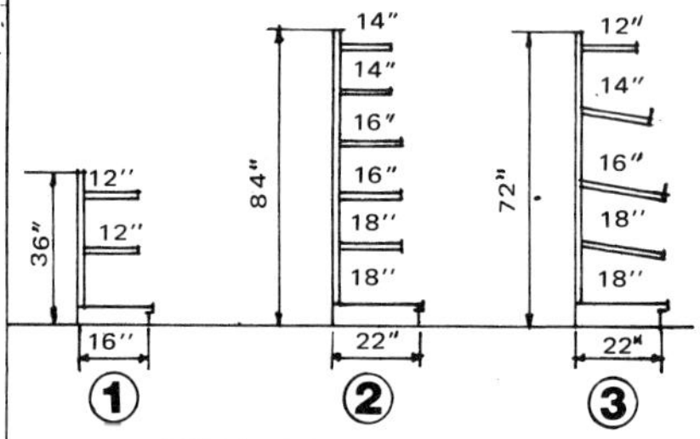
12"
12"
36"
16"
1
84"
14"
14"
16"
16"
18"
18"
22"
2
72"
12"
14"
16"
18"
18"
22"
3

SHELVING DISPLAY

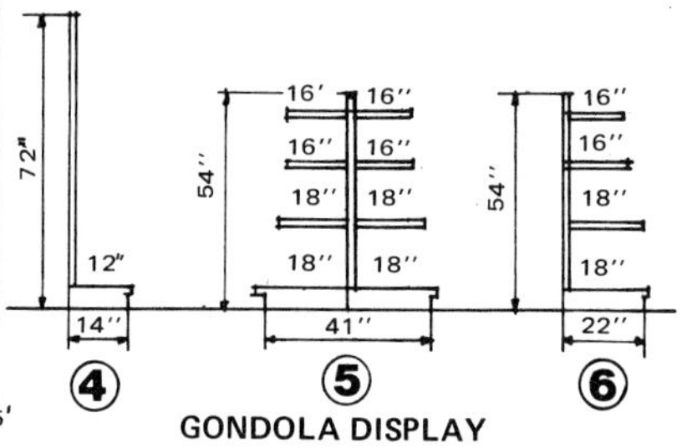
72"
12"
14"
4
54"
16'
16"
16"
16"
18"
18"
18"
18"
41"
5
54"
16"
16"
18"
18"
22"
6

GONDOLA DISPLAY

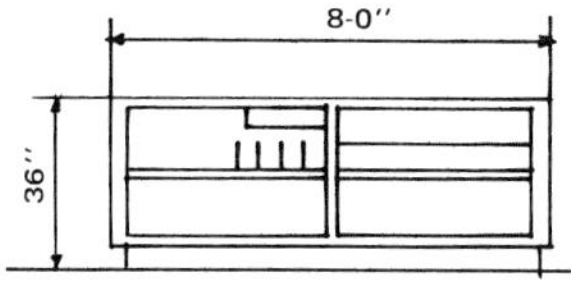
8-0"
36"

CASHIER'S COUNTER (8)

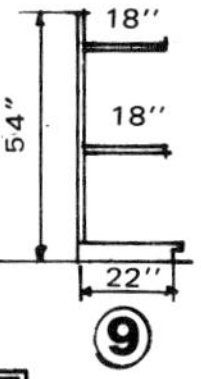
54"
18"
18"
22"
9

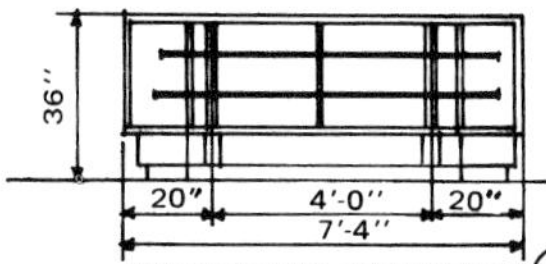
36"
20"
4'-0"
20"
7'-4"

CASHIER'S COUNTER (7)

People are in a rush today. Many do not like to take time to shop; therefore, the businessman who has a drive-in window for his liquor store will increase the sales potential. Most consumers know the particular brand they want to buy and are not interested in browsing around a store. A conventional three bay service station can be converted to a liquor store with a covered drive-in area as illustrated. A Spanish motif will attract attention and can be accomplished without great expense.

The additional expense for a second employee to serve the drive-in customers will be offset by increased income. A busy, well-located liquor store room, together with a walk-in cooler, is an important addition that will permit a larger inventory of fast moving items. The removal of the store room and sales walls improves the layout of the entire building. The spacious plan provides control of display areas. Self-service coolers are positioned for the convenience of the walk-in and drive-in customers.

The work to be performed on a three bay porcelain box building can be accomplished for a nominal expenditure. The porcelain must be removed and wing walls installed at both ends of the building, and the exterior walls will have stucco applied. A wall for the drive-in facility must be erected; new openings at the rear for an exit are to be completed; and an area in the front wall to house front-load coolers. This latter element will be part of the facade treatment. A pre-engineered roof system with metal tile roof can be completed, along with the work outlined at a nominal cost.

Equipment, Shelving & Displays and Fixtures Courtesy Universal-Nolin, Conway, Arkansas

LIQUOR STORE, WITH DRIVE-UP WINDOW

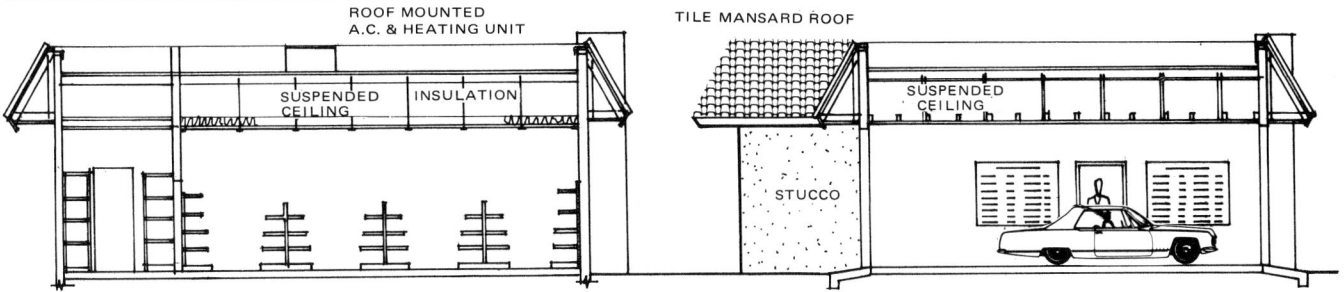

CROSS SECTION

CROSS SECTION

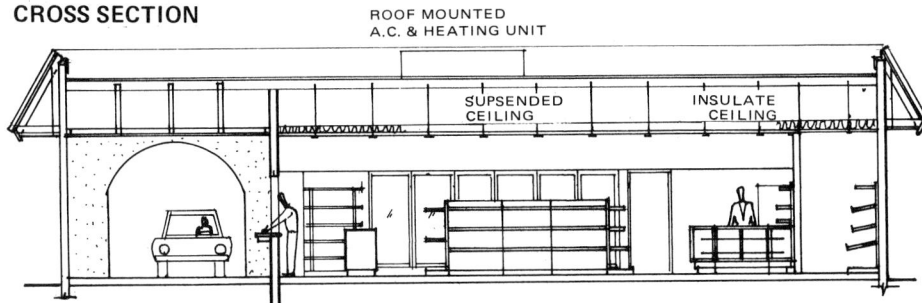

LONGITUDINAL SECTION

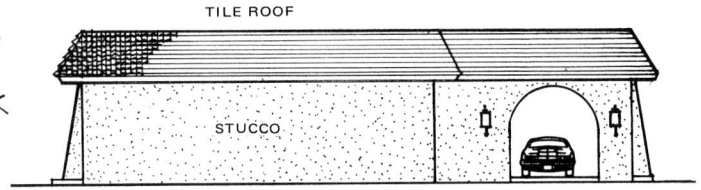

ELEVATION

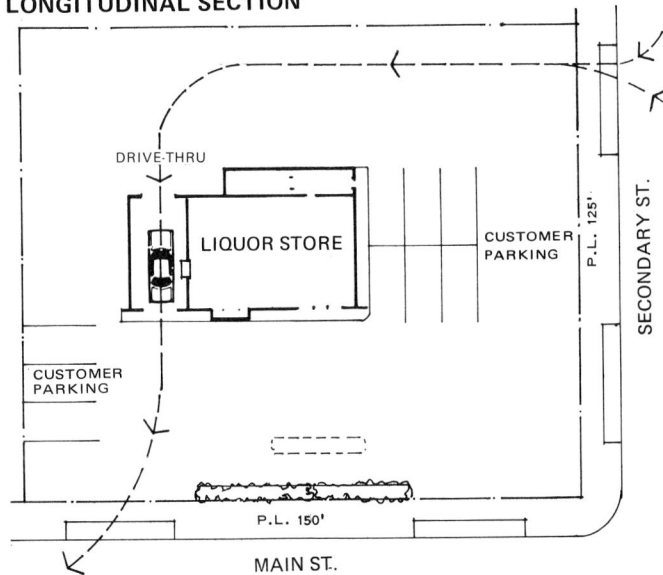

PLOT PLAN

SCALE 1″ = 50′

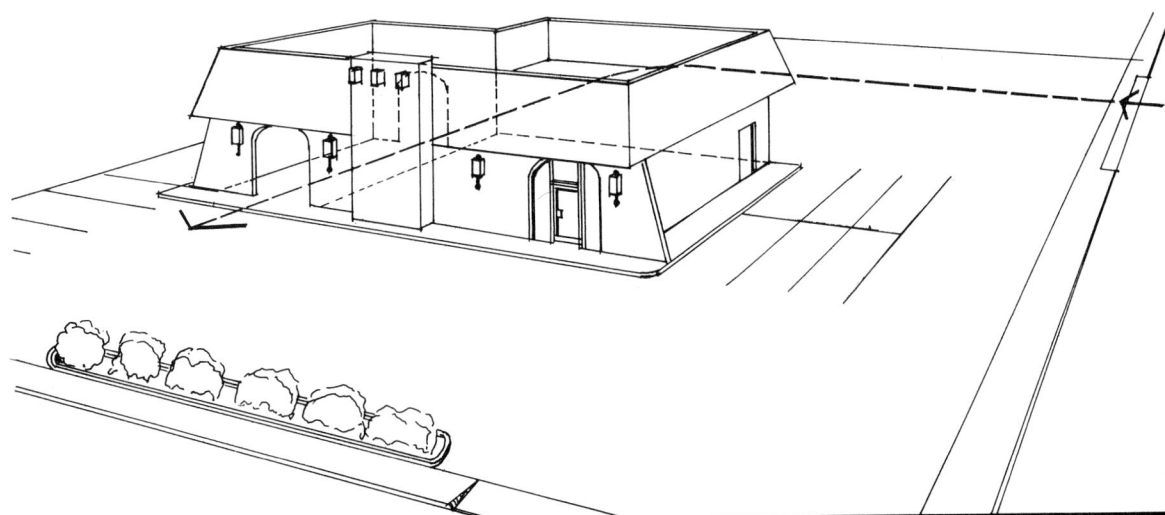

Equipment, Shelving & Displays and Fixtures Courtesy Universal-Nolin, Conway, Arkansas

LIQUOR STORE

The liquor store business is not usually adversely affected by a poor economy and is, therefore, considered to be recession-proof. Generally sales tend to increase during a poor business cycle because consumers find it less expensive to drink at home. Across the nation, liquor consumption continues to rise. "Michael's" is an example of an interesting conversion of a former three bay service station into a successful

OVERHEAD DOORS ARE CLOSED IN WITH CONCRETE BLOCK

THE REMOVAL OF PORCELAIN PANELS AND OVERHANG WAS DIFFICULT

AN ATTRACTIVE REMODELING

MICHAEL'S LIQUOR STORE

wine and liquor store. It is important that the interior and exterior of the store look uncluttered. The owner has accomplished this with vertical wood siding and excellent interior merchandising displays, easily iden-tifiable products, a limited supply of snack items and coolers for beer, wine and mixes. The actual liquor store occupies no more than 850 square feet. This does not include the storage area in the former end lubritory bay.

Courtesy Michael's Liquor Store, Hasbrouck Heights, N.J.

LIQUOR STORE

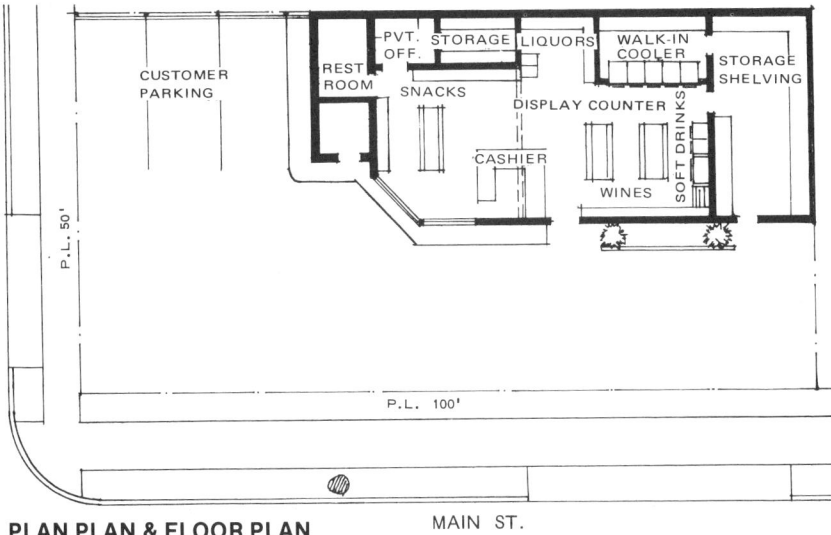

PLAN PLAN & FLOOR PLAN

The following is a list of the items deemed important by the owner in the selection of this site:

1. Located on a busy corner

2. Ample parking area

3. Near a traffic light

4. Storage gained by converting end bay

5. Larger sales area, allowing for the display of a variety of stock.

6. Good visibility of a corner location

The location is situated in a middle income residential area on one of the main thoroughfares. The land, measuring 100' x 50', and the abandoned three bay building, was purchased for $75,000. After resolving zoning objections, the owner obtained a permit and converted the building to a new business at a cost of $28,000 in 1974. Filling the gasoline tanks with sand and resurfacing the parking area amounted to an additional expenditure of $15,000. This is another example of why thousands of abandoned service stations offer excellent money saving opportunities for seasoned as well as inexperienced new business entrepreneurs.

The owner estimated that he had saved $15,000 by purchasing the property and converting the abandoned service station.

The owner's major problem was the removal of the porcelain enamel facing and structural members supporting the marquee. Redwood siding now covers the old masonry walls and provides a quality appearance at an economical cost. It will also hold a finish longer with minimum maintenance. The wood cedar shake roof shingled mansard is the finishing touch which transforms the former "white ice box" into a business establishment with character.

THE WINE CORNER

GONDOLAS DISPLAY PRODUCTS IN A WELL DESIGNED STORE

THE CASHIER'S COUNTER

Courtesy Michael's Liquor Store, Hasbrouck Heights, N.J.

LIQUOR STORE

GLENDALE LIQUORS

A GOOD LOOKING 40 YEAR OLD BUILDING

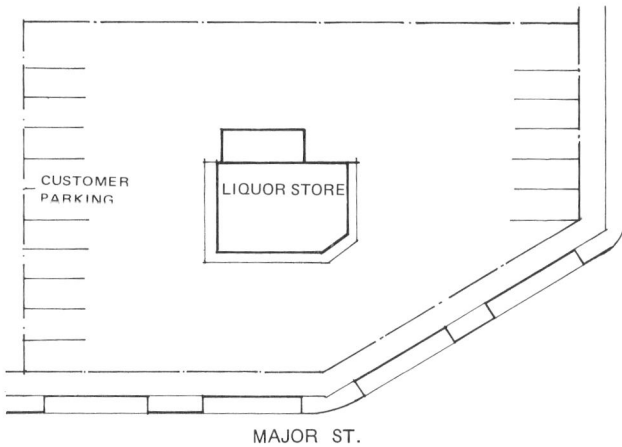
PLOT PLAN

CUSTOMER PARKING

LIQUOR STORE

MAJOR ST.

A STORAGE COOLER ADDITION

LONG GONDOLAS PROVIDE A FINE DISPLAY

In 1972, a former real estate agent purchased this twenty-five year old two bay service station, situated on an irregular 200' x 120' parcel for $20,000. The owner converted the building into a liquor store at a cost of $20,000, installing exterior redwood siding, a new store front and a suspended acoustical ceiling. Fixtures for the store included shelving gondolas, a counter and a large walk-in cooler which cost another $20,000. The arrangement of the display gondolas and counters allows the cashier to control the interior of the store. The store was designed to allow for a 25% growth factor. Although the sales area is limited to approximately 900 square feet, the owner estimates that it has the potential of $500,000 per year in gross sales.

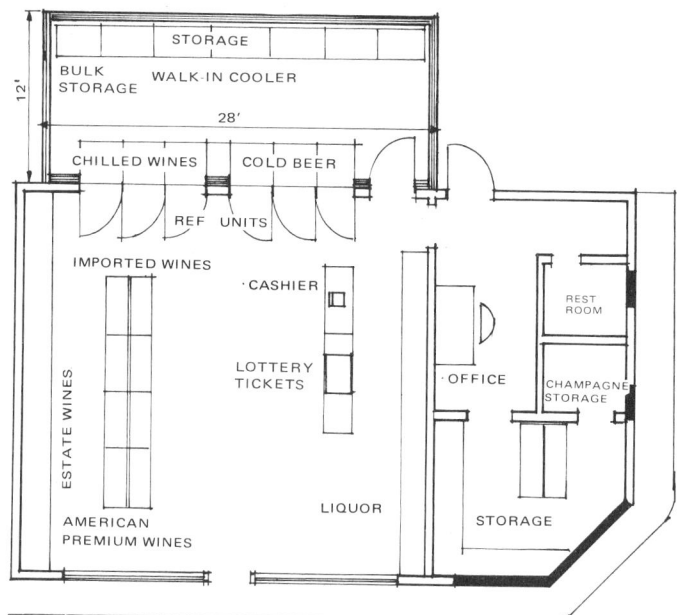
FLOOR PLAN

STORAGE
BULK STORAGE
WALK-IN COOLER
12'
28'
CHILLED WINES
COLD BEER
REF UNITS
IMPORTED WINES
CASHIER
REST ROOM
ESTATE WINES
LOTTERY TICKETS
OFFICE
CHAMPAGNE STORAGE
AMERICAN PREMIUM WINES
LIQUOR
STORAGE

LIQUOR STORE

THE INTERIOR OF THE COOLER

AN ORDERLY SALES DISPLAY

Courtesy Glendale Liquors, Trenton, N. J.

LIQUOR STORE

The two most important features of a liquor store are the location and the advertising and promotion of the business. This 1800 square foot three bay building provides the opportunity for a store layout that will help achieve high sales. The equipment layout is based on a successful design of one of the leading manufacturers of a complete line of refrigerated cabinets, walk-ins, sales counters and shelving.

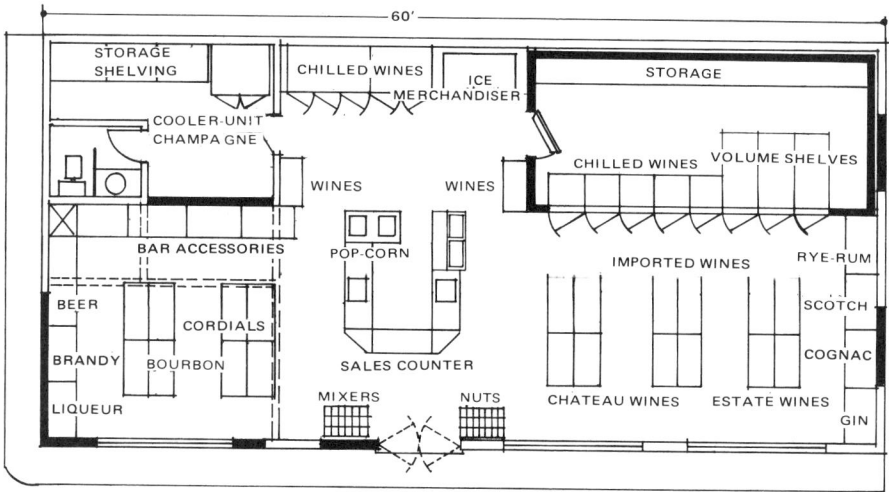

FLOOR PLAN

SCALE 0' 5' 10' 15'

FRONT ELEVATION

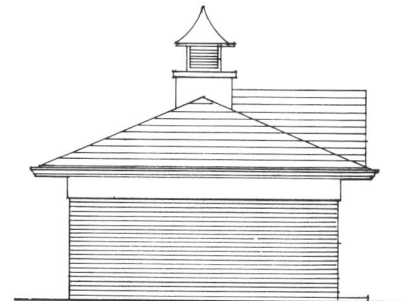

LEFT SIDE ELEVATION

Equipment, Shelving & Displays and Fixtures Courtesy Universal-Nolin, Conway, Arkansas

LAW AND REAL ESTATE OFFICE

Abandoned service stations make excellent offices for compatible professionals, such as real estate and law firms. A three bay building, with a floor area of 1800 square feet, can be subdivided into two equal offices. The hard lines of a flat top service station can be removed with a dramatic new mansard roof. A unique design is achieved with the versatility of the many patterns of redwood lumber siding. The siding may be installed vertically and diagonally to accent the building.

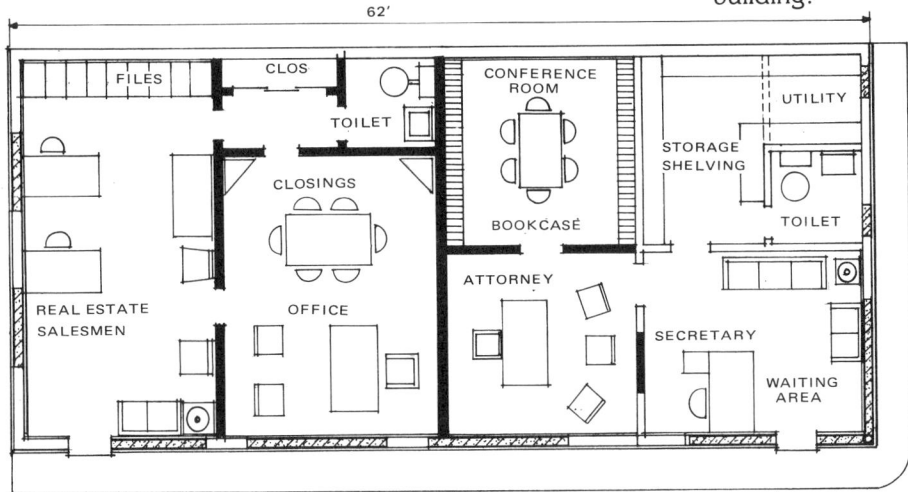

FLOOR PLAN

62'

FILES · CLOS · TOILET · CLOSINGS · CONFERENCE ROOM · BOOKCASE · UTILITY · STORAGE SHELVING · TOILET · ATTORNEY · REAL ESTATE SALESMEN · OFFICE · SECRETARY · WAITING AREA

SCALE 0' 5' 10' 15'

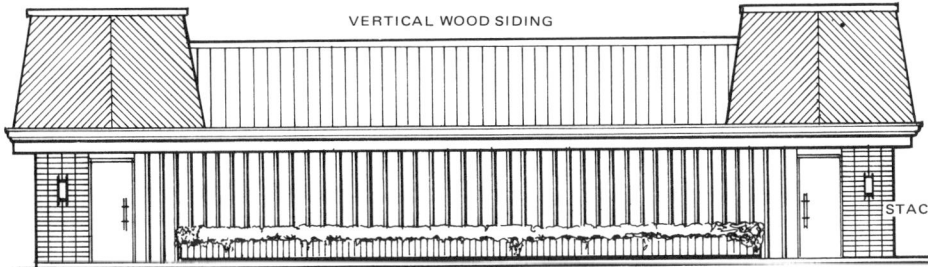

FRONT ELEVATION

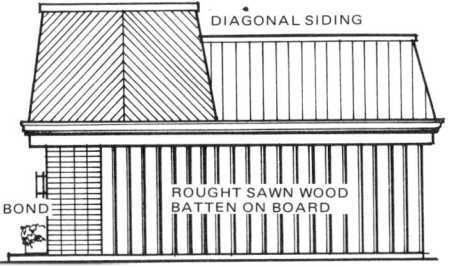

VERTICAL WOOD SIDING

LEFT SIDE ELEVATION

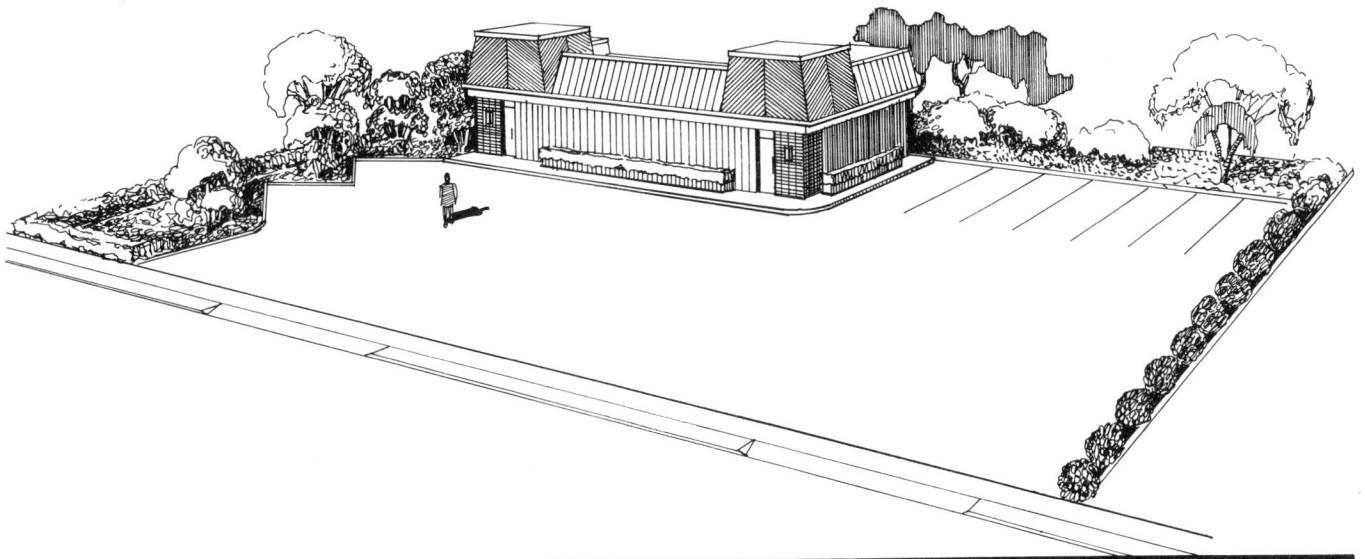

DIAGONAL SIDING

STACK BOND

ROUGHT SAWN WOOD BATTEN ON BOARD

140

MATTRESS STORE

This large corner property with a three bay canopy building is ideally suited for a box spring and mattress store. This type of a business will do well financially if it is located in a city with a population in excess of 150,000. The site should be located at a well travelled intersection and a good parking area is important. Utilization of the canopy for an additional 900 square foot of floor area is essential.

The enclosure of this structure with glass provides a showroom that can be viewed from vehicles waiting for a traffic light to change--a very big plus for the success of this business venture. The open floor area permits the display of a full line of beds, mattresses and box springs, such as the King, Queen, round and Waterbed, which can be displayed in the new wing.

A large storeroom with deep shelves is required for headboards, bedspreads, mattress covers, sheets and accessories. The interior of the store can be arranged into bedroom areas, complete with curtains or drapes to provide an attractive atomsphere.

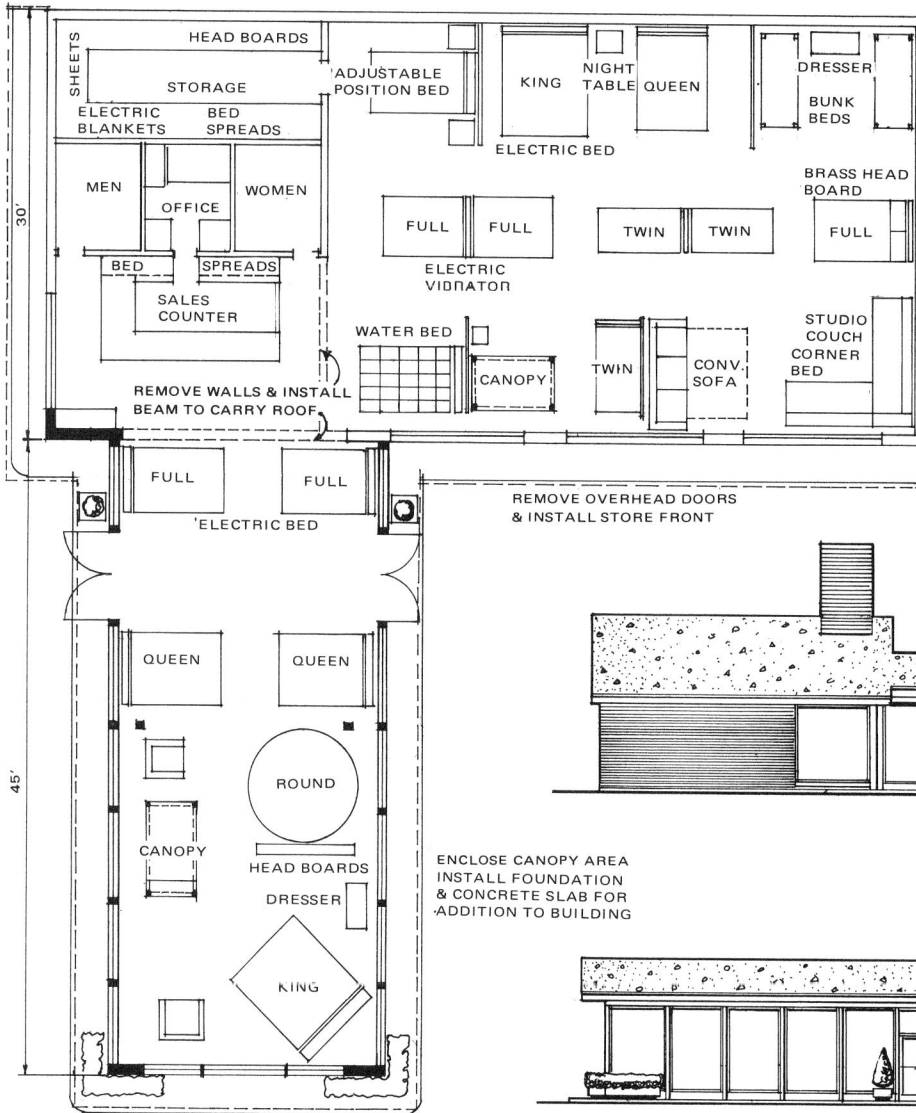

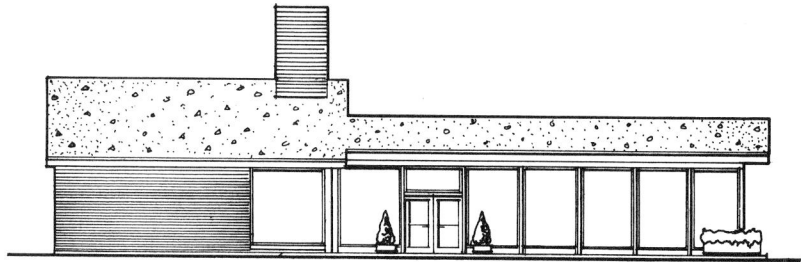

FLOOR PLAN

SCALE 0' 10'

LEFT SIDE ELEVATION

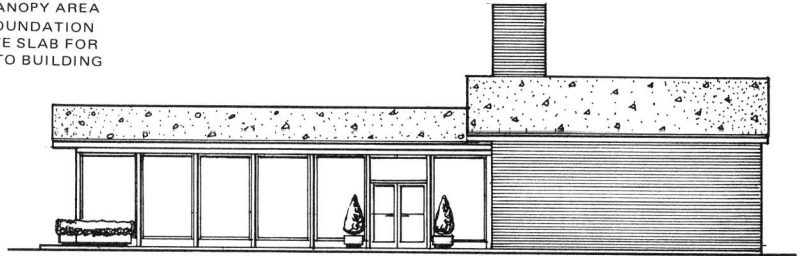

RIGHT SIDE ELEVATION

Bedtime Story

FRONT ELEVATION

MATTRESS STORE

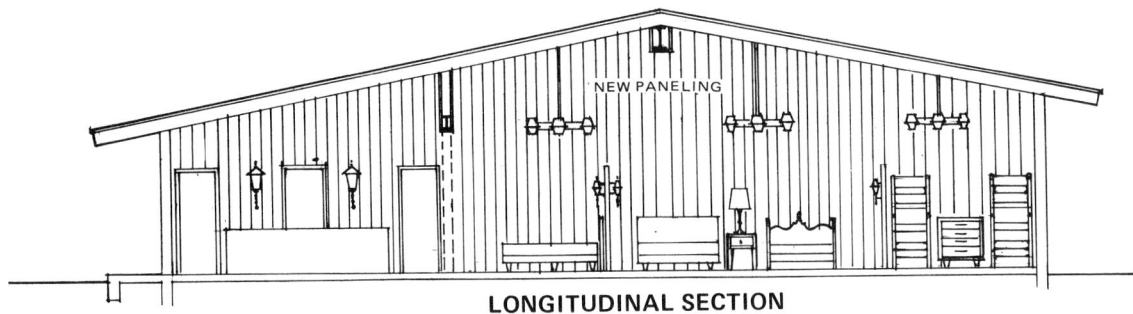

LONGITUDINAL SECTION

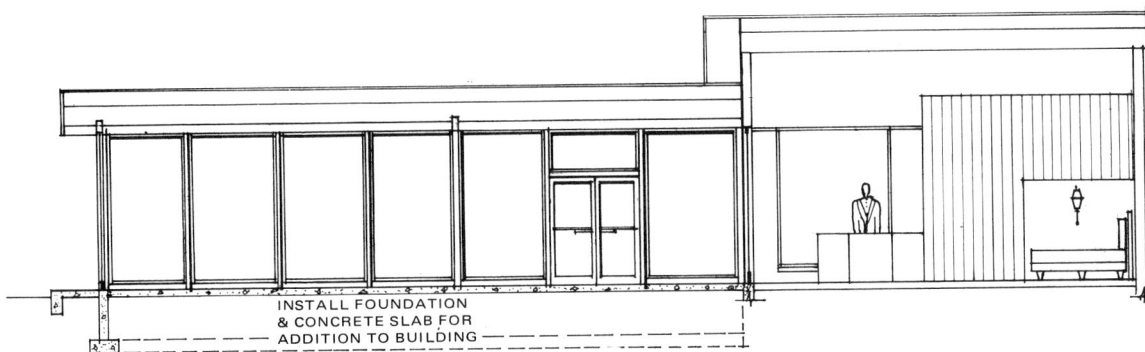

INSTALL FOUNDATION
& CONCRETE SLAB FOR
ADDITION TO BUILDING

CROSS SECTION

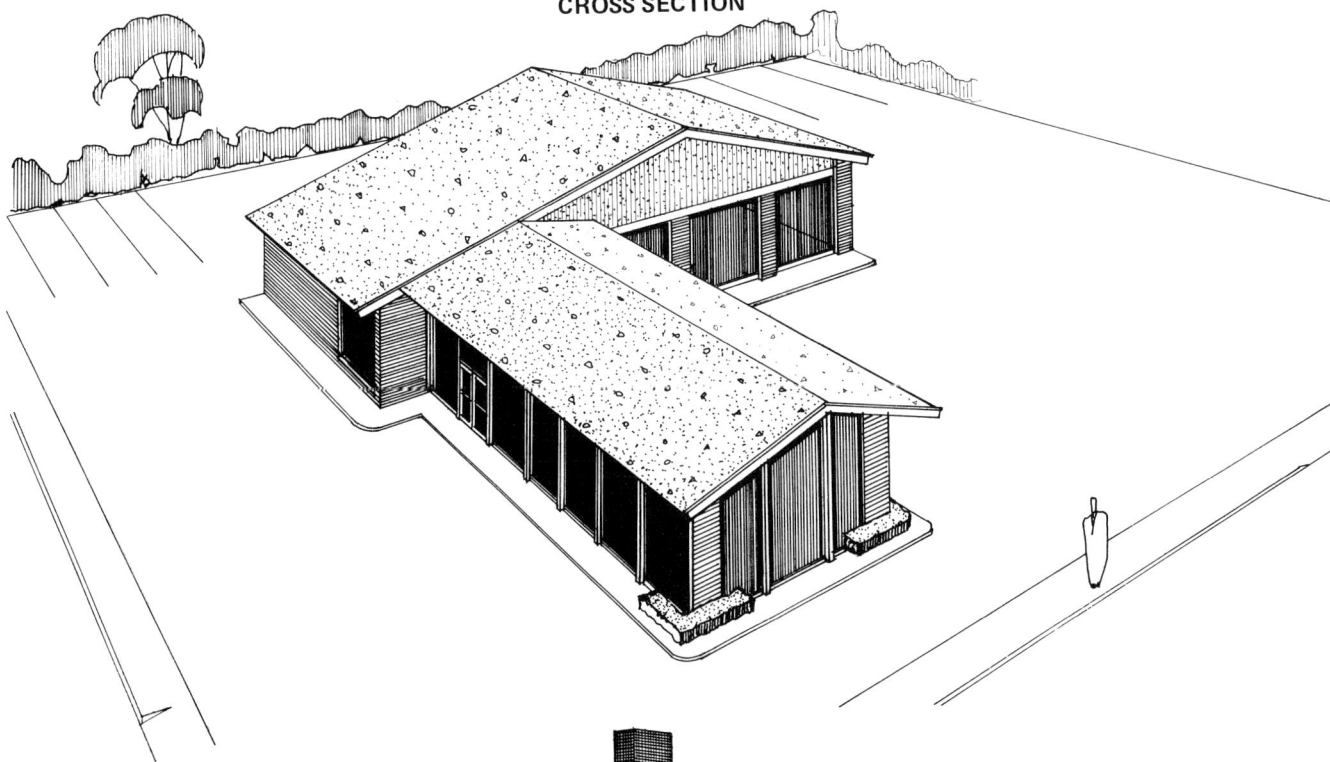

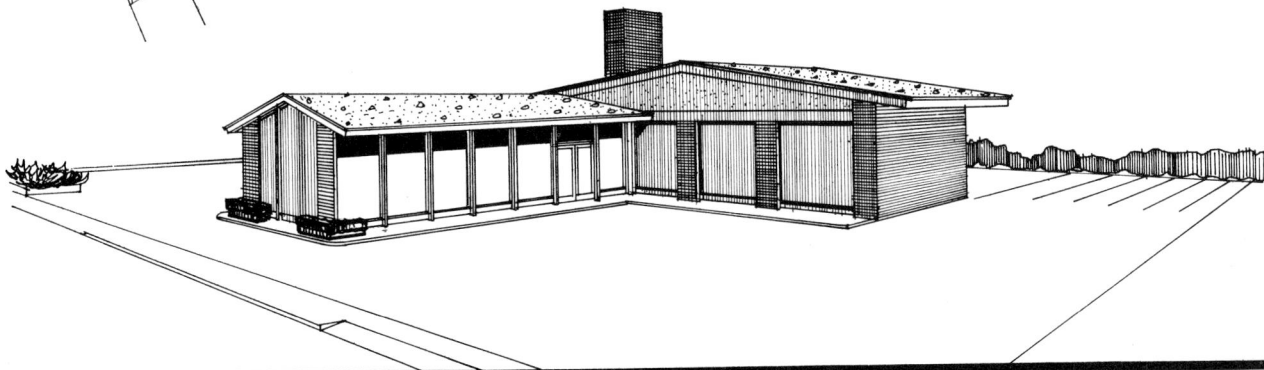

NEW PANELING

142

MULTI - BUSINESS USES, HARDWARE, GARDEN AND RENTAL CENTER

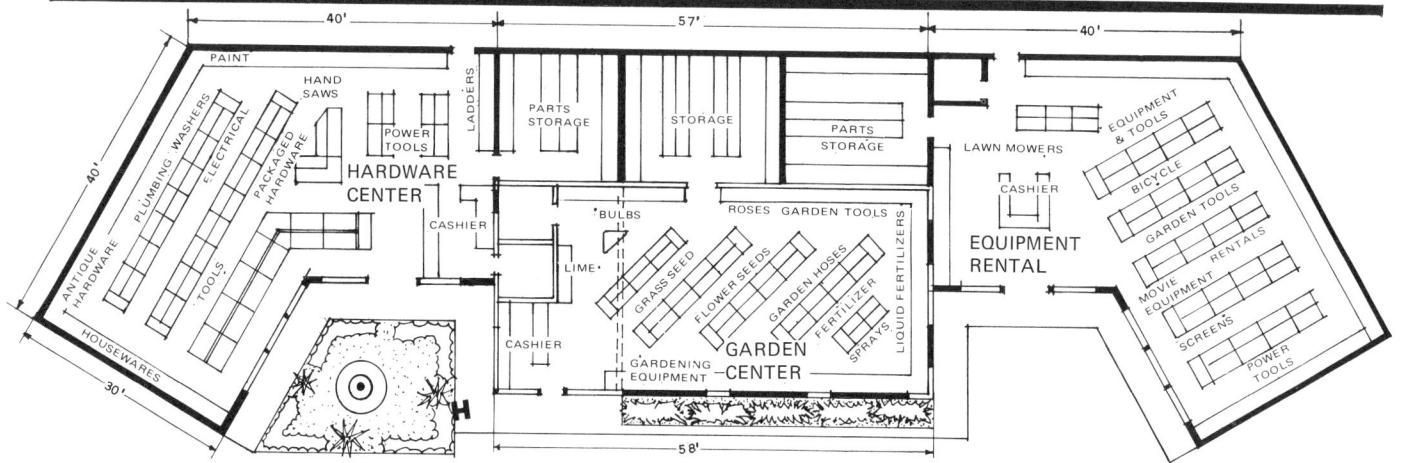

FLOOR PLAN

Labels in floor plan: PAINT, HAND SAWS, PLUMBING WASHERS, ELECTRICAL, PACKAGED HARDWARE, POWER TOOLS, HARDWARE CENTER, ANTIQUE HARDWARE, TOOLS, HOUSEWARES, CASHIER, LADDERS, PARTS STORAGE, STORAGE, PARTS STORAGE, BULBS, LIME, CASHIER, GARDENING EQUIPMENT, GRASS SEED, FLOWER SEEDS, ROSES, GARDEN TOOLS, GARDEN HOSES, FERTILIZER, SPRAYS, LIQUID FERTILIZERS, GARDEN CENTER, LAWN MOWERS, CASHIER, EQUIPMENT RENTAL, EQUIPMENT & TOOLS, BICYCLE, GARDEN TOOLS, RENTALS, MOVIE EQUIPMENT, SCREENS, POWER TOOLS

Dimensions: 40', 57', 40', 40', 30', 58'

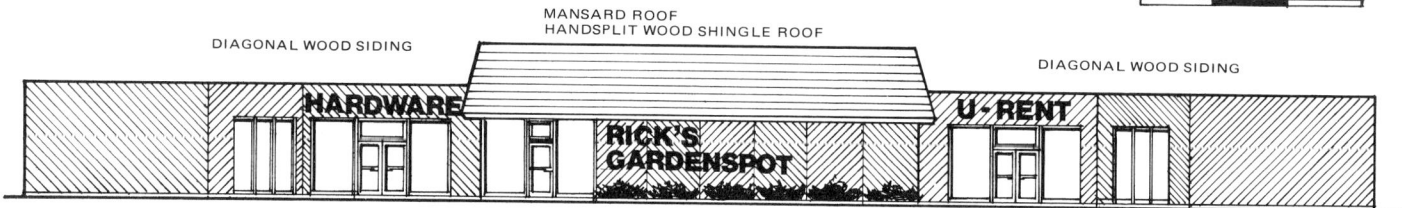

FRONT ELEVATION

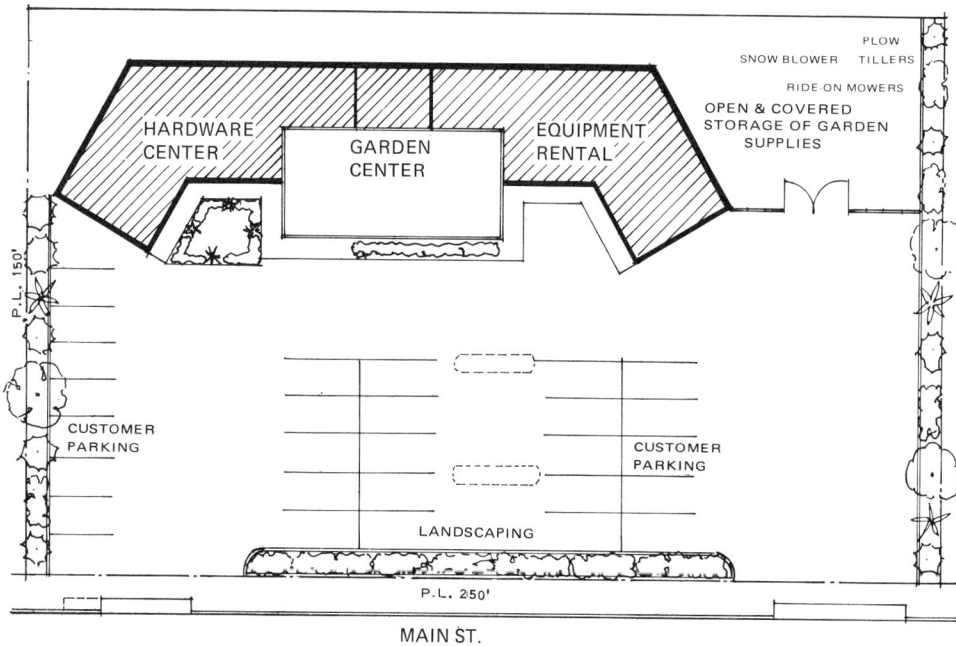

DIAGONAL WOOD SIDING · MANSARD ROOF · HANDSPLIT WOOD SHINGLE ROOF · DIAGONAL WOOD SIDING

HARDWARE · RICK'S GARDENSPOT · U - RENT

PLOT PLAN

HARDWARE CENTER, GARDEN CENTER, EQUIPMENT RENTAL, PLOW, SNOW BLOWER, TILLERS, RIDE-ON MOWERS, OPEN & COVERED STORAGE OF GARDEN SUPPLIES, P.L. 150', CUSTOMER PARKING, CUSTOMER PARKING, LANDSCAPING, P.L. 250', MAIN ST.

SCALE 1" = 50'

Abandoned service stations, situated on large properties in suburban communities, offer excellent opportunities for the development of multi-business facilities. The three uses illustrated complement each other. The primary business use is a general hardware store, flanked by a graden center and an equipment rental center. An important feature for each business is the need for ample storage areas. The hardware store should have an energy conservation center. The garden center must sell and/or rent a complete line of lawn mowers, and the equipment rental center must also have an outside open area for leasing trucks and heavy equipment.

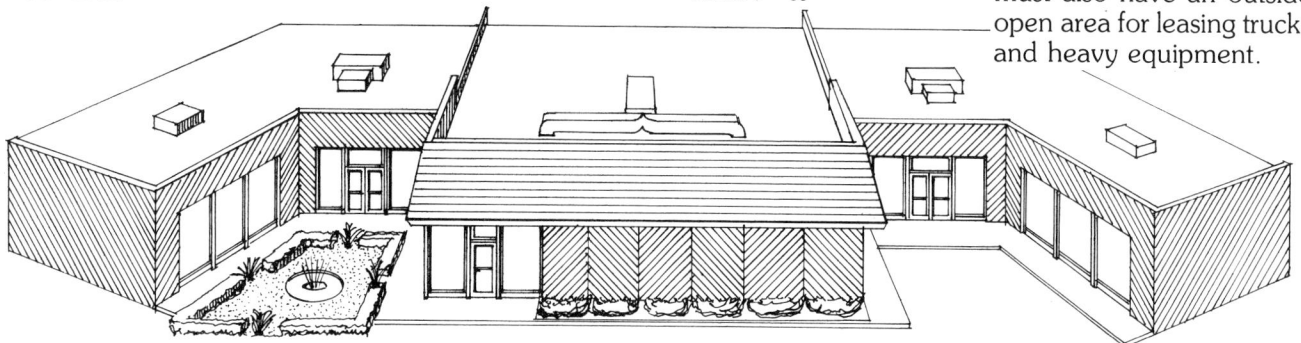

MULTI-USE STORES

EARLY STAGES OF THE CONVERSION

THE NEWLY REMODELED DONUT SHOP, CONVENIENCE STORE AND MEAT STORE

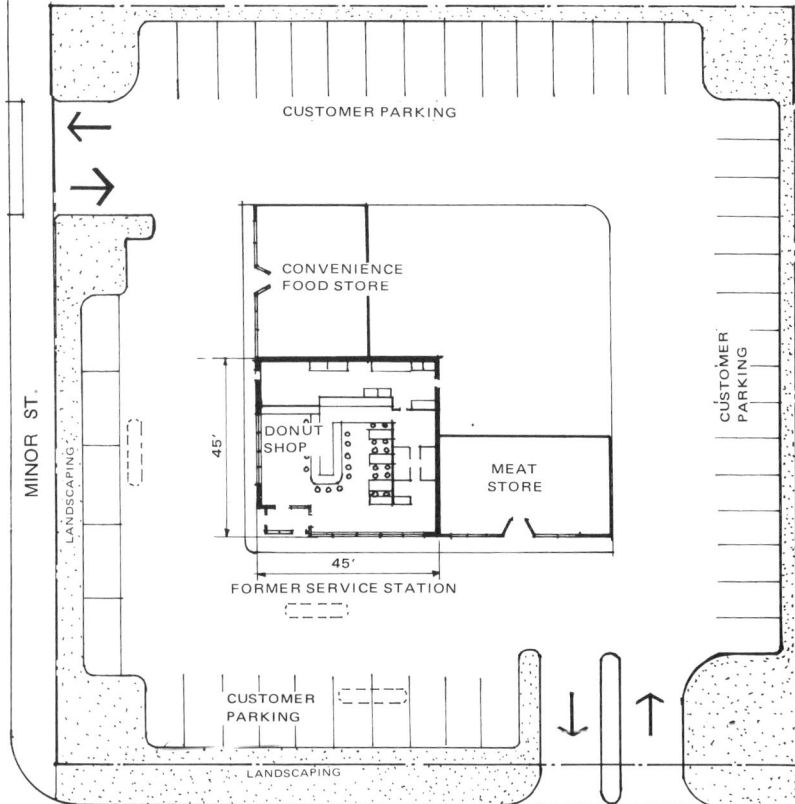

PLOT PLAN

A 200' x 175' property, situated on a busy intersection, was purchased by a real estate corporation for development for three business uses. The three bay side entry building is basically a square structure ideal for a use requiring 1800 square feet. It was situated in the center of the parcel and served as the base for expansion. An attractive donut shop was developed in the former service station portion. The butcher shop and convenience food store are retail operations that will attract customers and benefit the small complex.

COMFORTABLE BOOTHS ARE PROVIDED

DONUT SHOP SERVING COUNTER

DONUT RACKS AND PREPARATION AREA

Courtesy Cohen, Schatz & Assoc., Lakewood, N.J.
Courtesy Donut Hut of Howell, Howell, N.J.

MOTORCYCLE SUPPLY STORE

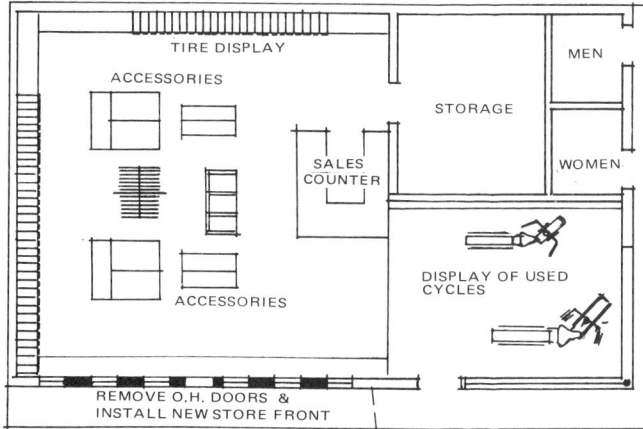

FLOOR PLAN

(Floor plan labels: TIRE DISPLAY, ACCESSORIES, MEN, STORAGE, WOMEN, SALES COUNTER, ACCESSORIES, DISPLAY OF USED CYCLES, REMOVE O.H. DOORS & INSTALL NEW STORE FRONT, LINE OF CANOPY)

An entrepreneur and her two sons, motor cycle enthusiasts, leased this two bay service station and converted it into a moped and motorcycle supply store. An inexpensive but practical conversion was accomplished by the owners. The overhead door openings were closed up and the sections of the doors were installed vertically, now serving as fixed windows. Wall to wall carpeting was installed throughout the interior. The walls were attractively covered with sales display items.

The store features cycle accessories, such as exhaust systems, windshields, tires, custom seats, and a wide variety of all weather clothes and footwear. Eight to ten used cycles are displayed in front of the building and under the canopy and are sold as an accommodation to customers.

The owners advise that ultimate expansion will be required to provide sales and services to this growing business.

A MODEST REMODELING OF A TWO BAY ABANDONED SERVICE STATION

DISPLAY OF CYCLES FOR SALE

Courtesy Tucker Cycle Supplies, Inc., Tucker, Georgia

MOTORCYCLE SUPPLY STORE

ALL-WEATHER CLOTHING

A WIDE RANGE OF ACCESSORIES

Courtesy Tucker Cycle Supplies, Inc., Tucker, Georgia

DOCTOR'S OFFICE

Two bay service station buildings lend themselves to excellent conversions for professional offices. A typical layout includes the waiting room, receptionist and doctor's offices, two examining rooms, and storage and rest rooms. The exterior of a twenty-year old colonial building can be restyled inexpensively to give it new distinction. The roof is extended over the door and windows to give a change in depth to the front of the building.

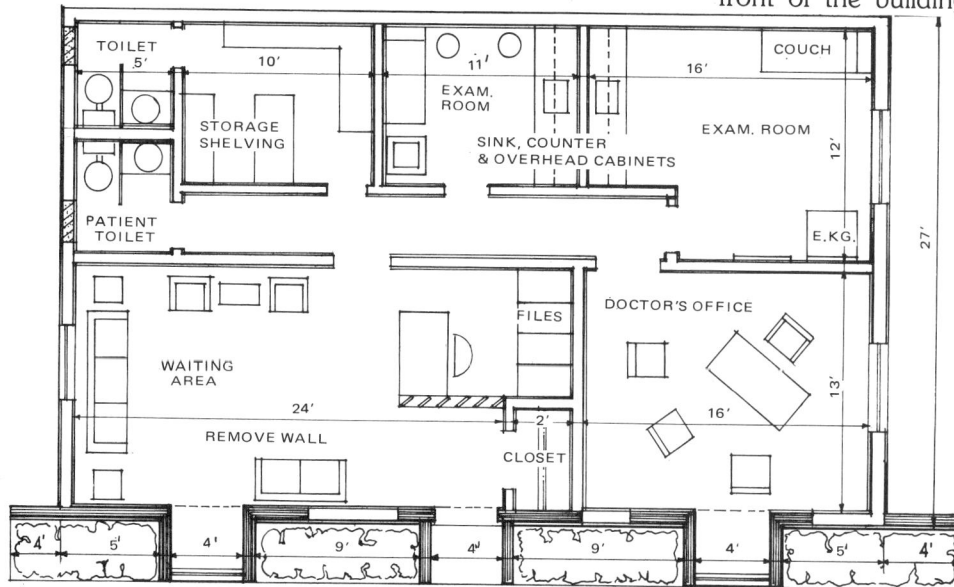

FLOOR PLAN

TOILET 5'
10'
STORAGE SHELVING
PATIENT TOILET
EXAM. ROOM 11'
SINK, COUNTER & OVERHEAD CABINETS
COUCH
16'
EXAM. ROOM
12'
E.KG.
27'
WAITING AREA
24'
REMOVE WALL
FILES
DOCTOR'S OFFICE
16'
13'
2'
CLOSET
4' 5' 4' 9' 4' 9' 4' 5' 4'

SCALE   3/32" = 1'-0"

LEFT SIDE ELEVATION

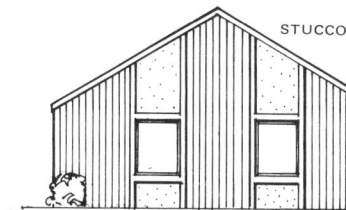

ROUGH SAWN WOOD BATTEN ON BOARD

RIGHT SIDE ELEVATION

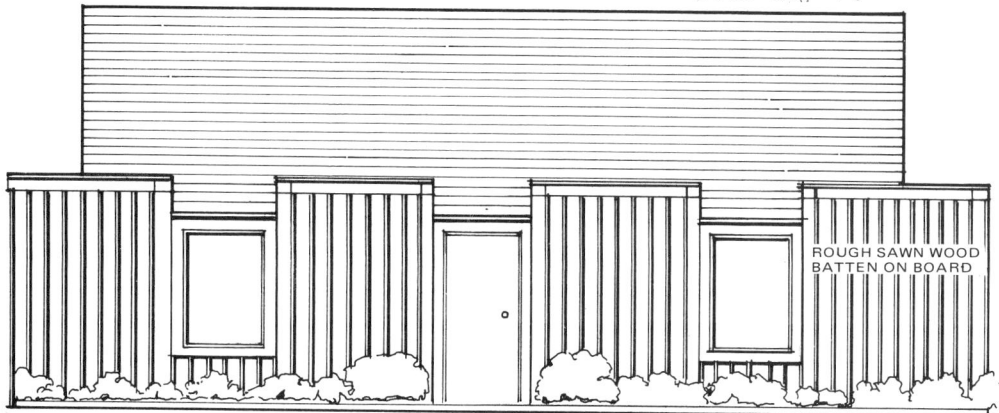

STUCCO

FRONT ELEVATION

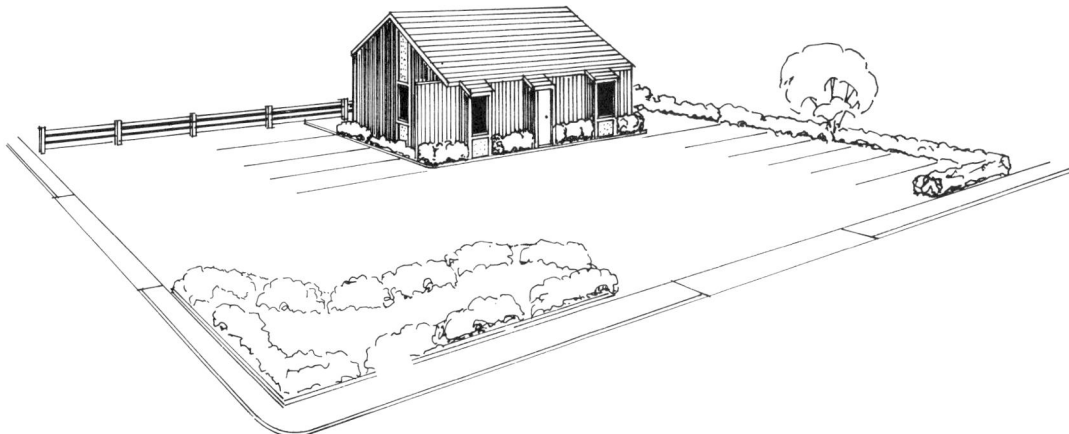

ROUGH SAWN WOOD BATTEN ON BOARD

OFFICE, PETROLEUM MARKETER

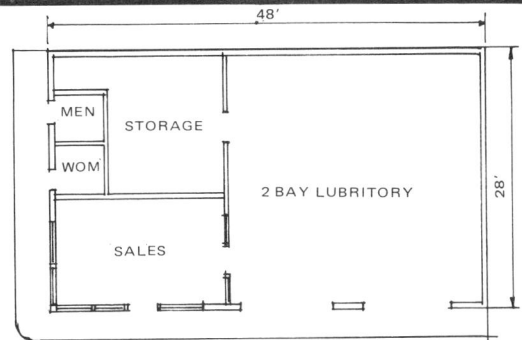

FLOOR PLAN OF BUILDING

THE FORMER TWO BAY SERVICE STATION

FOUR VIEWS OF THE CONVERSION TO AN OIL JOB BERS OFFICE

In 1977 an enterprising petroleum marketer from Butler, Georgia, was planning to remodel a two bay service station. He desired a dramatic change in the architectural treatment that would be an eye catcher to attract attention and gain more than his normal share of the transient and local traffic. He selected a design of an animal hospital found on page 49 of the author's first book, "A New Life for the Abandoned Service Station."

The jobber felt that this design provided the good looks and clean lines for a high volume outlet. The building was remodeled to operate as a distributor warehouse and retail self service station. A canopy was installed over a new and longer pump island. The converted building is colorful, attractive and different. As a result of the remodeling the retail business has increased six fold. The marketer served as a general contractor and supervised the entire project. The cost to accomplish the improvement was $33,800.

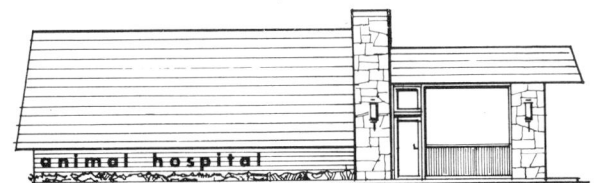

FRONT ELEVATION ANIMAL HOSPITAL

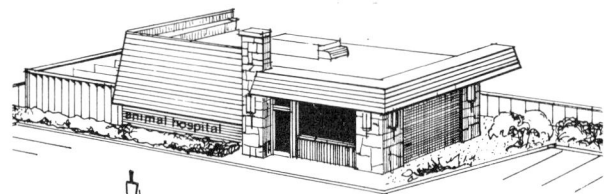

DRAWINGS THAT INFLUENCED THE OIL JOBBER'S CONVERSION

Courtesy Wilson Oil Co., Butler, Georgia

INSURANCE OFFICE

AN EXCELLENT TWO STORY CONVERSION

A COMMUNITY ASSET

AN ATTRACTIVE FACADE

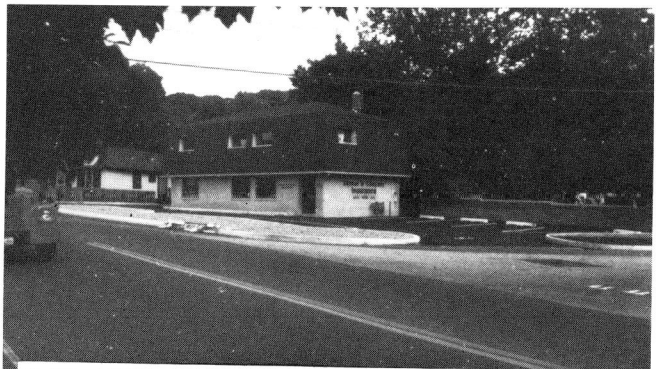

THIS VIEW ILLUSTRATES THE IMPORTANCE OF BUILDING VISIBILITY

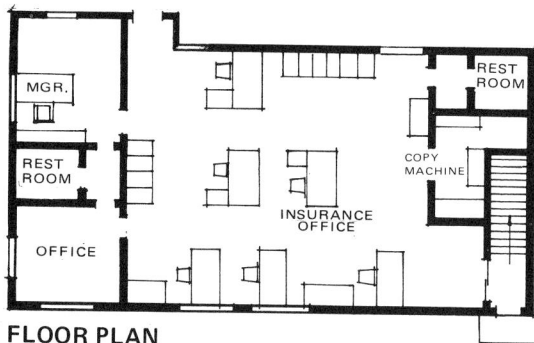

FLOOR PLAN

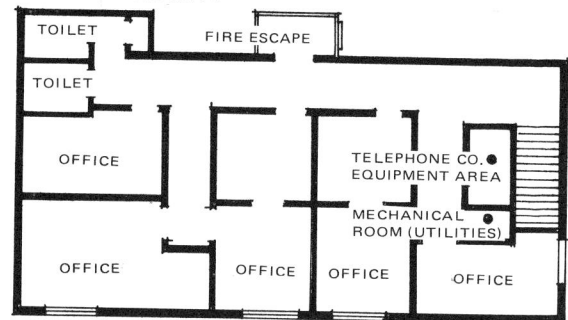

SECOND FLOOR PLAN

The owner of an insurance agency formerly rented an "intown" office of 900 square feet at a cost of approximately $560 per month. A new lease would have raised the monthly rent to a total of approximately $800. In his search for new office space, the insurance man found an abandoned service station in the suburbs of his community and purchased a 25' x 50' thirty year old building for $67,500. The original asking price had been $120,000. The new owner was aware that the town zoning board had twice denied previous requests for a variance to remodel the building to another use. The original service station was erected prior to the enactment of the current zoning laws of the community. By remodeling the old service station, the owner was able to retain the residence and light industrial buildings on the property as additional rental income.

Estimates, in the amount of $75,000, were obtained to convert the service station to a two story office facility. Due to the high costs, the owner decided to act as general contractor and sublet the construction work to various building trades. The renovation ultimately cost $55,000. The scope of work included:

1. Removal of interior partitions and the roof of the building. Increased thickness of existing-walls from 8" to 12" by adding 4" of concrete block. (The building department required 12" wall thickness for the second floor addition.)

2. Erection of prefab panel trusses for the second story structure. (The air conditioning and heating for the first floor were installed between the top and bottom chords of the truss.)

3. Installation of separate air conditioning and heating units on each floor.

Courtesy Salman & Company, Insurance, Ridgewood, N.J.

INSURANCE OFFICE

4. Installation of a built-up wood floor over the concrete floor of the old service station. Wood sleepers were shimmed level and two layers of plywood subflooring were installed for the wall to wall carpeting. This floor system permits the installation of electrical and telephone outlets under the flooring and provides maximum flexibility for desk arrangement in the insurance office.

5. Installation of attractive metal windows with thermal insulated glass.

The owner anticipated renting the second floor to one tenant; however, after four months, the area was unrented. The owner then decided to finish seven small offices on the second floor and completely furnished them with desks, chairs, cabinets and carpeting. Within a short period of time all the offices were rented, yielding $16 per square foot. Total monthly income from the tenants on the second floor is $900. Add this to the $850 paid by the insurance company and you can see how profitable it can be to convert an abandoned service station. The building is situated in a quiet community, ideal for local businessmen, sales and manufacturers representatives, etc. who require an office with parking facilities, for their activity. The owner provides an answering service for his tenants at a nominal charge.

The owner is one of many professionals and businessmen we have interviewed who "bit the bullet" and acted as general contractor, sublet the work to subcontractors and saved on conversion costs.

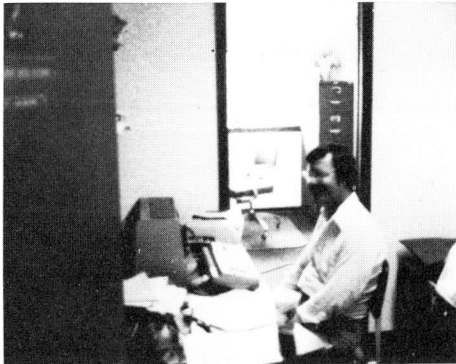

SECOND FLOOR RENTED OFFICE SPACE

STORE ROOM AND COPYING MACHINE

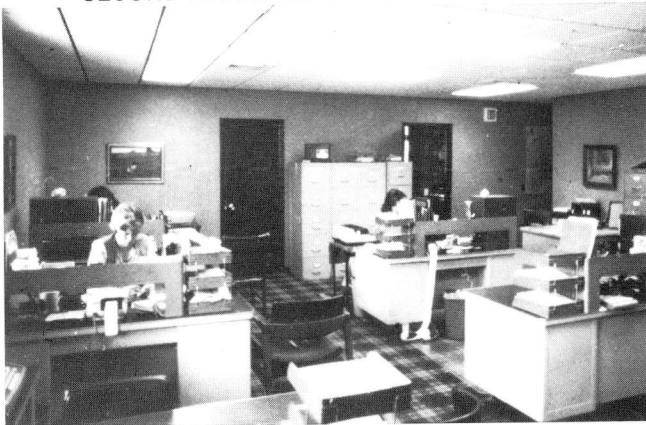

FOUR VIEWS OF THE ADMINISTRATIVE OFFICE

LAW OFFICE

The conversion of service stations with canopies generally result in an excellent remodeling. The recycling of a two bay porcelain box into a law office is another example of the maximum utilization of the available structure. The building was remodeled in September 1979 at a cost of $60,000 by an attorney who, with his contractor, designed the interior and exterior improvements. The property measures approximately 135' x 165' and has ample parking space for employees and clients. The interior parcel is situated on a heavily travelled main highway in the suburbs of a large city.

The original building contained approximately 1100 square feet of floor space. The floor area was increased an additional 700 square feet by enclosing the perimeter of the canopy. The building provides offices for four attorneys, a law library conference room, storage room, two rest rooms, a spacious waiting room and desk space for three secretaries. A brown brick exterior and a new four foot high mansard roof covered with brown shingles presents a warm inviting building to the public. Expensive paneling was used throughout the interior. The parking area was upgraded and landscaping installed at a cost of $4,000. The conversion of the building cost the owner $33 per square foot a reasonable price for construction of this quality. In the author's opinion the building has a market value of $90,000.

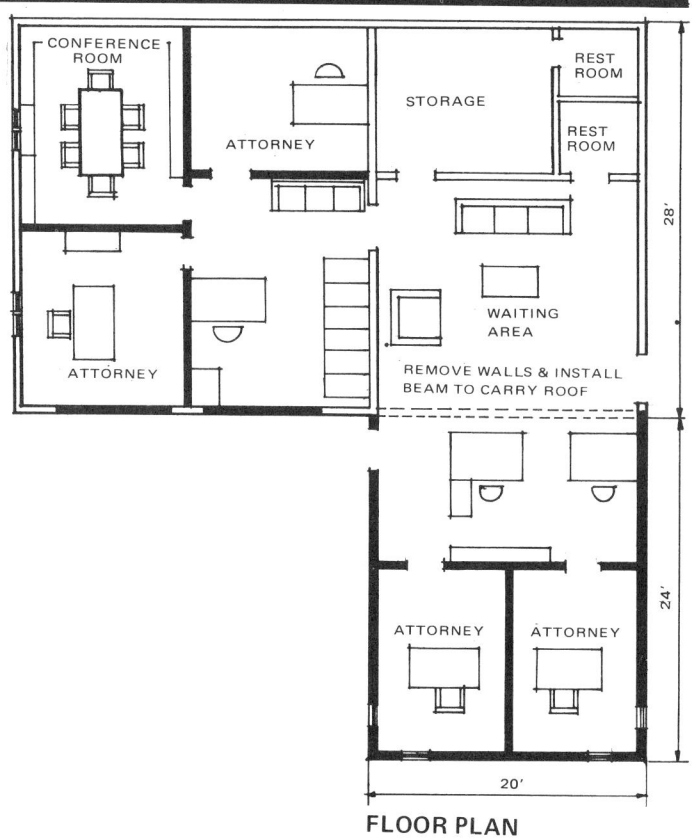

FLOOR PLAN

PREPARING THE CANOPY FOR A NEW ROOF

THE ENCLOSED CANOPY PROVIDES ADDED OFFICE SPACE

Courtesy Snyder, Leonard, Bigger & Dodd, Asheville, N. C.

LAW OFFICE

WAITING AREA

LAW LIBRARY AND CONFERENCE ROOM

SECRETARIES AND RECEPTION AREA

ATTORNEY'S OFFICE

ATTORNEY'S OFFICE

An important advantage to the owner is the benefit of the investment tax credit. The major disadvantages encountered were the abnormally high cost to complete the plumbing and electrical work as well as the excavation and removal of the underground storage tanks. The owner is to be congratulated for his efforts that have resulted in the removal of a blight on a highway in a popular resort community.

Courtesy Snyder, Leonard, Biggers & Dodd, Asheville, N. C.

MUNICIPAL BUILDING

In 1979, a municipal utility distributing natural gas and electricity under a franchise agreement to a small midwestern city was looking for a new branch business office in the downtown area. However, because there was no vacant land available, it was necessary to buy existing developed property. A former three bay canopied service station became available and was purchased for $40,000. Approval of the Planning Commission, the City Council and the Regional Building Department was required to change the business use of the abandoned service station.

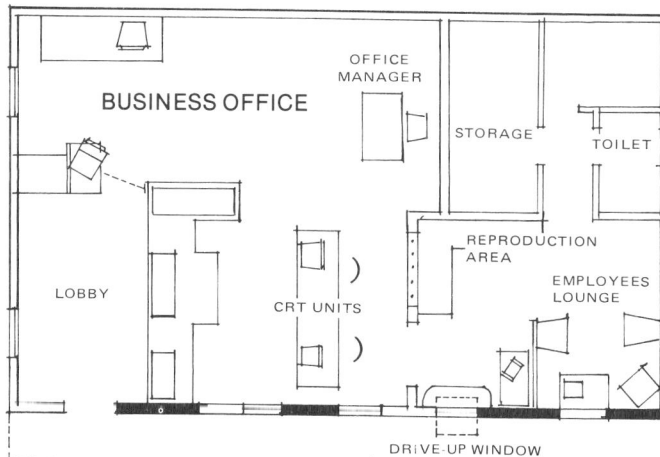

A contractor was retained to block up the overhead doors, modify other openings and apply a new face brick exterior. Canopy columns were enclosed with brick and an attractive planter was constructed covering the old pump island area. A new concrete yard was installed and fencing erected around the perimeter of the property.

The interior work was performed by both the contractor and the personnel of the utility company. The paneling, new acoustical ceiling, counters, cabinetry, and general woodwork was well done. The former service station heating system was retained and one of the rest rooms was converted into a deluxe facility for the employees. The cost to accomplish this work was approximately $55,000.

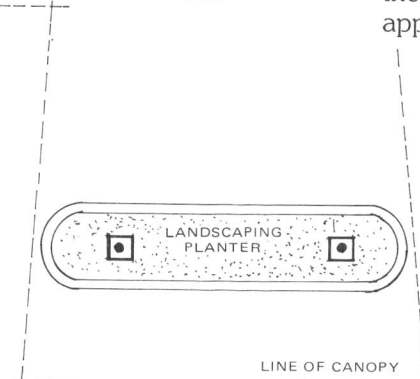

FLOOR PLAN

Courtesy City of Colorado Springs, Department of Public Utilities, Manitou Division Office, Colorado.

MUNICIPAL BUILDING

A BRIGHT ATTRACTIVE OFFICE

THE ENTRANCE LOBBY

THE CRT UNITS ARE IMPORTANT TO PROVIDE
CURRENT DATA AT ALL TIMES

REPRODUCTION EQUIPMENT

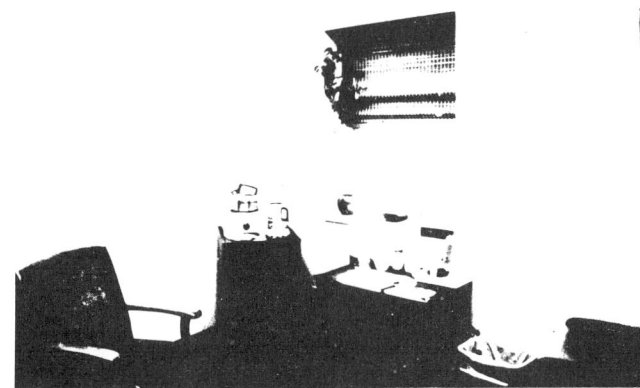

EMPLOYEE LOUNGE AREA

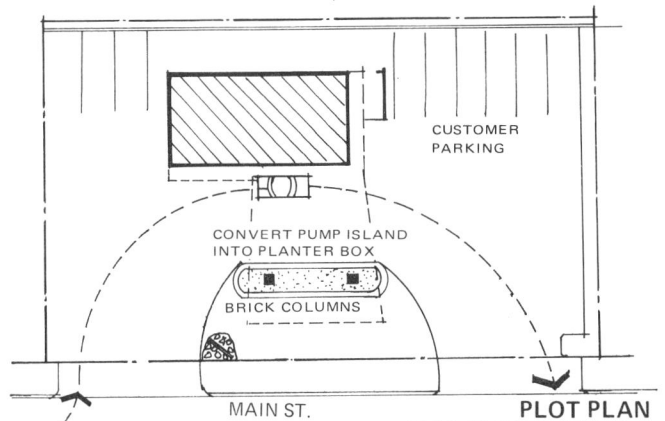

CUSTOMER
PARKING

CONVERT PUMP ISLAND
INTO PLANTER BOX

BRICK COLUMNS

MAIN ST.

PLOT PLAN

Courtesy City of Colorado Springs, Department of Public Utilities, Manitou Division Office, Colorado.

OPTICIAN'S OFFICE

OPTICIANS OFFICE

AN OWNER CREATED DESIGN

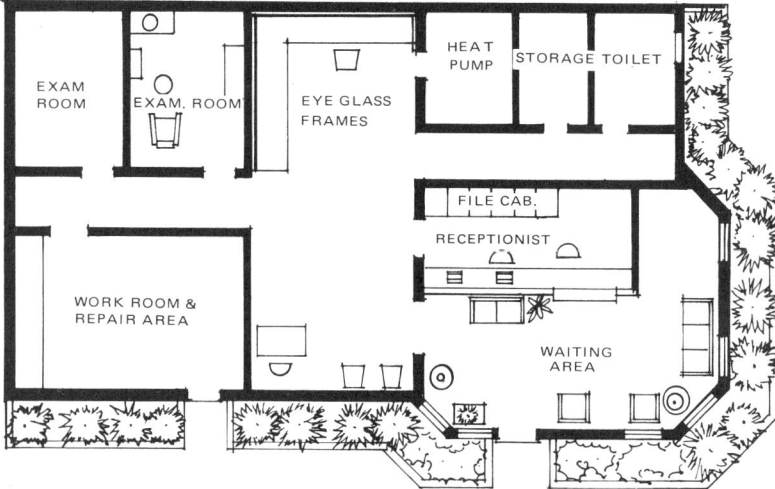

FLOOR PLAN

An optometrist purchased a former two bay service station in 1976 for $43,000. The property, measuring 100' x 100', is situated on a heavily travelled main highway passing through the center of the city. The owner and a local contractor consulted on the scope of work and the materials to be used. The interior is panelled with cherry wood; a suspended acoustical ceiling was installed; and the sales wing was converted to a waiting room, receptionist's office and a room for the heat pump. The lubritory area was converted into two examining rooms, a work room and a fitting area adjacent to the mounted displays of eye glass frames. The exterior is attractively treated in a contemporary motif with a tinted blue stained vertical siding. As a result of careful planning, using selective materials and a good contractor, the project was completed in thirty days at a cost of $27,000.

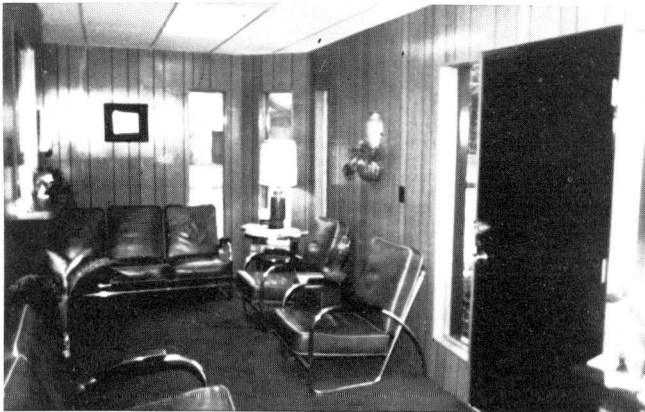

THE WAITING ROOM

THE RECEPTIONIST AND FILES

EXAMINATION ROOM

DISPLAY OF EYEGLASS FRAMES

Courtesy Dr. Colin Brown, Johnson City, Tennessee

OPTICIAN'S OFFICE

THE RESTYLED THREE BAY COLONIAL BUILDING

MULTI-BUSINESS USE

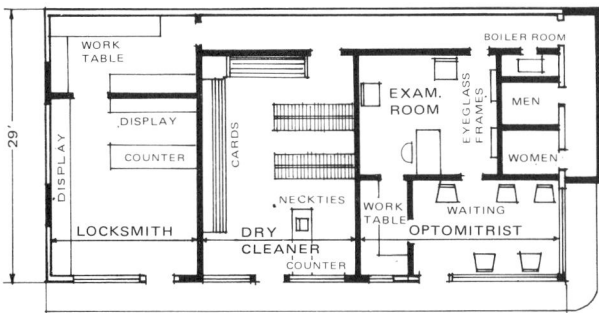

FLOOR PLAN

A three bay colonial building in the center of town, opposite the municipal building had been vacant for eighteen months until an optician bought the 150′ x 150′ property for $51,000. The original asking price had been $65,000. The new owner required only 430 square feet of the 1,470 square foot building and subdivided the remainder of the floor area into two rental areas. The two tenants are a dry cleaner and a locksmith.

The owner sublet the work to the various building trades to complete the following:

1. Raise the floor throughout the lubritory bay area.

2. Perform plumbing, electrical, carpentry and painting to install a new suspended acoustical ceiling, floor mounted warm air heating system and storefronts.

The cost to accomplish this work was $24,000.

The only disadvantage the owner encountered is the lack of storage area. Ultimately he will expand toward the rear of the property.

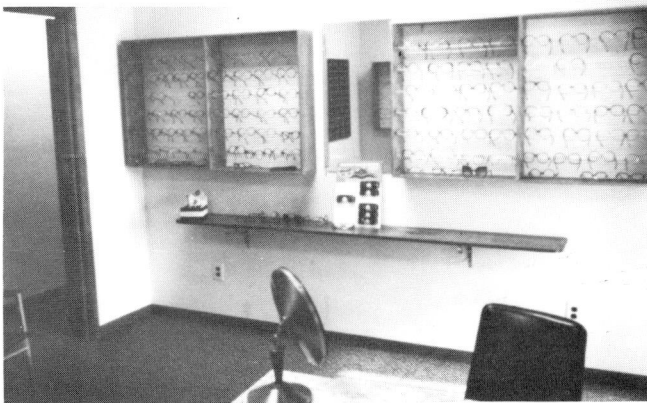

EXAMINING ROOM AND DISPLAY CASES FOR EYEGLASS FRAMES

THE DRYCLEANER

THE LOCKSMITH SHOP

Courtesy Deptford Eyeglass Center, Deptford, N.J.

REAL ESTATE OFFICE

In September, 1974, a realtor joined the "Gallery of Homes" and started looking for a new headquarters for his real estate firm. He purchased for $48,500 this three bay canopied service station that had been closed for more than eighteen months. The building is situated on a full block front, (approximately 175′ x 100′) on a heavily travelled main artery. The major requirement of the new realty network is a large reception area with a living room atmosphere. The owner and his architect decided to utilize the canopy to enlarge the building floor area by 300 square feet and enclosed the canopy and retain the 13′ high ceiling.

FLOOR PLAN

The side walls were constructed of stone to a height of 7′ and glass was installed to the underside of the roof. The 13′ front wall is made entirely of glass. A fireplace, two attractive chandeliers and a round table, sofa and chairs create a warm atmosphere not seen in many real estate offices. An impressive feature of the office is the area occupied by the sales representatives. The room measures approximately 13′ x 45′ with one wall of partitioned booths as illustrated. The overhead doors were covered with vertical cedar siding. The parking area for customers was landscaped in a professional manner. The building conversion was accomplished at a cost of $33 per square foot, ($66,000).

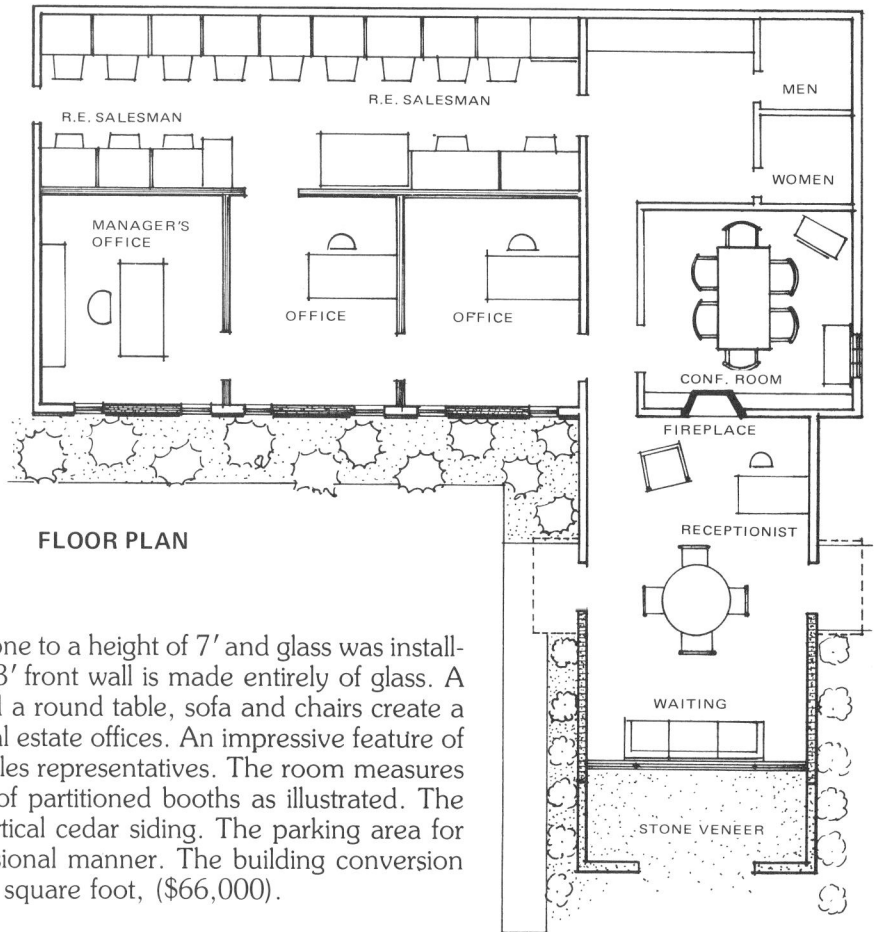

THE SIGN AND LOGO IS MOUNTED OVER THE FOCAL POINT OF THE BUILDING

NATURAL WOOD SIDING CONCEALS THE OLD BAY AREA

THE CONVERSION OF A THREE BAY CANOPIED SERVICE STATION

Courtesy J. D. Jackson Associates, Asheville, N. C.

REAL ESTATE

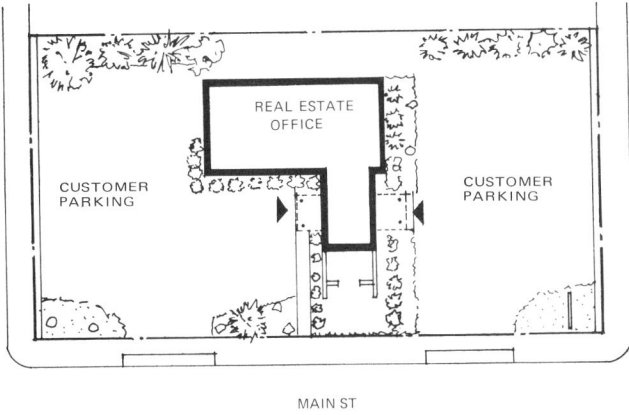

PLOT PLAN

MAIN ST

THE BUSY REAL ESTATE SALES ROOM

A WARM LIVING ROOM ATMOSPHERE

A MULTI - PURPOSE ROOM FOR CLOSINGS AND CON-
FERENCES

FURNITURE, CHANDELIERS AND FIREPLACE
DECORATED THE RECEPTION ROOM

Courtesy J. D. Jackson Associates, Asheville, N. C.

OFFICE, REAL ESTATE INVESTMENT COUNSELOR

ATWOOD'S, REAL ESTATE COUNSELING

A UNIQUE CONVERSION OF SMALL BUILDING AND PLOT

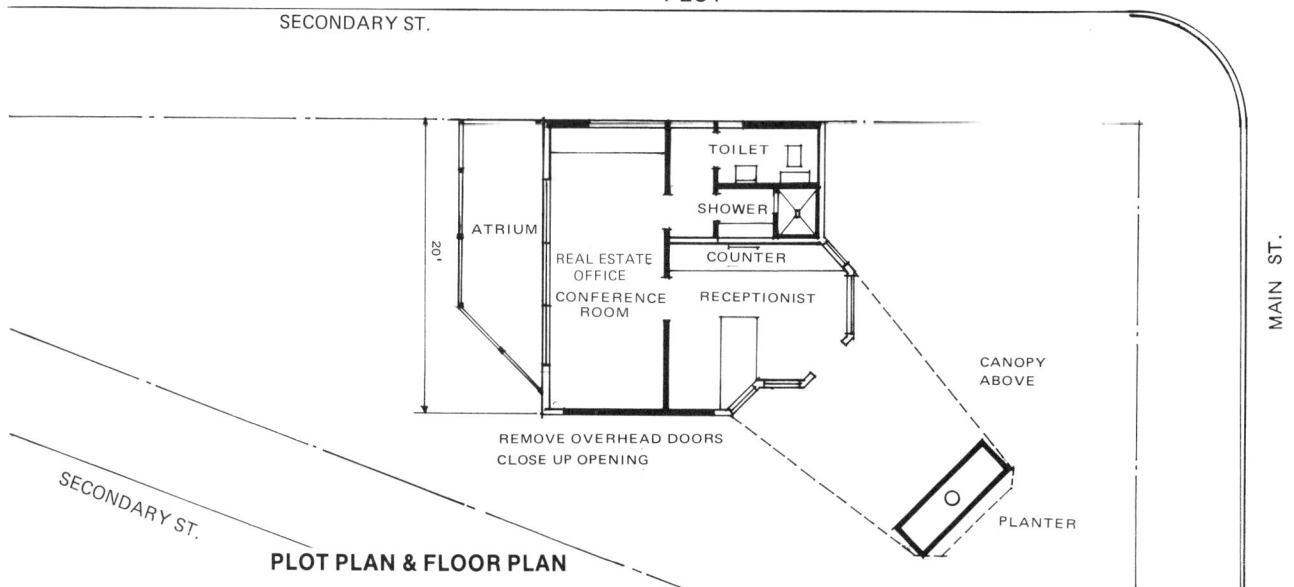

PLOT PLAN & FLOOR PLAN

A real estate investment counselor from Portland, Oregon, seeking an office for his business, purchased an abandoned service station and plot in 1974. The service station was a one bay building containing 450 square feet of floor area, situated on a 2800 square foot triangular parcel. No local zoning problems were encountered in the change of the business use of the building. A local architect was retained to prepare the alteration plans.

The old sales room is now a functional office for the counselor's secretary, complete with base and wall cabinets for storage. The old lubritory has been converted into an attractive 10' x 20' office and conference room.

The following scope of work was performed to convert the former service station;

1. Removal of the overhead doors and enclosure of the opening with 2" x 4" studs, 1/2" gypsum board inside, with 3/8" exterior plywood on outside wall. Existing windows were retained.

2. Removal of the existing lift and installation of a level concrete top over exisiting floor. Installation of new carpeting throughout building.

3. Installation of new interior 2" x 4" stud partitions with 1/2" gypsum board.

4. Installation of suspended acoustical ceiling with 6" batt insulation throughout building, with recessed lighting fixtures.

5. Conversion of one rest room into a large fiberglass lined shower. Remodeling of one rest room with new fixtures, accessories and the installation of vinyl floor and electric unit heater.

6. Installation of built-in counter top in office and conference room; new base and wall cabinets and new counter top desk with formica finish.

7. Installation of new entrance door and side light panels, and complete painting of the interior and exterior.

Courtesy J. A. Atwood, Portland, Oregon

OFFICE, REAL ESTATE INVESTMENT COUNSELOR

THE INTERIOR OF THE INVESTMENT COUNSELOR'S OFFICE

THE SECRETARY AND RECEPTION ROOM

A FIBER GLASS LINED SHOWER

8. Installation of air conditioner on roof with ducts to offices. Air conditioning unit contains add on electric heat section.

9. Installation of rough sawn cedar fence around landscaped area at rear of building; planter around old pump island and matching cedar boards around building.

The above work was completed at a cost of $17,000, (1974 dollars).

The owner has not made any structural changes to the building in order that he may retain the right to operate the facility as a service station at some future date if he so desires. The building department was advised that the remodeling was performed for the owner's own personal office on an interim basis only. The reason for converting the service station in this manner is that the property will appreciate in value while it is being used as an office by the owner. The greatest advantage to the owner has been the increased name familiarity and identity this high traffic central business district location has afforded.

Courtesy J. A. Atwood, Portland, Oregon

LOAN OFFICE & OFFICE CENTER

STAGE I - A NEW FINANCE OFFICE

STAGE II - A NEW ADDITION FOR RETAIL RENTAL

STAGE III - ANOTHER ADDITION
THE COMPLETED BUILDING WITH SPACE FOR SEVEN TENANTS

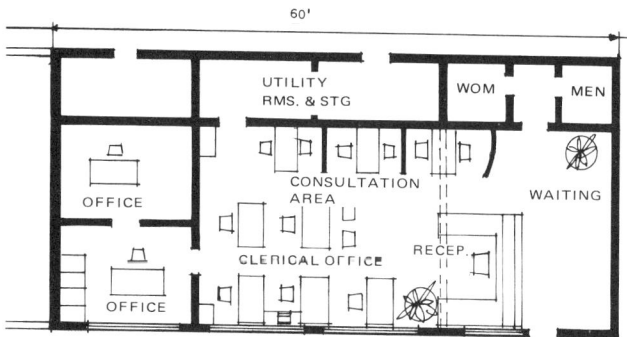

FLOOR PLAN

In 1973 a three bay service station, situated on a large parcel on a major highway, was converted to a financial business office. A modest but attractive remodeling of the building was accomplished by replacing the overhead doors with a new storefront, applying brick veneer to the front wall and installing a mansard roof covered with handsplit wood shingles. The interior alteration included removal of old walls and provisions for an open space plan for the administration department of the loan office. In 1977 an addition of approximately 100' was completed, and before the paint dried another 75' addition was started. The imaginative owner has developed a series of rental spaces in an attractive manner. The simplicity of the conversion and subsequent additions has permitted the owner to develop a series of stores to better serve the community.

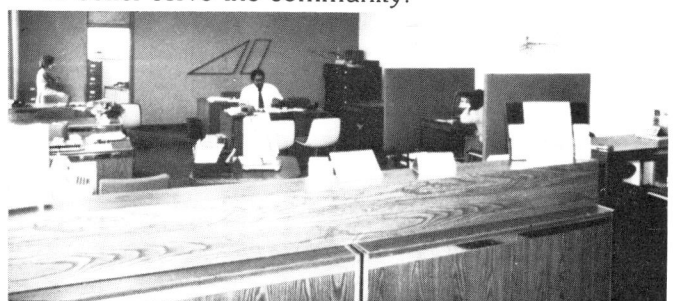

THE ADMINISTRATIVE OFFICE FOR FINANCIAL SERVICES

Courtesy Avco Financial, Trenton, N.J.

THE BOOMING PET FOOD BUSINESS

In 1980 for the first time in many years, the dog population in the United States declined. However, pet food sales have continued to boom because Fido is no longer getting table scraps. The weight of the average dog has increased from thirty-five to thirty-seven pounds, and canned and dry pet foods have gained in popularity. An indication of the increasing growth of this business is the average 210 lineal feet of shelving space devoted to dog and cat food in supermarkets. More space is devoted to pet foods than any other grocery item. The Wall Street Journal said that sales have increased at nearly 4% per year during the past five years. Pet food sales were $2.5 billion in 1976; $3.1 billion in 1978; and $3.7 billion in 1979.

It costs approximately $100 a year to feed the average dog and $65 to feed the average cat. Approximately 40% of all American households own at least one dog and 20% own at least one cat. This amounts to forty-one to forty-nine million pet dogs and twenty-three to twenty-six million pet cats.

In the author's opinion, the best business opportunity is the conversion of an abandoned service station to a pet food store. To be successful, it should be located in a business district of a densely populated suburban community. Pet products typically yield profit margins ranging between thirty-five and fifty per cent. (This includes the very high profit margins obtainable from pet accessories). The average pet owner is careful to make certain that his animal is well fed and cared for.

This is a business that has been captured almost exclusively by the supermarkets and small grocery stores. Many convenience food stores now devote thirty to fifty lineal feet of shelving for pet foods and accessories. The cost to convert a lubritory to a sales area or warehouse is nominal. This is a "no frills" store with the customers serving themselves. Shelving, bins and counters can be fabricated by a carpenter-contractor.

PET FOOD STORE

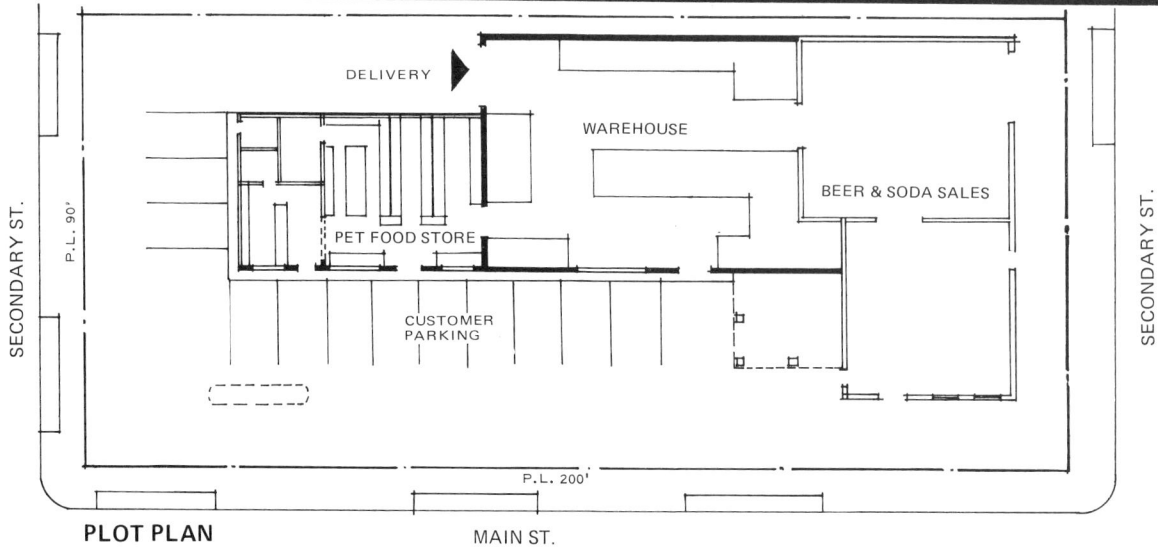

DELIVERY

WAREHOUSE

BEER & SODA SALES

PET FOOD STORE

CUSTOMER
PARKING

SECONDARY ST.

SECONDARY ST.

P.L. 90'

P.L. 200'

PLOT PLAN

MAIN ST.

THE CLOSED SERVICE STATION

STAGE 1 COMPLETED 1976

A UNIQUE LOGO

THE MAJOR CORNER IN TOWN

Courtesy Mutts & Butts, Merrick, N. Y.
Drawing Courtesy, Leon Rosenthal, A.I.A., Architect, Babylon, N. Y.

PET FOOD STORE

A VIEW OF THE ENTIRE FRONTAGE

THE REMODELING UNDERWAY

STAGE II COMPLETED 1980

PET STORE, WAREHOUSE AND BEER AND SODA STORE

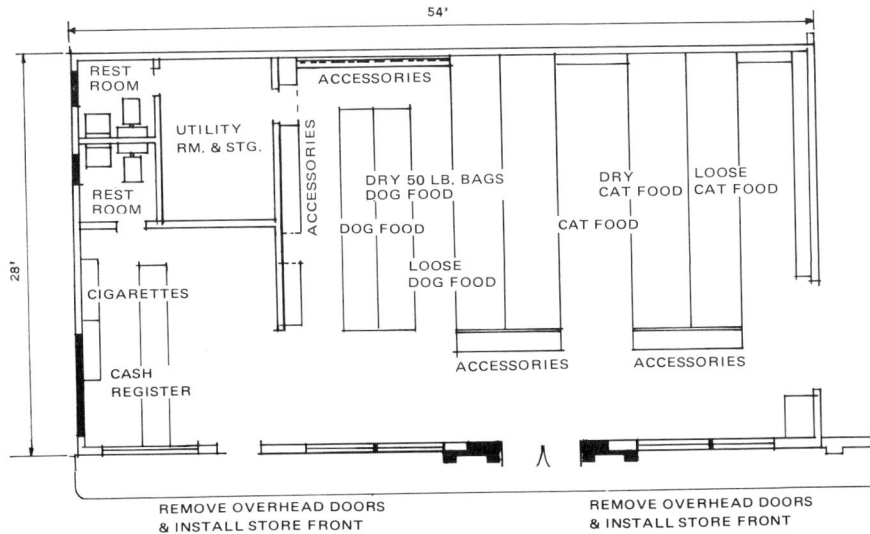

FLOOR PLAN

Courtesy Mutts & Butts, Merrick, N.Y.
Drawing Courtesy, Leon Rosenthal, A.I.A., Architect, Babylon, N.Y.

PET FOOD STORE

The owner of a beer and soda distributorship, situated on a heavily travelled commercial artery in Long Island, purchased an adjacent abandoned service station in 1976. The principal, a man of vision, decided to test a new business concept and converted the three bay service station into a discount pet food store for a minimal investment of $20,000. The only improvement that was made at that time was the installation of metal shelving to the 12′ high ceiling. Four foot aisles were left for the fork lift truck to stack the product. The overhead doors remained for convenient trailer truck deliveries. Sales were by the case only; i.e., 12, 24, and 48 cans to the case of one type and brand of pet food. Customers could not mix the case cans and obtain a variety of products. Most pet foods were discounted 5¢ per can below supermarket prices. The majority of customers purchased one or two months' needs for their pets and store sales were brisk, even though initially the store had a limited inventory and seven out of ten customers did not buy anything. After two years, the steady increase in business convinced the owner that he must expand.

In 1980 the service station was remodeled with added shelving and display aisles, and a twenty foot high warehouse addition was erected. The new addition extends from the pet food store to the beer and soda store. The warehouse addition enabled the owner to carry a larger variety of pet food brands than supermarkets at a substantial discount, and business has tripled as a result of the expanded facilities and additional weekly newspaper advertising. Currently only one out of twenty customers will leave the store without buying any products. It is interesting to note that a number of customers will shop in both stores.

The owner continues to discount in the 5¢ per can range even if he loses money on a few items. He feels

BISCUIT SALES BY THE POUND

CANNED AND DRY PET FOOD

ACCESSORIES DEPARTMENT

Courtesy Mutts & Butts, Merrick, N. Y.
Drawings Courtesy Leon Rosenthal, A. I. A., Architect, Babylon, N. Y.

PET FOOD STORE

it is essential to maintain a discount image with the consumer. The store sells for the same price a grocer pays for the product when he buys from a pet food warehouse. Customers now have the opportunity to mix cans of pet foods as well as purchase the products by the case. A complete line of accessories is also available. Cigarettes by the pack or carton are sold in the sales area at the cashier's counter.

The owner recommends a $100,000 inventory be maintained for a complete line of pet foods. This permits the store to devote a section to gourmet pet foods. According to the Wall Street Journal, a pet owner can select special diets for old dogs, young dogs, fat dogs, urban and suburban dogs. The Wall Street Journal also states that the average supermarket devotes approximately 210 lineal feet of shelf space to dog and cat meals, an increase of approximately 30% in five years.

The average sale exceeds $10 per customer and many other sales range from $50 to $100. Pet owners are extremely particular about the quality of the food they buy. The owner states that one of the reasons for his success is that he carries only the best brands of pet foods. At Christmas time, pet owners will buy expensive accessories and treats for their pets as presents. Accessories can yield as much as 100% profit. There are also many pet owners who may have from five to ten cats to feed. Their pet food expenditures can run as high as $150 monthly. It is essential that they obtain a discount and buy by the case.

The owner has worked hard to establish his business. He made contact with all manufacturers of pet food products and ultimately established a line of credit in the industry. He will purchase only trailer load quantities. It is interesting to note that two years ago the most profitable of the two businesses was the beer and soda distributorship. Currently, with the completion of the new warehouse addition and the remodeling of the service station, pet foods are more profitable.

As the writer stated on the opening page, most of the successfully businesses have outgrown their original service stations and have added storage or sales areas if possible. It is an important fact that very few entrepreneurs will relocate a successful business to a new location. They would rather live with the hardship of limited space than gamble on a new site.

The success of the pet food store is a credit to the owner's vision, hard work and perserverance. He is ably assisted by his wife and two hardworking sons who help manage the two businesses. It was also determined that the owners are considering franchising their business concept.

CASHIER COUNTER AND CIGARETTE SALES

PET CARRIERS, ACCESSORIES, CANNED AND DRY DOG FOOD

GOURMET PET FOODS CASE AND SALE OF PET FOODS BY THE CAN

Courtesy Mutts & Butts, Merrick, N.Y.

Drawings Courtesy Leon Rosenthal, A. I. A., Architect, Babylon, N. Y.

PET FOOD STORE

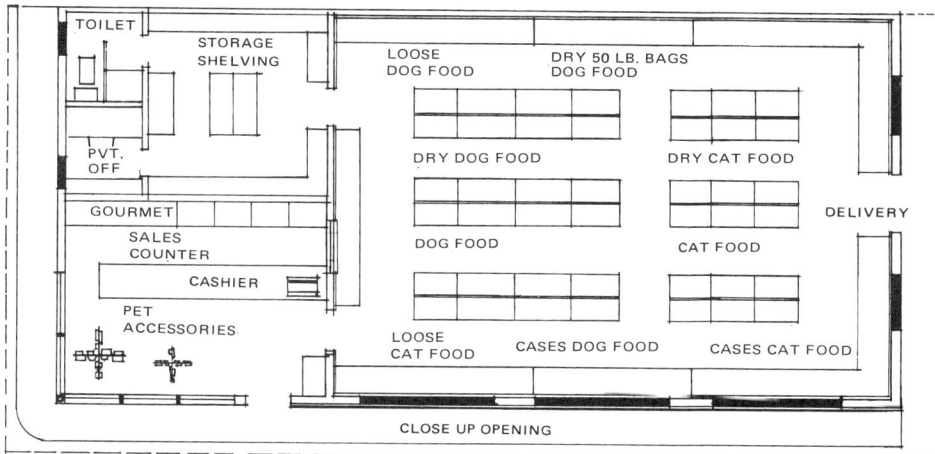

Unlike the warehouse concept, with sales by the case or carton, this floor plan provides gondola and shelf areas for self-service customers. Two-thirds of the store has been devoted to pet food sales. To meet the needs of the consumer, a variety of pet food products are available at discount prices. Higher priced pet foods are becoming popular; therefore, selection of brands is important. Sufficient counter space must be dedicated to dry dog food, which accounts for one-third of all pet food sales for the economny-minded consumer. Sufficient area should be available for warehouse and storage expansion. The conversion of the old three bay building to a profitable new business venture can be accomplished with a limited budget as illustrated. A diagonal redwood siding exterior is attractive and requires a minimum of maintenance.

FLOOR PLAN

SCALE 0' 5' 10' 15'

FRONT ELEVATION

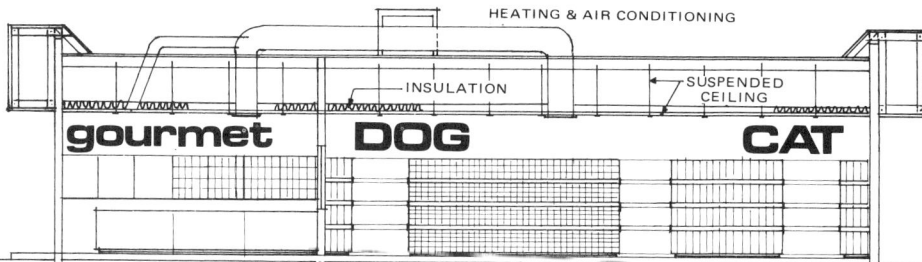

PET PLACE

DIAGONAL WOOD SIDING

LONGITUDINAL SECTION

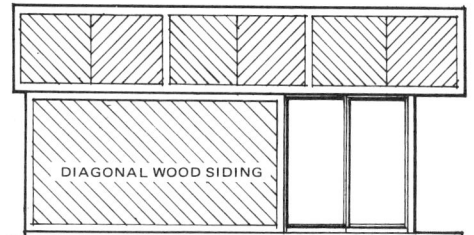

HEATING & AIR CONDITIONING

INSULATION

SUSPENDED CEILING

gourmet   DOG   CAT

LEFT SIDE ELEVATION

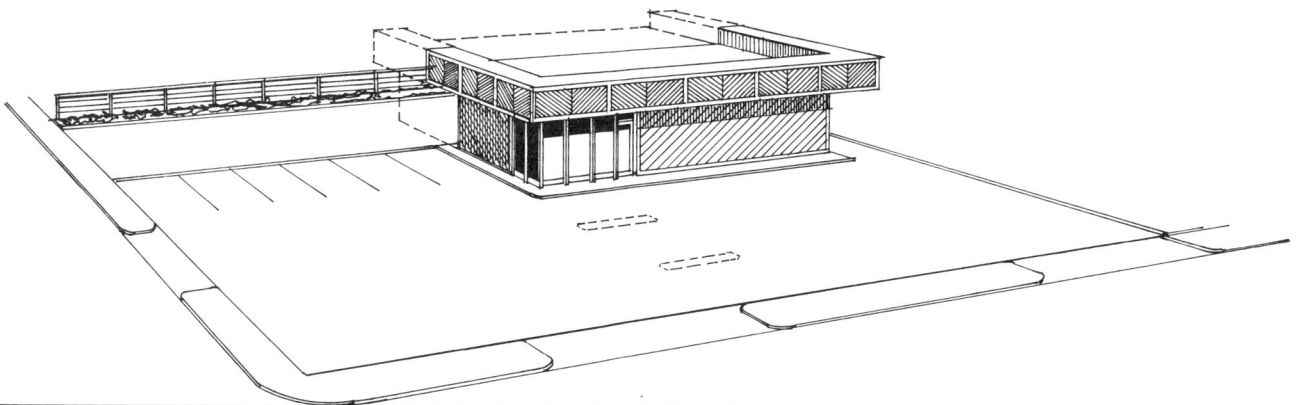

DIAGONAL WOOD SIDING

PET GROOMING CENTER

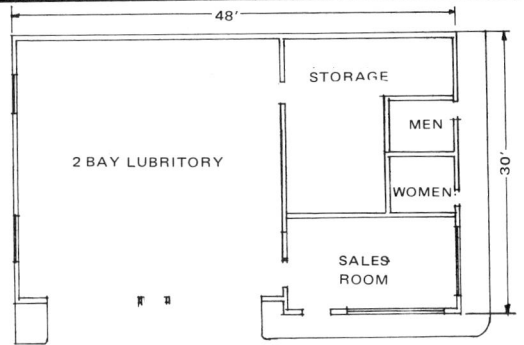

FLOOR PLAN OF ABANDONED BUILDING

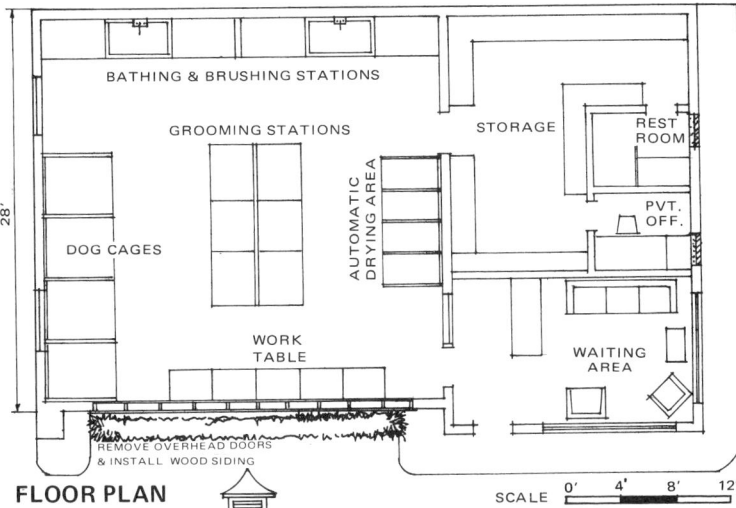

FLOOR PLAN

SCALE 0' 4' 8' 12'

Americans spend more than $10 billion annually on their dogs. Substantial sums are spent on dog grooming even during periods of a sagging economy. This former two bay colonial service station can be converted into an animal grooming center at minimal cost. The overhead door openings may be covered with batten on board and an attractive planter. The sloping lubritory floor and floor drains are an asset to the new business. Each night the floors must be washed thoroughly after animals have been washed and groomed. White epoxy painted walls are essential for a sterile appearance.

FRONT ELEVATION

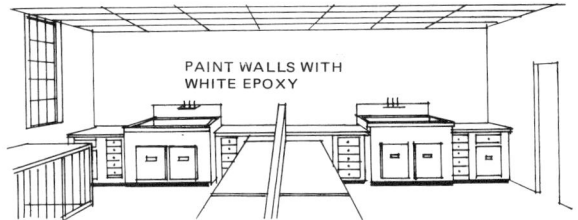

INTERIOR VIEW

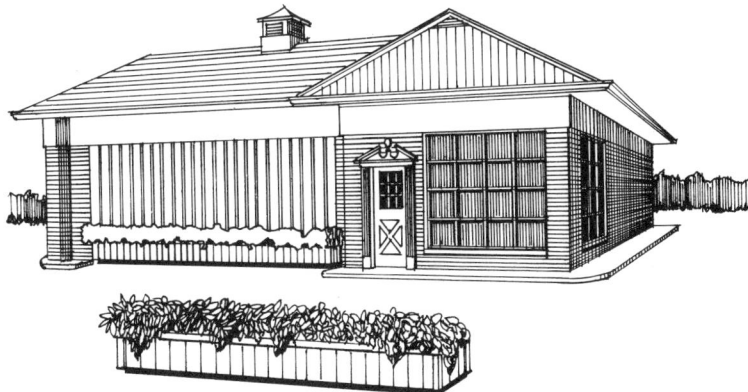

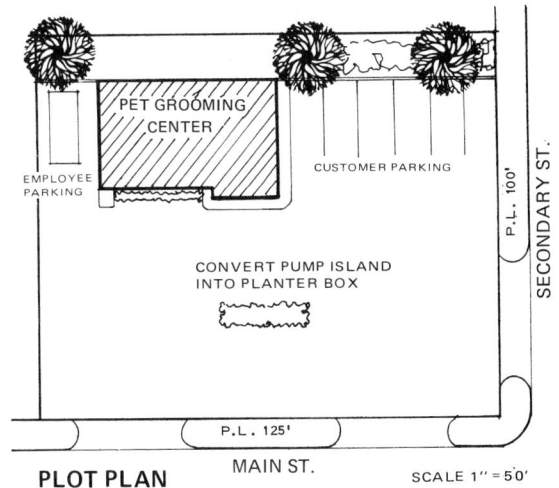

PLOT PLAN   MAIN ST.   SCALE 1" = 50'

PLANT STORE

THE ABANDONED TWO BAY BUILDING

THE INTERIOR OF THE LUBRITORY PRIOR TO REMODELING

THE INTERIOR OF THE LUBRITORY AS THE REMODEL-ING IS COMPLETED

THE FIRST STAGE OF THE PLANT STATION

In 1973 a college graduate with an excellent business background, who wanted to work for himself and had $1500 to invest, leased an abandoned two bay service station located in a densely populated suburban area on a heavily travelled major artery. The lease was for $400 per month for a five year term with a five year option and an inflationary clause tied to the Consumer Price Index. The young entrepreneur obtained a $5000 loan from a local branch bank to renovate this forty to fifty year old solid brick service station, containing 1100 square feet, located on a small 6,750 square foot lot. The interior had peeled; and the sheetrock was in poor condition. The local zoning board was in favor of the conversion to another business because it was a higher and better commercial use and eliminated a blight in the community.

Courtesy The Plant Station, Abington, Pa.
Courtesy Regis Mc Cann, Abington, Pa.

PLANT STORE

A RUST COLORED RED BRICK WAS COVERED WITH PROCELAIN ENAMEL

THE EXTERIOR IS REMODELED

The conversion was a monumental task for a young businessman, but he was not discouraged. He became his own general contractor and sublet mechanical work that required a licensed professional. Much of the work was performed by the new tenant and one helper. He also designed and specified every detail of the remodeling. The work was done in two stages. The first stage included the following work:

1. The installation of two skylights for plant light.
2. The sandblasting of interior walls.
3. The installation of 1″ x 6″ pine flooring, laid ½″ apart and painted with epoxy varnish (provided drainage after plants were watered).
4. The installation of rough sawn hanging arbors and wood plant benches.
5. The installation of plumbing lines to plant areas.

The business was an immediate success, and sales increased constantly. The store was closed and the second stage was started. This included:

NEW BAY WINDOWS

REMODELING STAGE 2

1. The erection of a 15′ x 24′6″ greenhouse.
2. The addition of a slat house.
3. The removal of porcelain from the exterior walls.
4. The sandblasting or cleaning of the masonry walls.
5. The installation of bay windows at the old overhead door openings.
6. The installation of a mansard roof with handsplit wood shingles.
7. The paving and striping of the entire parking area.
8. The installation of 100 amp electric service and a business sign.

Courtesy The Plant Station, Abington, Pa.
Courtesy Regis McCann, Abington, Pa.

PLANT STORE

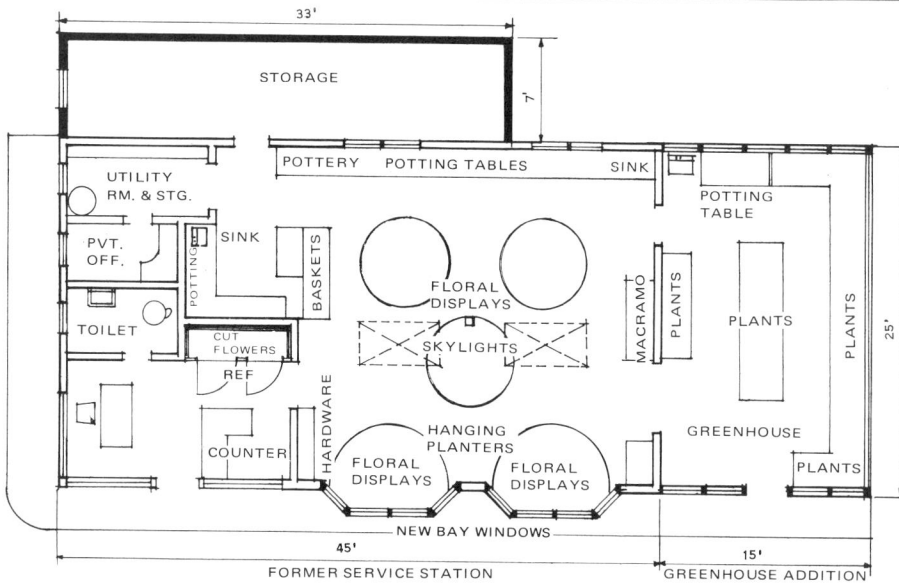

FLOOR PLAN

POTTED PLANT RACK DISPLAYS

GIFT DISPLAYS

THE SLATTED FLOOR IS FINISHED WITH AN EPOXY

A WATER SPRINKLER SPRAYS PLANTS DAILY

FLORAL AND PLANT DISPLAYS

Courtesy The Plant Station, Abington, Pa.
Courtesy Regis McCann, Abington, Pa.

PLANT STORE

The cost to accomplish the conversion work was $27,000.

Subsequently the store was sold to a former employee who is having continued success with the operation. An important feature of the store is to provide plant service for all plants in certain price ranges. This continued service results in repeat sales because customers are guaranteed satisfaction. Average sales range from $1 to $5, with some sales as high as $300.

On July 30, 1975, the owner was awarded a "Business for Beauty" certificate of commendation by the Council of the local township.

The successful young businessman made several comments of great interest.

1. The owner of a plant store must stay on the premises. Small businesses require personal contact.

2. He had been approached by investors to franchise his business but declined. He has this advice for those interested in franchising their business. "A good franchise is one that continues to research merchandising methods, has educational seminars passing on new data to franchisees, and has a constant advertising program to promote activity in a city for all outlets".

AN AWARD BY THE TOWN COUNCIL, "BUSINESS FOR BEAUTY"

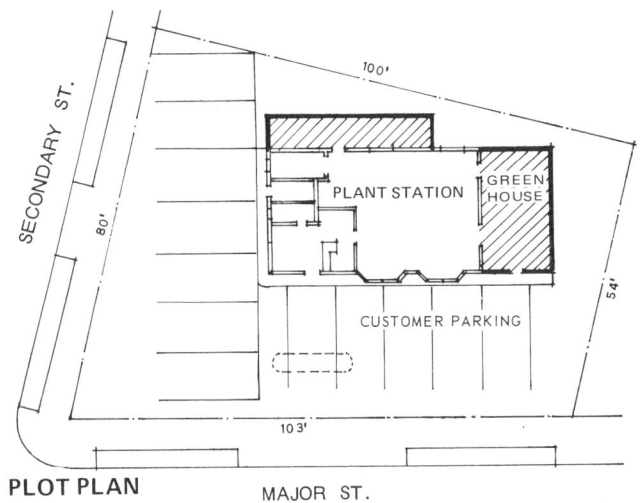

PLOT PLAN   MAJOR ST.

Courtesy The Plant Station, Abington, Pa.
Courtesy Regis Mc Cann, Abington, Pa.

PLANT STORE

REMOVE

FOR LEASE

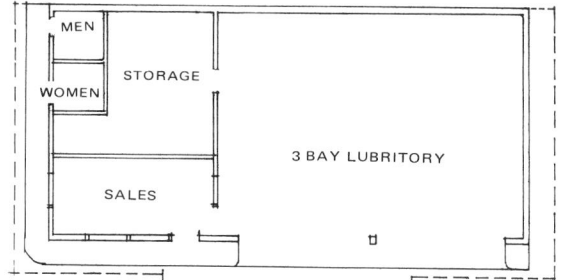

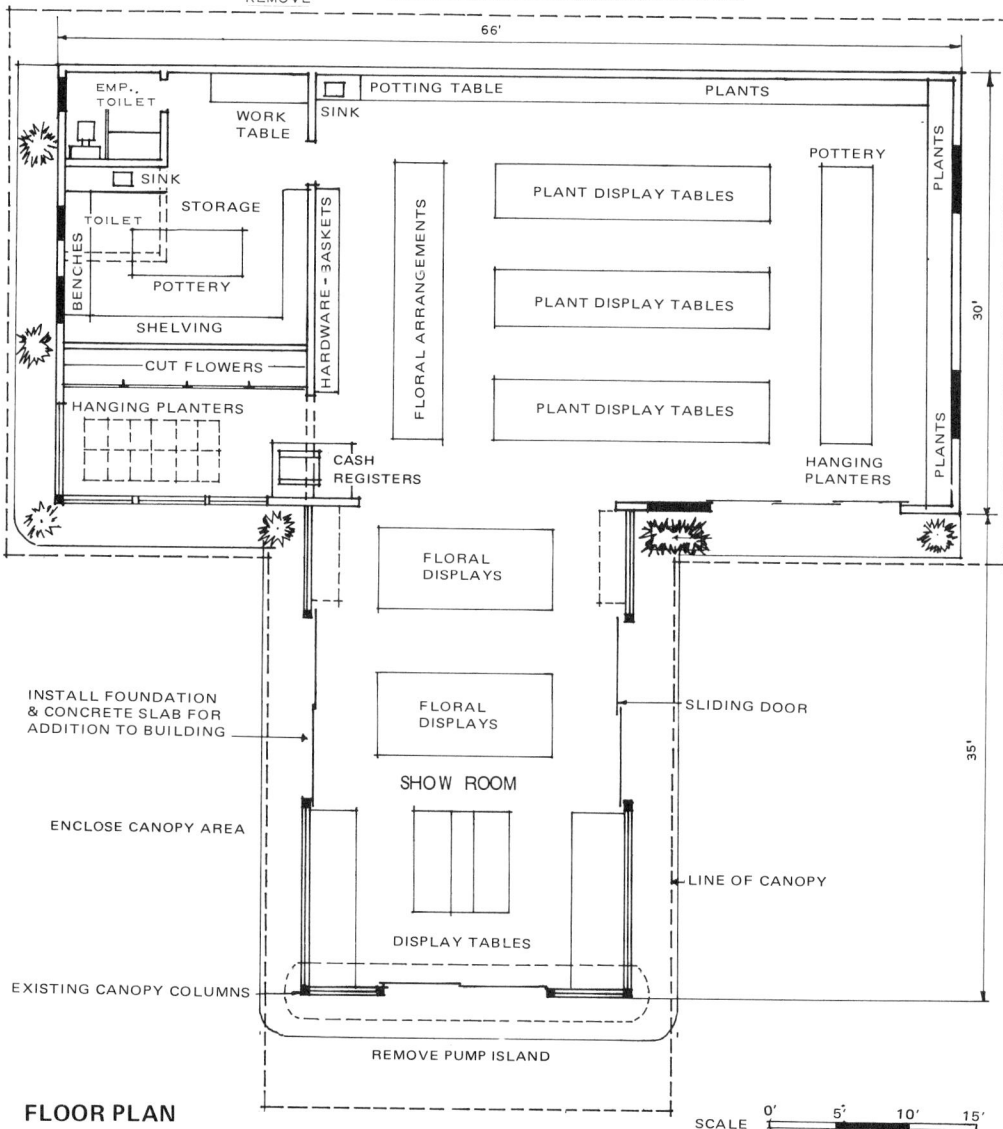

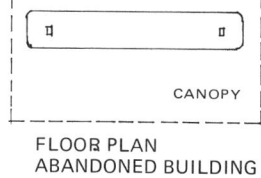

FLOOR PLAN ABANDONED BUILDING

MEN

WOMEN

STORAGE

SALES

3 BAY LUBRITORY

CANOPY

66'

EMP. TOILET

WORK TABLE

SINK

POTTING TABLE

PLANTS

SINK

POTTERY

PLANTS

TOILET

STORAGE

BENCHES

POTTERY

SHELVING

HARDWARE - BASKETS

FLORAL ARRANGEMENTS

PLANT DISPLAY TABLES

PLANT DISPLAY TABLES

PLANT DISPLAY TABLES

PLANTS

CUT FLOWERS

HANGING PLANTERS

HANGING PLANTERS

CASH REGISTERS

30'

FLORAL DISPLAYS

SLIDING DOOR

INSTALL FOUNDATION & CONCRETE SLAB FOR ADDITION TO BUILDING

FLORAL DISPLAYS

35'

ENCLOSE CANOPY AREA

SHOW ROOM

LINE OF CANOPY

EXISTING CANOPY COLUMNS

DISPLAY TABLES

REMOVE PUMP ISLAND

FLOOR PLAN

SCALE 0' 5' 10' 15'

The abandoned service station, illustrated in the photograph, is an attractive building which requires a minimum amount of work to complete an appropriate coversion to a new business use. A plant store requires a building for the display of floral arrangements, hanging planters, and a variety of potted house plants. A greenhouse or a covered area, such as a canopy, will provide space for these house plants.

HANGING PLANTERS

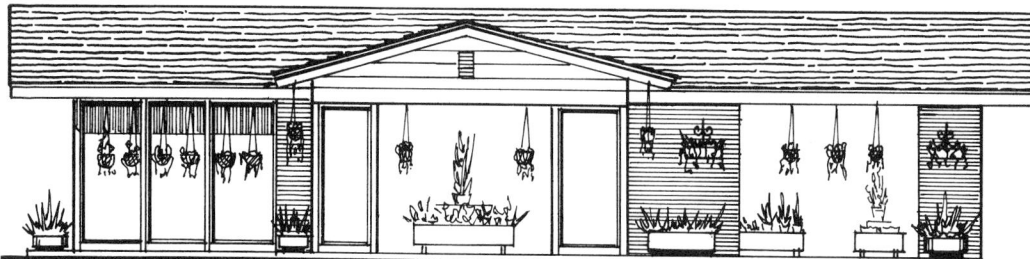

FRONT ELEVATION

PLANT STORE

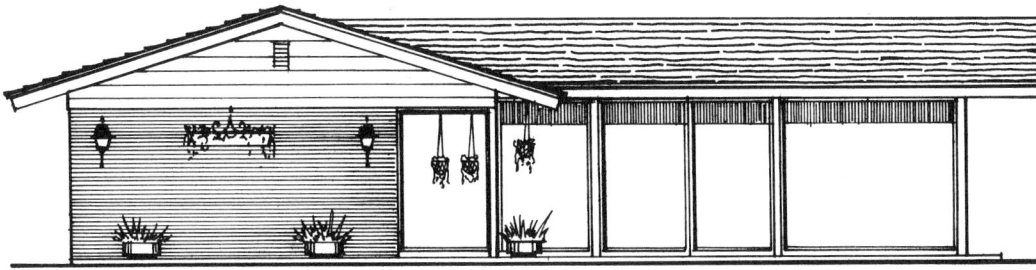

LEFT SIDE ELEVATION

HANGING PLANTERS

DISPLAY TABLES

CUT FLOWERS

LONGITUDINAL SECTION

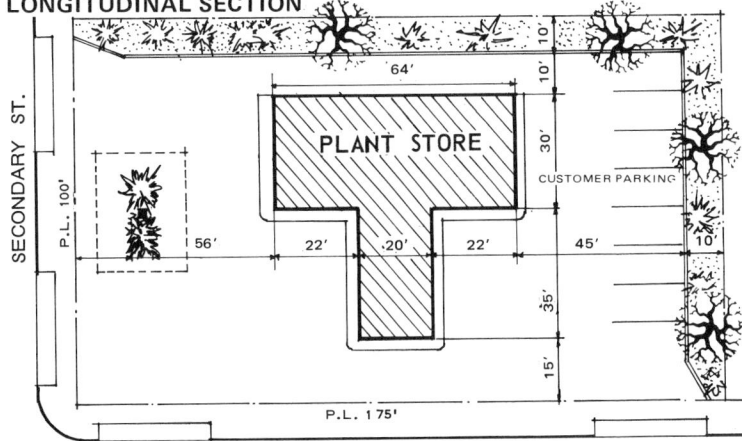

The canopy should be enclosed with fixed glass and sliding glass doors and a concrete floor, and a foundation around the perimeter of the structure should be installed. The lubritory floor drains should be retained and additional drains installed in the floor of the new showroom. The installation of a slatted redwood floor on 2″ x 4″ wood sleepers will allow the watering of potted plants and the drainage of any excess runoff. The building overhang, as well as the freestanding canopy, provides an excellent area for the display of hanging plants which will attract the motoring public.

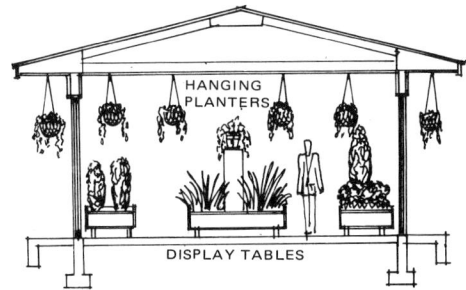

SECONDARY ST.

P.L. 100'

PLANT STORE

64'

10' 10'

30'

CUSTOMER PARKING

56'

22'   20'   22'

45'

35'

15'

10'

P.L. 175'

PLOT PLAN

MAIN ST.

SCALE 1″ = 40'

HANGING PLANTERS

DISPLAY TABLES

CROSS SECTION

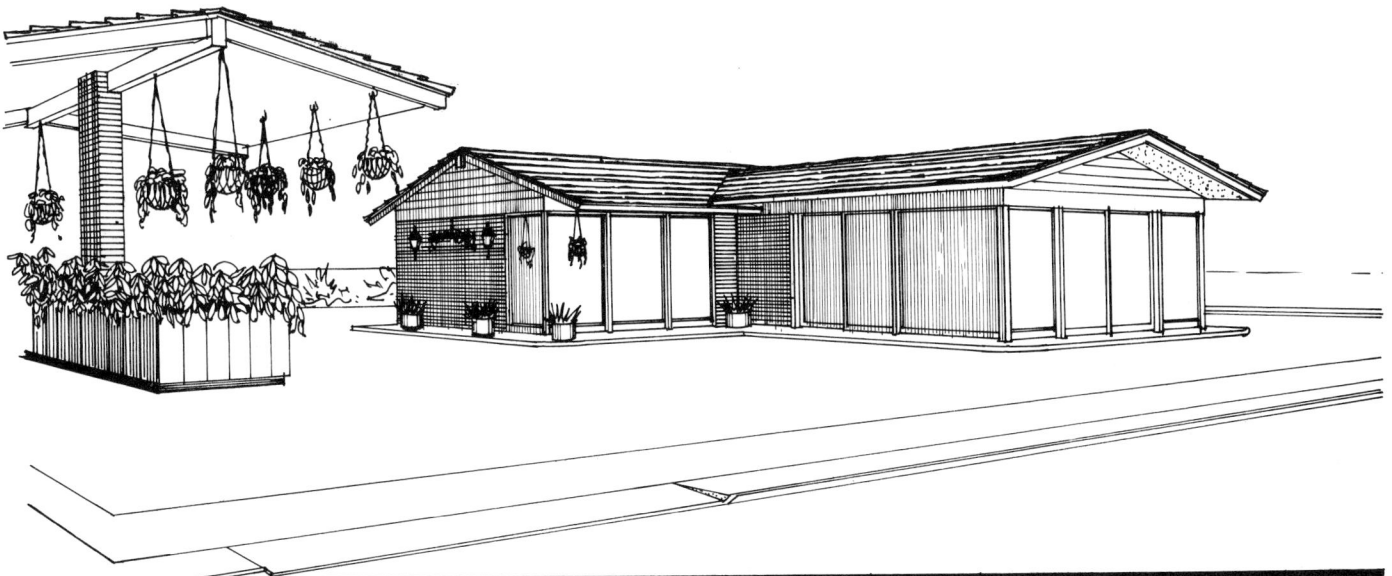

PLASTER CRAFT STORE

THE MAIN STREET ELEVATION

PLASTERCRAFTS

AN ADDITION IS REQUIRED FOR DISPLAY AREA

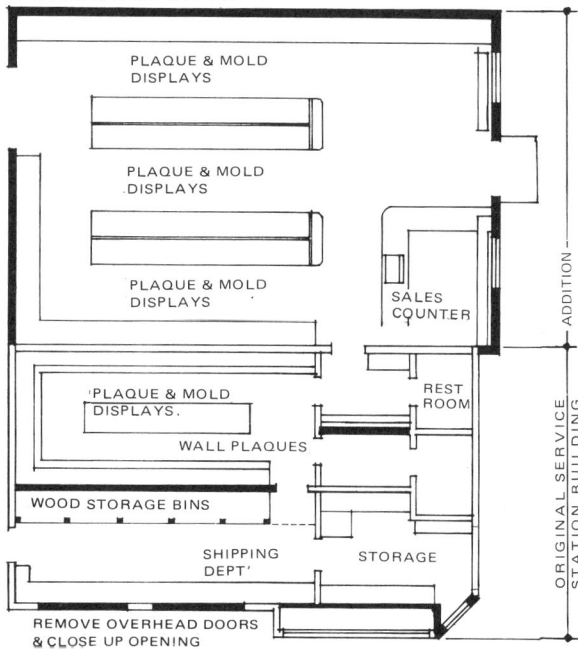

FLOOR PLAN

The acquisition of an abandoned two bay service station on this very busy intersection was a very wise decision by an investor. He remodeled the building and erected a 30' x 45' addition for a new tenant. The new mansard roof, stone veneer and brick addition add to the exterior appearance. The new tenant panelled the interior and installed shelving and counters for the display of plastercraft products. The old lubritory area was subdivided into a display area for plain and unfinished plaques and figures.

The present owner of the store has experienced an increase in business and is faced with a shortage of storage space. Although this may limit future growth the tenant never considered moving from this excellent corner location where there is a superior traffic flow on both arteries.

Courtesy Hodge Podge, Allentown, Pa.

PLASTER CRAFT STORE

THE ORNAMENTAL AND WALL PLAQUE DISPLAY

THE PLAQUE AND MOLD DISPLAY ROOM

THE SALES COUNTER

Courtesy Hodge Podge, Allentown, Pa.

PICTURE FRAMING STORE

In recent years, custom framing and "do-it-yourself" stores have flourished. The shop for custom framing includes paintings, original prints and graphics, laminating, mirror frames, mirror glass, gold leaf, hand carved and natural wood moulding, antique engravings, bird prints, needle art, limited edition prints, guns, medals, coins, diplomas, certificates and restoration work.

An area must be devoted to the display of unframed original paintings, both oil and water color, prints and engravings.

After a customer selects his print, he may frame it himself or have it custom framed. The "do-it-yourself" area is set up so that customers can make their own frames and seek help and advice from an experienced manager. The customer makes his selection from a wide range of mouldings and mats, then assembles them and frames the picture. The ultimate savings realized by the customer can be as high as 50%. The building facade should be representative of the business. The removal of all interior walls is important to provide an open space plan.

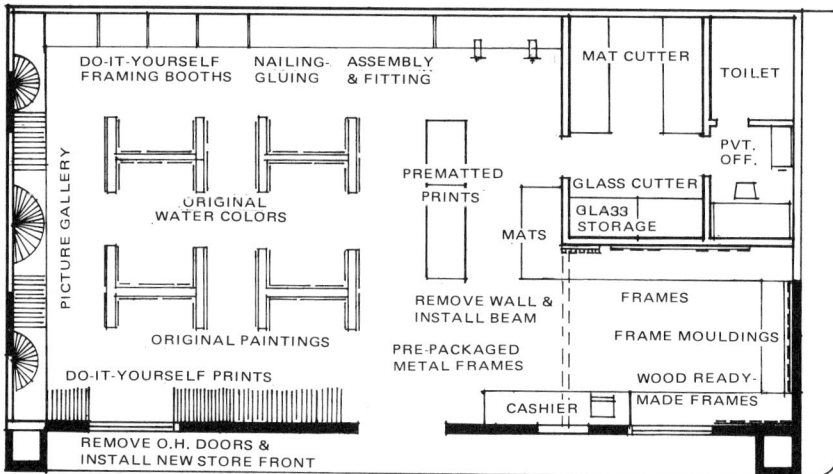

FLOOR PLAN

DO-IT-YOURSELF FRAMING BOOTHS
NAILING-GLUING
ASSEMBLY & FITTING
MAT CUTTER
TOILET
PICTURE GALLERY
ORIGINAL WATER COLORS
PREMATTED PRINTS
PVT. OFF.
GLASS CUTTER
GLA33 STORAGE
MATS
ORIGINAL PAINTINGS
REMOVE WALL & INSTALL BEAM
FRAMES
FRAME MOULDINGS
DO-IT-YOURSELF PRINTS
PRE-PACKAGED METAL FRAMES
WOOD READY-MADE FRAMES
CASHIER
REMOVE O.H. DOORS & INSTALL NEW STORE FRONT

SCALE 0' 5' 10' 15'

FRONT ELEVATION

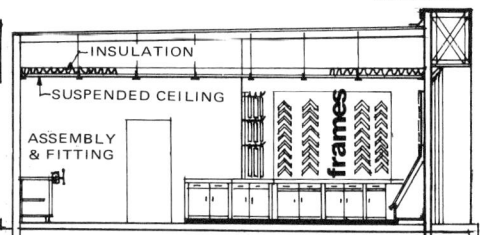

PRE-ENGINEERED ROOF
U FRAME IT
STUCCO FINISH
PLANTERS

CROSS SECTION

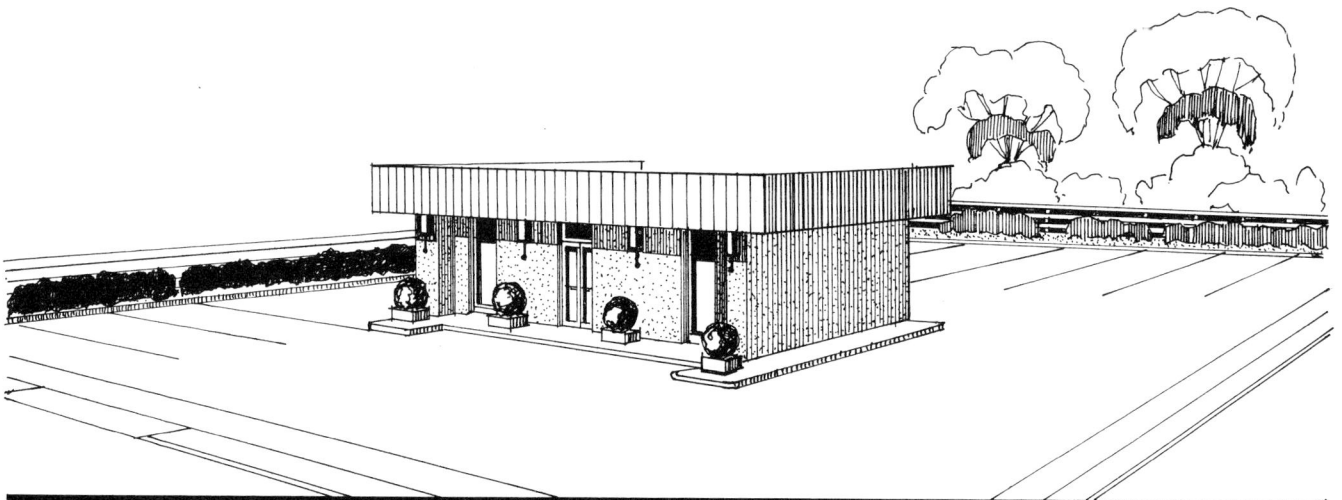

INSTALL NEW PRE-ENGINEERED MARQUEE
INSULATION
SUSPENDED CEILING
ASSEMBLY & FITTING
frames
WOOD READY-MADE FRAMES

POSTAL BOX RENTAL CENTER

Post Office box rentals are in great demand in large cities. The waiting period, in many instances, may be as long as nine to twelve months. Abandoned service stations in metropolitan areas offer opportunities for the businessman searching for room to house a rental box center. The two bay service station illustrated contains 1200 square feet. It also has a customer parking area, not normally found in business districts. This building will accommodate 500 to 600 rental boxes and provide a monthly rental income ranging from $5,500 to $6,500. Income from the operation is related to the number of rental boxes and the location of the facility. Monthly box rentals vary from a low of $2. to a high of $15. in business sections of a city. The application of natural cedar siding to the exterior walls and overhang converts the porcelain box to an attractive new business use.

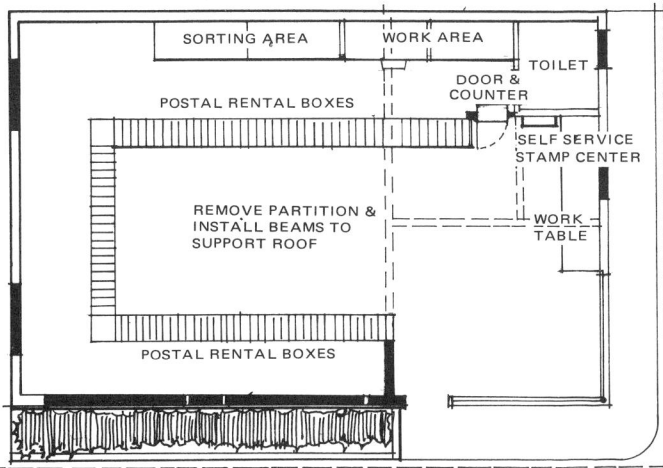

FLOOR PLAN

SORTING AREA   WORK AREA   TOILET

POSTAL RENTAL BOXES   DOOR & COUNTER

SELF SERVICE STAMP CENTER

REMOVE PARTITION & INSTALL BEAMS TO SUPPORT ROOF

WORK TABLE

POSTAL RENTAL BOXES

SCALE 0' 5' 10' 15'

RIGHT SIDE ELEVATION

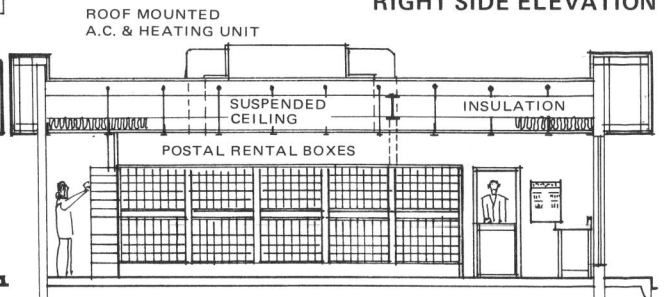

DIAGONAL WOOD SIDING

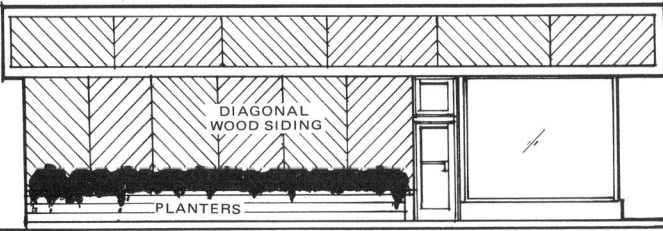

FRONT ELEVATION

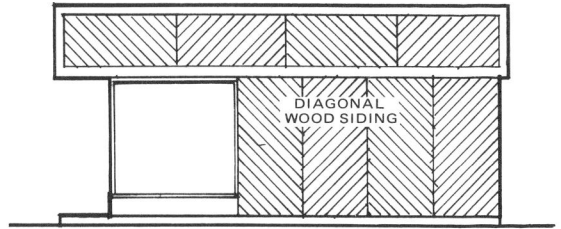

DIAGONAL WOOD SIDING

PLANTERS

LONGITUDINAL SECTION

ROOF MOUNTED A.C. & HEATING UNIT

SUSPENDED CEILING   INSULATION

POSTAL RENTAL BOXES

PRINT AND COPYING CENTER

The print and copying franchise operation is one of the fastest growing in the country. According to the Department of Commerce publication "Franchising in the Economy", printing and copying services have grown from 2578 stores in 1979 to an estimated 3230 in 1981. Franchise sales were $302.8 million in 1979 and are estimated to increase to $428.6 million in 1981. The average annual sales in 1979 were $117,500 per store and are forecast to increase to $132,700 in 1981.

Enclosing the area under the canopy of this abandoned three bay station increases the floor space from 1800 to 2600 square feet and provides a sales center for art and drafting supplies. The self-service photo copier at the entrance is an important profit center to consider. This business will prosper in commercial areas near municipal buildings and adjacent to a college campus or a high school.

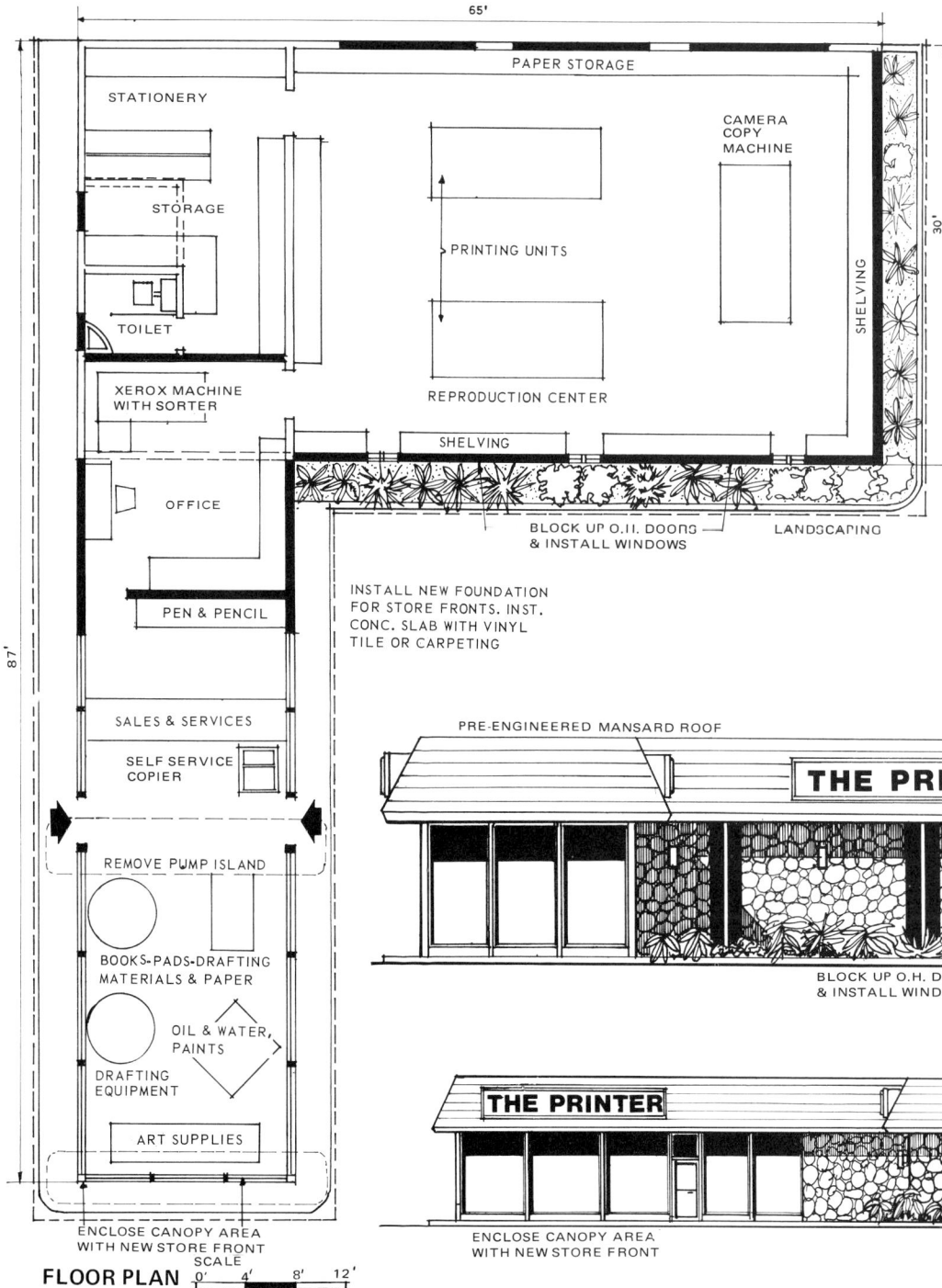

FLOOR PLAN

65'

30'

87'

PAPER STORAGE

STATIONERY

CAMERA COPY MACHINE

STORAGE

PRINTING UNITS

SHELVING

TOILET

REPRODUCTION CENTER

XEROX MACHINE WITH SORTER

SHELVING

OFFICE

BLOCK UP O.H. DOORS & INSTALL WINDOWS

LANDSCAPING

PEN & PENCIL

INSTALL NEW FOUNDATION FOR STORE FRONTS. INST. CONC. SLAB WITH VINYL TILE OR CARPETING

SALES & SERVICES

SELF SERVICE COPIER

REMOVE PUMP ISLAND

BOOKS-PADS-DRAFTING MATERIALS & PAPER

OIL & WATER PAINTS

DRAFTING EQUIPMENT

ART SUPPLIES

ENCLOSE CANOPY AREA WITH NEW STORE FRONT

SCALE 0' 4' 8' 12'

PRE-ENGINEERED MANSARD ROOF

THE PRINTER

STONE VENEER

BLOCK UP O.H. DOORS & INSTALL WINDOWS

FRONT ELEVATION

THE PRINTER

MANSARD ROOF

STONE VENEER

ENCLOSE CANOPY AREA WITH NEW STORE FRONT

LEFT SIDE ELEVATION

PRINT AND COPYING CENTER

MANSARD ROOF

THE PRINTER

AGGREGATE PANELS

ENCLOSE CANOPY AREA
WITH NEW STORE FRONT

LEFT SIDE ELEVATION

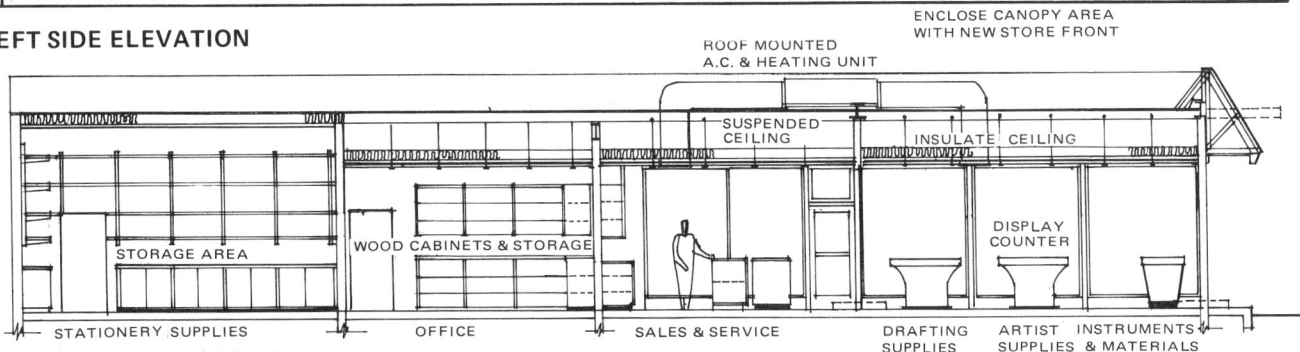

ROOF MOUNTED
A.C. & HEATING UNIT

SUSPENDED
CEILING

INSULATE CEILING

DISPLAY
COUNTER

STORAGE AREA

WOOD CABINETS & STORAGE

STATIONERY SUPPLIES

OFFICE

SALES & SERVICE

DRAFTING
SUPPLIES

ARTIST
SUPPLIES

INSTRUMENTS
& MATERIALS

LONGITUDINAL SECTION

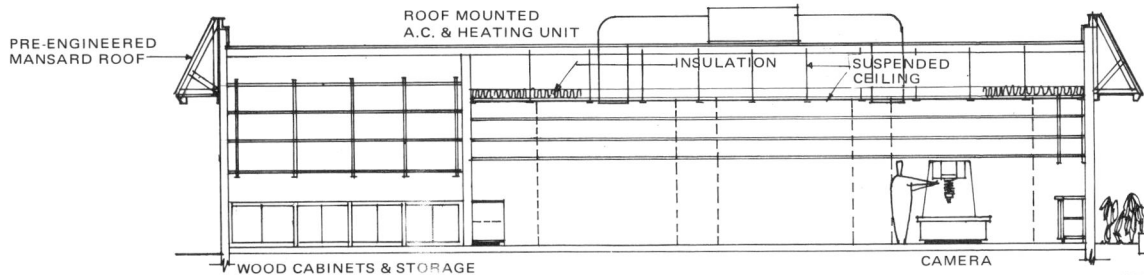

ROOF MOUNTED
A.C. & HEATING UNIT

PRE-ENGINEERED
MANSARD ROOF

INSULATION

SUSPENDED
CEILING

WOOD CABINETS & STORAGE

CAMERA

CROSS SECTION

THE PRINTER

THE PRINTER

THE PRINTER
ARTIST SUPPLIES
REPRODUCTIONS

FAST FOOD INDUSTRY

Fast foods account for one-third of all customer eating out occasions. The food service industry received a tremendous boost when people determined that eating away from home was a form of entertainment.

Abandoned service stations provide inexpensive opportunities for many small and occasionally large chains to expand their programs. The conventional colonial and ranch service stations can be converted into attractive building images. The 1200 to 1800 square foot floor area of a two or three bay building is sufficient to provide a dining area with a capacity to seat 40 to 45 patrons; a kitchen with preparation tables, stoves, ovens, refrigerators, and counters for eat-in and take-out customers. The existing rest rooms can be utilized after minor fixture and door relocations have been completed. Quarry tile floors throughout the building are essential from a sanitary standpoint. Former service stations that have attached or free-standing canopies can be utilized for additional dining area. The drive-up window should be located along the side of the building near the cashier. The interviews that follow include a cross section of franchises and individually owned businesses.

Ethnic fast foods are popular in cities where the immediate community is of the same origin. The interiors and exteriors often reflect the origin of their customer's culture. The introduction of pictures, statues, sculpture, and wall paper create an old world atmosphere.

One of the newest and most successful entries into a crowded market is the "nutrition" fast food unit. Health foods are featured in the assortment of food served, including barbecued chicken, desserts, soup, hot dogs, pizzas, etc. at one of the new restaurants. If the businessman is interested in this concept he might wish to include a profit center containing display counters for herbs, vitamins and other health foods. A variety of fast moving items should be displayed in a compact area adjacent to the cashier.

Automation provides a non-labor intensive production system to the industry, and reasonable quality food at a modest price served in a clean atmosphere in a short period of time.

FAST FOOD STATISTICS

The best definition of fast foods was found in an edition of NATION'S RESTAURANT NEWS:

What is fast food?

1. No waiter or waitress service.

2. Food served within one to three minutes of the time a person orders it.

3. Food eaten by customers within ten to twenty minutes.

4. Average per person check of under $2.75.

Fast food accounts for one-third of all customer eating out occasions.

Growth of fast foods:

Year	No. Units	Sales	% Growth	Sales Growth
1967	54,492	$ 3,417,642	—	—
1972	78,301	8,893,938	43.7%	160.2
1977	100,493	20,476,111	28.3%	130.2
1981		31,507,990	53.9%	

(Note: Sales in thousands)

In 1970 the average initial investment per unit for land, building and equipment ranged from $75,000 to $200,000 as compared with 1980 investments ranging from $400,000 to $800,000.

In 1970 the average check was estimated as low as $1.00 and in 1981 it increased to $2.65.

(Source: NATION'S RESTAURANT NEWS)

FRIED CHICKEN STORE & RESTAURANT

HENNY-PENNY

A 13 YEAR OLD CONVERSION

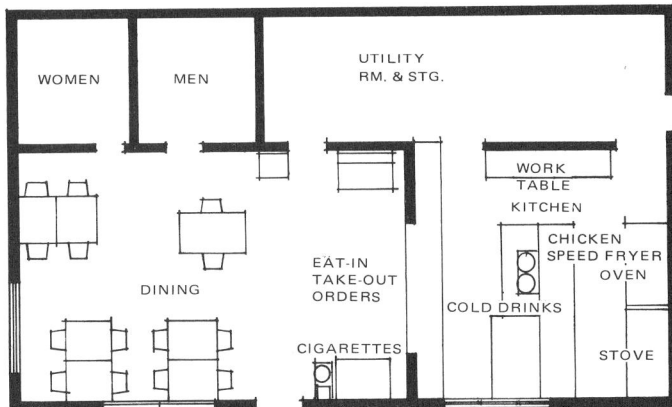

FLOOR PLAN

In 1968, an obsolete abandoned service station, situated at a major intersection of a large southern city, was converted to a fried chicken fast food restaurant. Although the floor space was considered substandard, the perceptive owner realized the potential of the location and proceeded with the remodeling. Overhead doors were removed, new storefronts were installed and the exterior was painted red and white. The manufacturer of the food-service equipment offered counseling with the layout equipment selection, a merchandising display, and an excellent brand name for the store. The majority of purchases are take-out orders; however, a small eat-in dining area has also been provided.

THE TAKE-OUT COUNTER

THE KITCHEN AND PREPARATION AREA

EAT-IN DINING AREA

Courtesy Bell's Henny Penny, Johnson City, Tenn.

MEXICAN RESTAURANT

THE FINAL STAGES OF THE BUILDING REMODELING

THE FINISHED PRODUCT

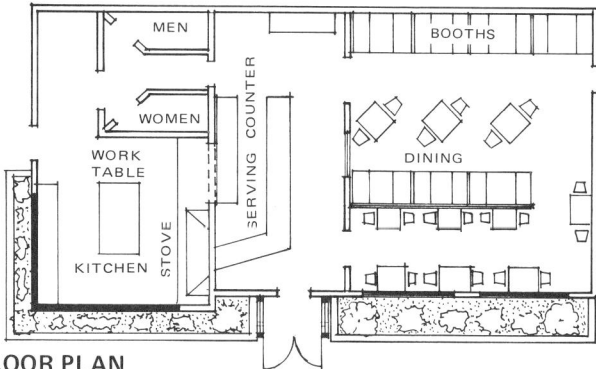

FLOOR PLAN

MEXICAN RESTAURANT

In 1978, the author met an energetic young man and his helper as they were remodeling a three bay service station that he had purchased for $165,000. Six months and $50,000 later, the building had been converted to a Mexican restaurant with authentic cooking by the owner. The restaurant is situated on a 250' x 200' plot on a heavily travelled artery with a 40,000 car per day traffic count. The interior has been remodeled with dark wood paneling and a stucco facade was applied to the exterior to create a Spanish atmosphere. It was necessary for the owner to appear before the local zoning board to obtain approval to change the business use. The new use was granted and one of the conditions required that the parking area be enlarged to accommodate maximum occupancy. It is unusual for the Zoning and or Planning Boards to deny the conversion of a nonconforming business to a higher and better use.

Courtesy Mexican Food Factory, Marlton, N.J.

SEAFOOD RESTAURANT

A NATURAL WOOD EXTERIOR

A HAND SPLIT WOOD SHINGLE MANSARD ROOF

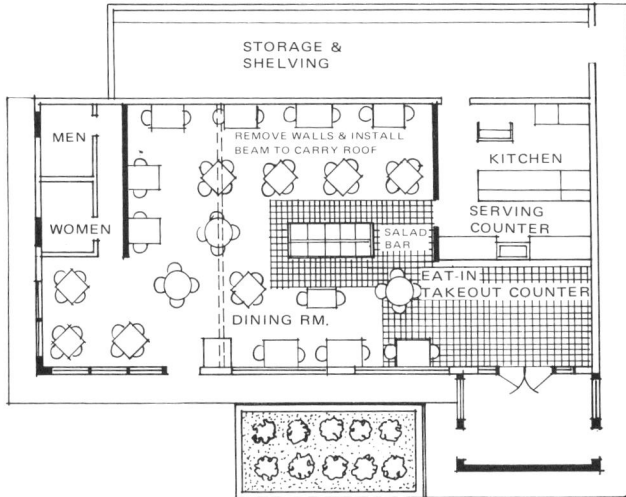

FLOOR PLAN

In 1979, a major seafood chain converted a three bay service station into one of their new marketing concepts. For $129,000 they turned an abandoned station into an attractive restaurant with an authentic Cape Cod atmosphere.

The interior of the building was gutted, and a kitchen, with a large serving counter, was installed. The lubritory and former sales room were remodeled into a dining area with room for a salad bar. Natural wood siding was installed on the exterior and then stained a rustic brown.

The big advantage in converting the abandoned service station was the retention of the existing building setback. The franchisor stated that the location of the property was the most important fact in the acquisition.

THE EAT-IN, TAKE-OUT COUNTER

THE DINING ROOM AND SALAD BAR

Courtesy Arthur Treacher's Fish & Chips, Towson, Md.

RESTAURANT

A NEW USE FOR AN ABANDONED TWO BAY SERVICE STATION

THE FORMER PUMP ISLAND HAS BEEN CONVERTED TO A PLANTER

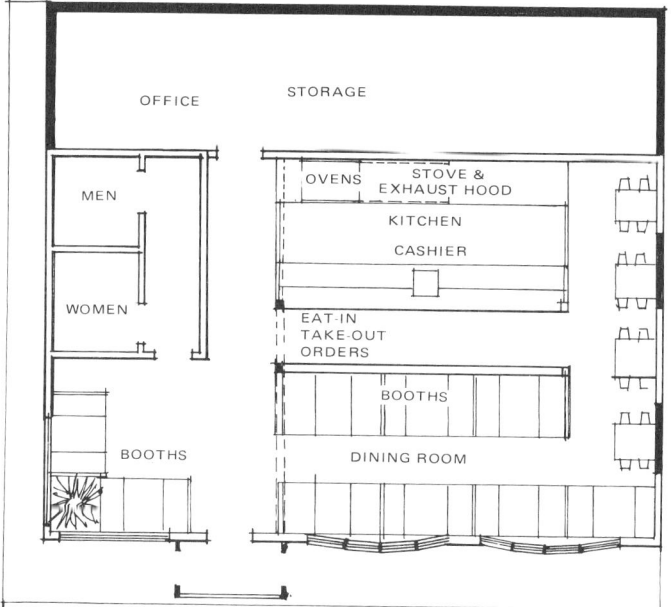

FLOOR PLAN

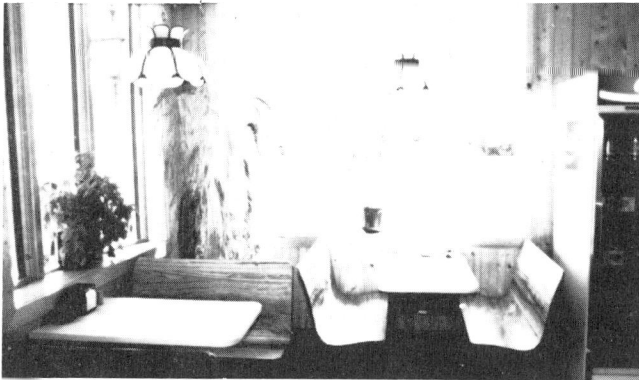

BAY WINDOWS WERE INSTALLED IN ALL SALES AND LUBRITORY OPENINGS

ORDER COUNTER AND PIZZA OVENS

Courtesy Leo's Pizza Palace, Pitman, N.J.

RESTAURANT

Some food experts estimate that business increases of approximately 15% annually makes the pizzeria the fastest growing restaurant business in today's economy. The stores that feature many varieties of pizza are generally the most popular. In 1981, the owner purchased an abandoned two bay service station situated on a large parcel on two busy streets and remodeled it into an attractive pizza restaurant. The interior walls were covered with knotty pine and the floor with quarry tile. The Tiffany lamps are placed throughout the restaurant and add to the warm atmosphere.

The kitchen, ovens and serving area are situated centrally to serve both eat-in and take-out customers. The booths accommodate more than fifty customers. The interior finishes require limited maintenance.

The large parking area available is more than ample for employees and customers. The prudent owner reused the former area and island lights and the colonial building lights. An addition to the rear houses the owner's office, storage and refrigeration units, and preparation area. The exterior walls were refinished in an attractive stucco.

A LARGE KITCHEN AND SERVING COUNTER

AN ATTRACTIVE BRIGHT CORNER OF THE RESTAURANT

TIFFANY LAMPS AND A KNOTTY PINE INTERIOR
PROVIDE A CHEERFUL ATMOSPHERE

Courtesy Leo's Pizza Palace, Pitman, N.J.

RESTAURANT

WINDOWS AND ENTRANCE FOYERS REPLACE THE OVERHEAD DOORS

A NEW SIGN ON THE TOP OF THE HIGH RISE POLES

A VIEW OF THE PROPERTY

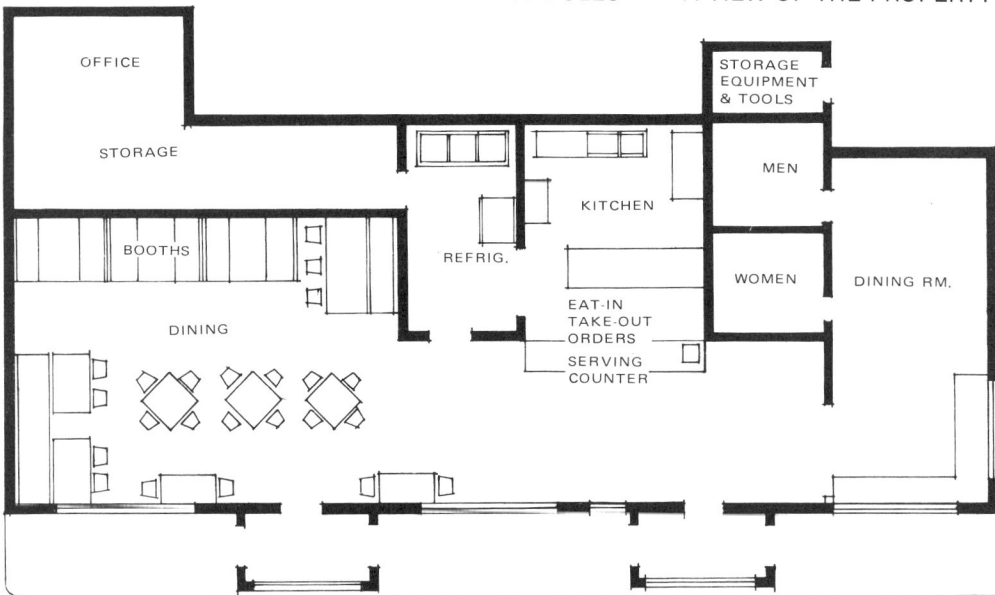

FLOOR PLAN

A six acre tract, with an abandoned three bay colonial service station, was purchased by a local businessman in 1978 for $75,000. The property is situated at the exit from an Interstate highway on the main road to one of the south's popular tourist attractions; an excellent location for a restaurant.

The interior remodeling included new partitions for the dining room, kitchen, storage room, counter area and private office. A 10′ addition was required to enlarge the rest rooms to accommodate the handicapped customers. This alteration work was accomplished for $30,000. Kitchen equipment, freezers, refrigerators, sinks, counters, etc. were purchased and installed for $35,000.

Courtesy Buddy Burger, Natural Bridge, Virginia
Photographs Courtesy of Gerard J. Bryce.

RESTAURANT

New internally illuminated business signs were placed on the former oil company identity and high rise poles which are visible from the Interstate highway. The signs are a tremendous asset that the average interstate restaurant does not enjoy.

The owner stated that the major advantages in buying this facility were: obtaining a well constructed colonial building that could easily be converted into a restaurant, an asphalt paved driveway and parking lot for customers, landscaping, the high rise poles for the sign and the six acres adjacent to the interstate which is ample for the development of a future motel complex. In the author's opinion the 1980 market value of the land and building is $175,000.

ENTRANCE TO TAKEOUT COUNTER

ORDER AND TAKEOUT COUNTER

THE DINING ROOM

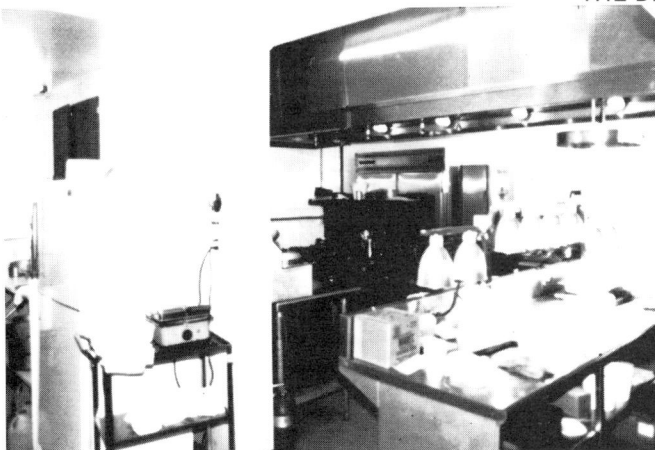
THE KITCHEN

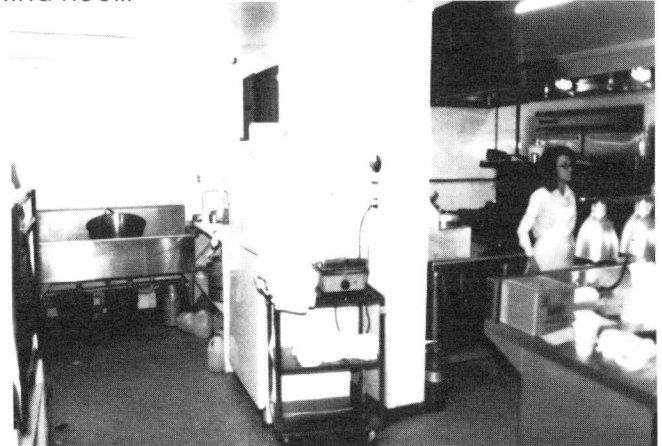
THE PANTRY

Courtesy Buddy Burger, Natural Bridge, Virginia

RESTAURANT-CATERER

A THREE BAY COLONIAL BUILDING CLOSED FOR TWO YEARS

THE OUTSIDE DINING AREA

THE BUILDING IS CONVERTED

PLOT PLAN

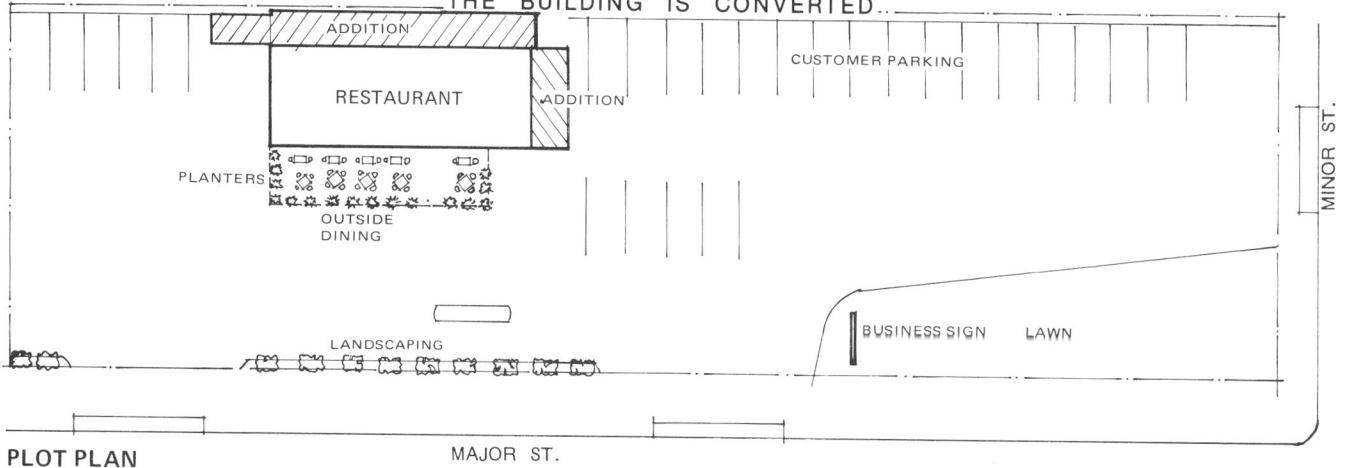

Courtesy Olde Tymes, Lawrenceville, N. J.

RESTAURANT-CATERER

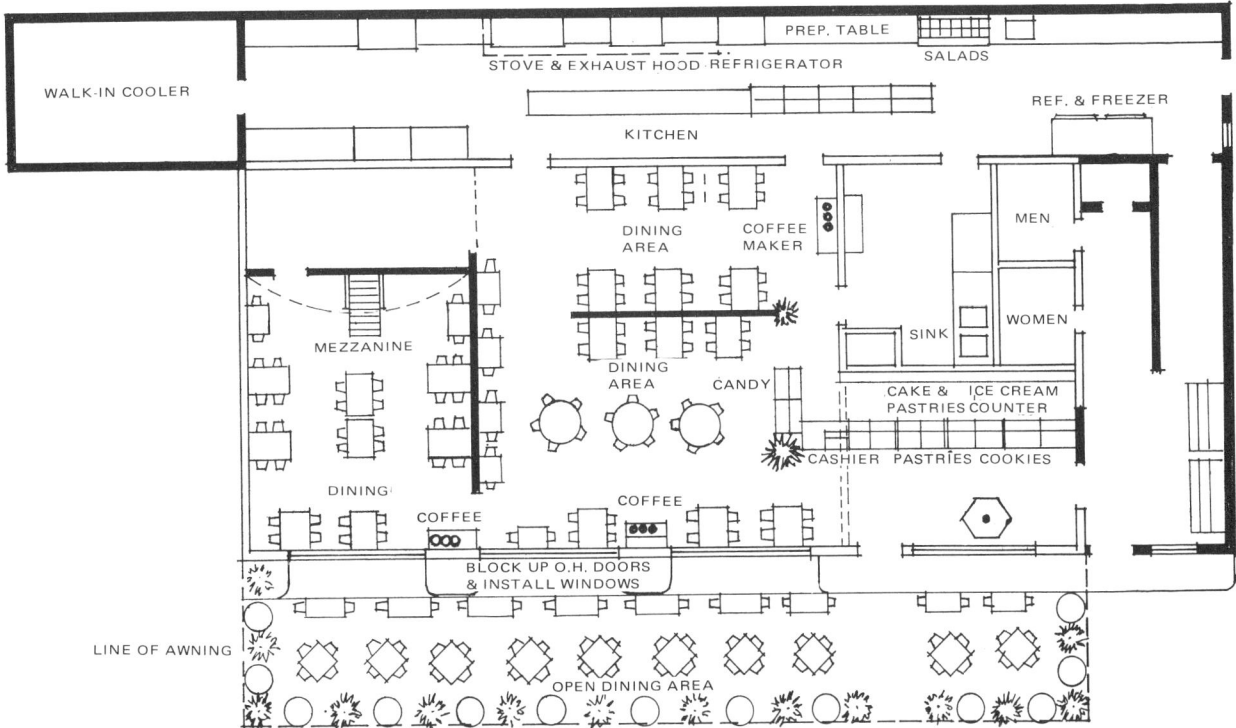

Plan labels:
WALK-IN COOLER
PREP. TABLE
STOVE & EXHAUST HOOD · REFRIGERATOR
SALADS
KITCHEN
REF. & FREEZER
DINING AREA
COFFEE MAKER
MEN
MEZZANINE
WOMEN
SINK
DINING AREA
CANDY
CAKE & ICE CREAM PASTRIES COUNTER
DINING
CASHIER PASTRIES COOKIES
COFFEE
COFFEE
BLOCK UP O.H. DOORS & INSTALL WINDOWS
LINE OF AWNING
OPEN DINING AREA

A NEW ADDITION FOR THE KITCHEN

A DINING AREA

THE LARGE DINING ROOM

Courtesy Olde Tymes, Lawrenceville, N. J.

RESTAURANT-CATERER

In 1978, an enterprising young man, who owned his own sandwich and deli shop in an urban area of a large city, purchased a three bay colonial service station that had been abandoned and boarded up with plywood for two years. The property, measuring 375' x 135' on the main artery entering the state capital, was purchased for $130,000. The young man, his father, and two brothers performed 75% of the building remodeling themselves. The interior was originally converted into a food and deli center, complete with an ice cream and soda counter, bakery and restaurant with a mezzanine. The owner removed the weathered siding and beams from two barns, situated in neighboring Pennsylvania and used them to decorate the interior. The remodeling and equipment for the kitchen, deli display counters, refrigerators and stoves, together with the tables, (formerly old sewing machines) and chairs in the restaurant area cost approximately $35,000.

COOKIES AND FINE PASTRIES

THE CHOCOLATE AND CANDY CASE

ICE CREAM SECTION

Courtesy Olde Tymes, Lawrenceville, N. J.

RESTAURANT - CATERER

In January, 1982, the owner remodeled the interior, enlarged the dining areas and added a large kitchen, complete with walk-in refrigeration facilities. The anticipated seating capacity in the dining room area will be for 135 patrons. The cost to complete the second conversion and provide additional equipment was approximately $50,000.

This is a busy family operated enterprise. The owner features catering in addition to the enlarged restaurant facilities.

During summer months, a covered outside dining area, concealed by a row of planters, is open for dining. An additional sixty customers can be served in this area.

THE MEZZANINE DINING AREA

LOOKING DOWN ON THE DINING ROOM FROM THE MEZZANINE

THE PIANO IN THE MEZZANINE DINING AREA

Courtesy Olde Tymes, Lawrenceville, N. J.

RESTAURANT

In 1976, a husband and wife team purchased a two bay service station from the former owner-operator for $35,000 for conversion to a new business use. The building, erected in the early 1930's, had a stucco exterior with red brick insets, a popular motif during this period in the midwest and southwest.

Ninety percent of the work was performed by the new owners. They removed the pump island with sledge hammers and used a jack hammer to remove the floor to install new plumbing lines to the relocated rest rooms. Two brick planters were constructed at opposite ends of the garden dining area. This outdoor area was then enclosed with a wrought iron fence.

Tables with umbrellas were placed in this attractive setting which seats 40 comfortably for dining. The restaurant seats an additional 66 in the dining room. The remodeling cost the owners $21,000. The property, measuring 90' x 150' provides parking for ten cars.

The conversion of this former service station is an asset to the restored area of this historic city. The facility now provides an inviting atmosphere for tourists and the community. The owners have developed a steady customer trade by serving fine foods at reasonable prices.

THE BRICK AND STUCCO FACADE HAS BEEN RESTORED

THE BRICK PLANTERS MATCH THE BUILDING EXTERIOR

THE GARDEN DINING AREA REPLACES THE FORMER PUMP ISLAND AREA

Courtesy of Rodden's Restaurant, Old Colorado City, Colorado Springs, Colorado.

RESTAURANT

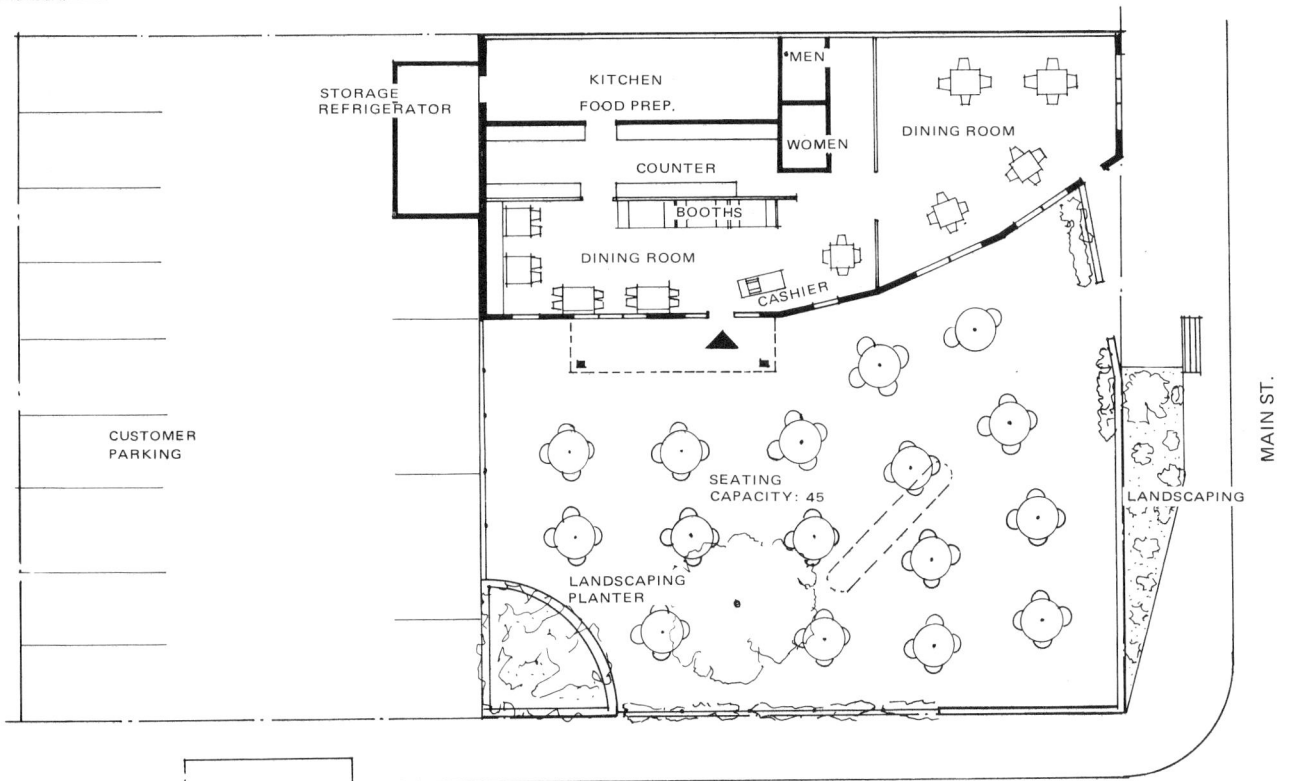

PLOT PLAN & FLOOR PLAN

THE RELOCATED REST ROOMS

THE DINING AREA

ATTRACTIVE CURTAINS COVER THE WINDOWS

THE SERVING COUNTER

Courtesy of Rodden's Restaurant, Old Colorado City, Colorado Springs, Colorado.

194

THE ABANDONED THREE BAY SERVICE STATION BEFORE REMODELING

A LAWN, ATTRACTIVE LANDSCAPING AND A SHUFFLE
BOARD COURT

A VITAL NEW CENTER FOR THE SENIOR CITIZENS OF
THE COMMUNITY

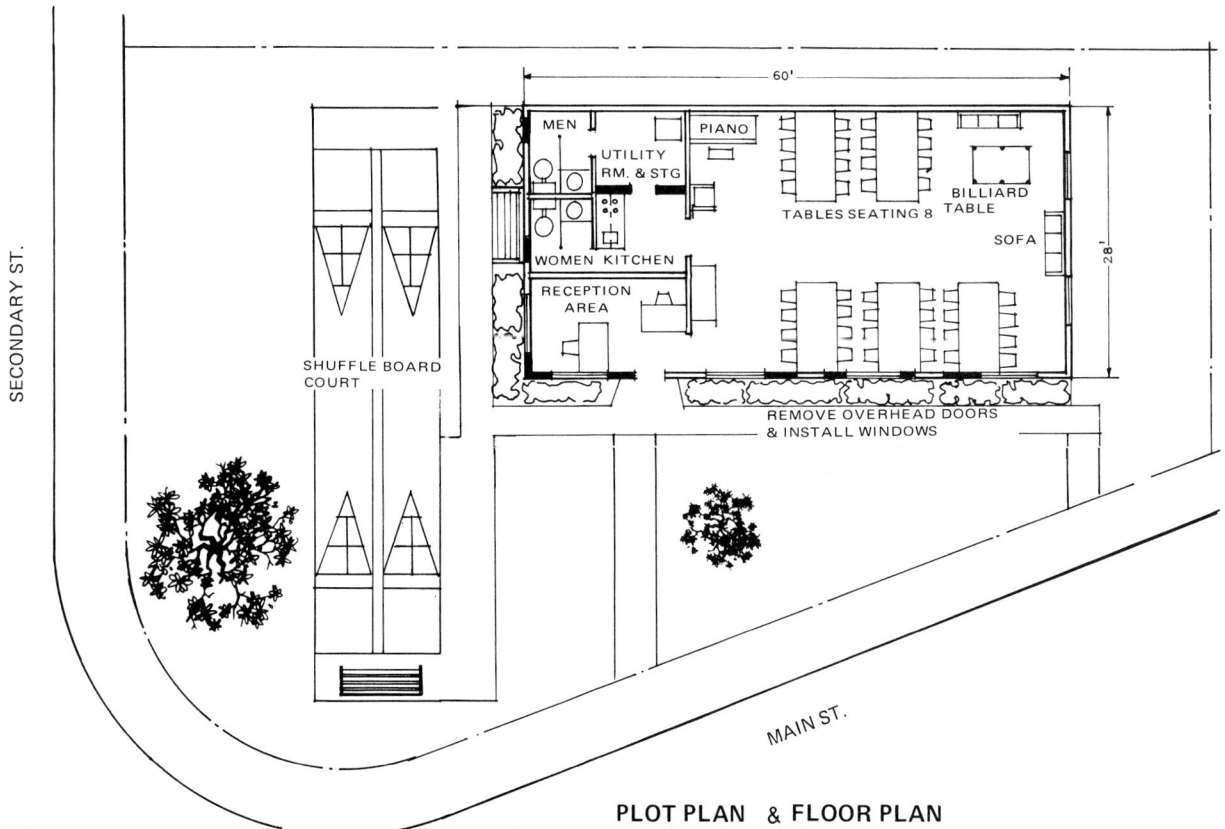

PLOT PLAN & FLOOR PLAN

Courtesy Village of Rockville Centre, N.Y.

SENIOR CITIZEN'S CENTER

In 1975, the widening of a main road in a large Long Island community left a three bay service station with insufficient yard area to continue its operation. The mayor and trustees of the village recommended acquisition of the property and demolition of the building. However, the village Recreation Director had other ideas. He had obtained a copy of the author's original book and reviewed the contents with the members of the village's governing body. The conversion concepts contained in the publication together with Recreation Director's presentation, convinced the village that recycling was not only economical but also practical. The building remodeling and the property improvements were accomplished for $32,000 in September 1976, a modest expenditure. The village reports a $58,000 savings was realized when compared to the demolition of the service station and new construction. The exterior of the building was faced with vertical siding and openings were enclosed with windows and a new entrance. A major advantage of the building was the entry level to the Center which is excellent for senior citizens. The toilet facilities, however were not large enough for the elderly. The old blacktop yard was removed and a park like area with benches and a shuffle board court was developed. A suspended acoustical ceiling was installed and the walls covered with sheetrock and then painted.

More than 400 senior citizen's enjoy the activities that are organized and conducted by an administrator with social worker at the Center. The Village Administrator advises that the Village recently purchased a 70' x 140' parcel situated to the rear of the Center and is now planning a $200,000 federally funded building extension. The exterior of the new addition will match the facade of the present building.

THE MEETING ROOM SEATS MORE THAN FORTY

A COMPACT KITCHEN OCCUPIES THE FORMER STORE ROOM

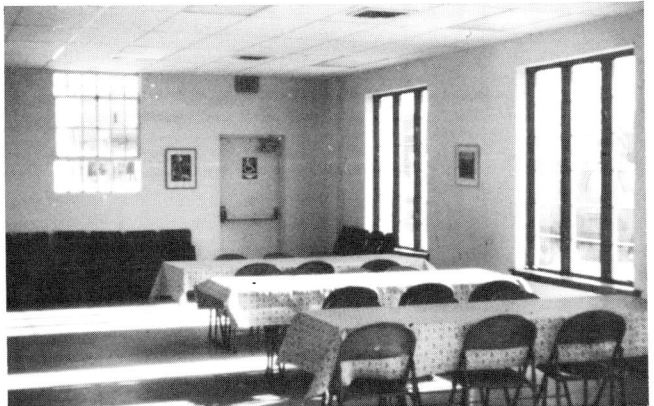

A PIANO AND BILLIARD TABLE FOR MANY HOURS OF PLEASURE

Courtesy Village of Rockville Centre, N.Y.

SHOE STORE

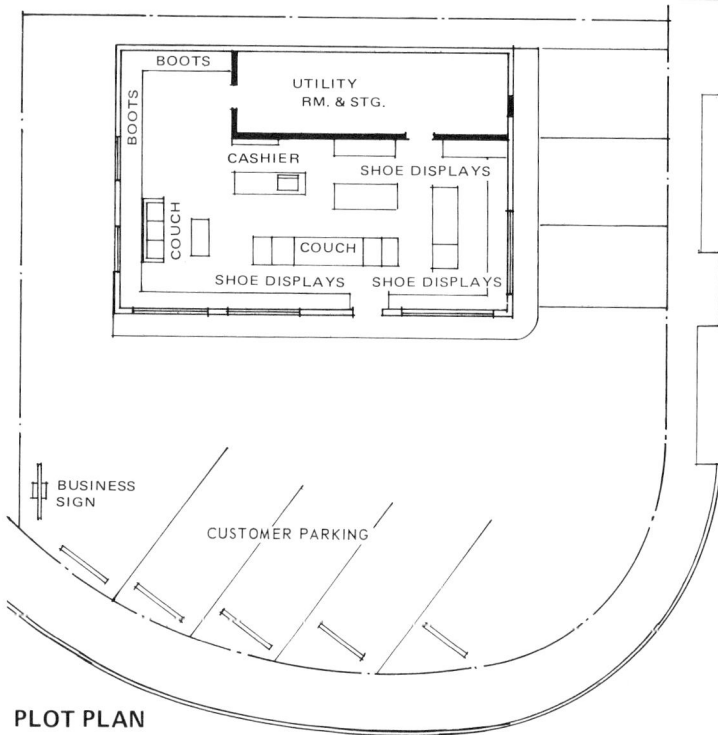

BOOTS
BOOTS
UTILITY RM. & STG.
CASHIER
SHOE DISPLAYS
COUCH
COUCH
SHOE DISPLAYS
SHOE DISPLAYS

BUSINESS SIGN

CUSTOMER PARKING

PLOT PLAN

The successful conversion of a twenty-year old two bay colonial service station to a shoe store was accomplished at a cost of $20,000 in 1977. This is an excellent example of the development of a small 70′ x 70′ property. The parcel is situated at a major intersection in a suburban community of a large city. Although limited parking area is available the store is also accessible to considerable pedestrian traffic.

The major portion of the conversion was performed by the owner. He retained an architect to prepare the remodeling plans. The renovation is an excellent example of a modest expenditure to develop an attractive store. The interior open space plan with shoe inventory displayed throughout the store illustrates a highly successful merchandising concept. Attractive open beamed ceiling, light fixtures and hanging planters create an excellent atmosphere for shoppers.

PARKING FOR EIGHT CUSTOMERS

A SUBSTANDARD CORNER BECOMES A SUCCESSFUL BUSINESS

Courtesy Shoe Bit, Balacynwd, Pa.

SHOE STORE

THE ATTRACTIVE REMODELING IS AN ASSET TO THE COMMUNITY

THE COLONIAL BUILDING LENDS ITSELF TO AN INEXPENSIVE REMODELING

NEATLY ARRANGED SHOE DISPLAYS LINE THE STORE

TRACK LIGHTING HIGHLIGHTS SPECIAL SALES

THE INTERIOR DECOR PRESENTS A PLEASANT PLACE TO SHOP

THE AREA DEVOTED TO BOOT SALES

SHOE DISPLAYS PERMIT EXAMINATION OF SALE ITEMS BY CUSTOMERS

Courtesy Shoe Bit, Balacynwd, Pa.

MINI - STORAGE

Many abandoned service stations in suburban areas are situated on large parcels of land. Properties containing three-quarters or more acres of land have the potential of being converted to a mini-storage concept. The high return per square foot of space makes this a potentially profitable business venture. It is important to have a resident manager's office and apartment on the premises to provide protection against vandalism. The former service station building has been converted to the office and residence. A cyclone fence should surround the entire complex. Monthly rental fees range from $15 to $20 for a small closet to $60 to $100 for a 300 to 400 square foot storage unit. The storage spaces are designed to permit the relocation of partitions and provide flexibility.

In the course of new constructions, the structure is generally built of concrete block walls with interior partitions of plywood or corrugated metal. The roof generally consists of flexicore, with built-up three ply roofing. Pre-engineered metal walls and roof structure are an inexpensive alternative. Construction costs range from $12 to $24 per square foot depending on the area. (1981 Costs)

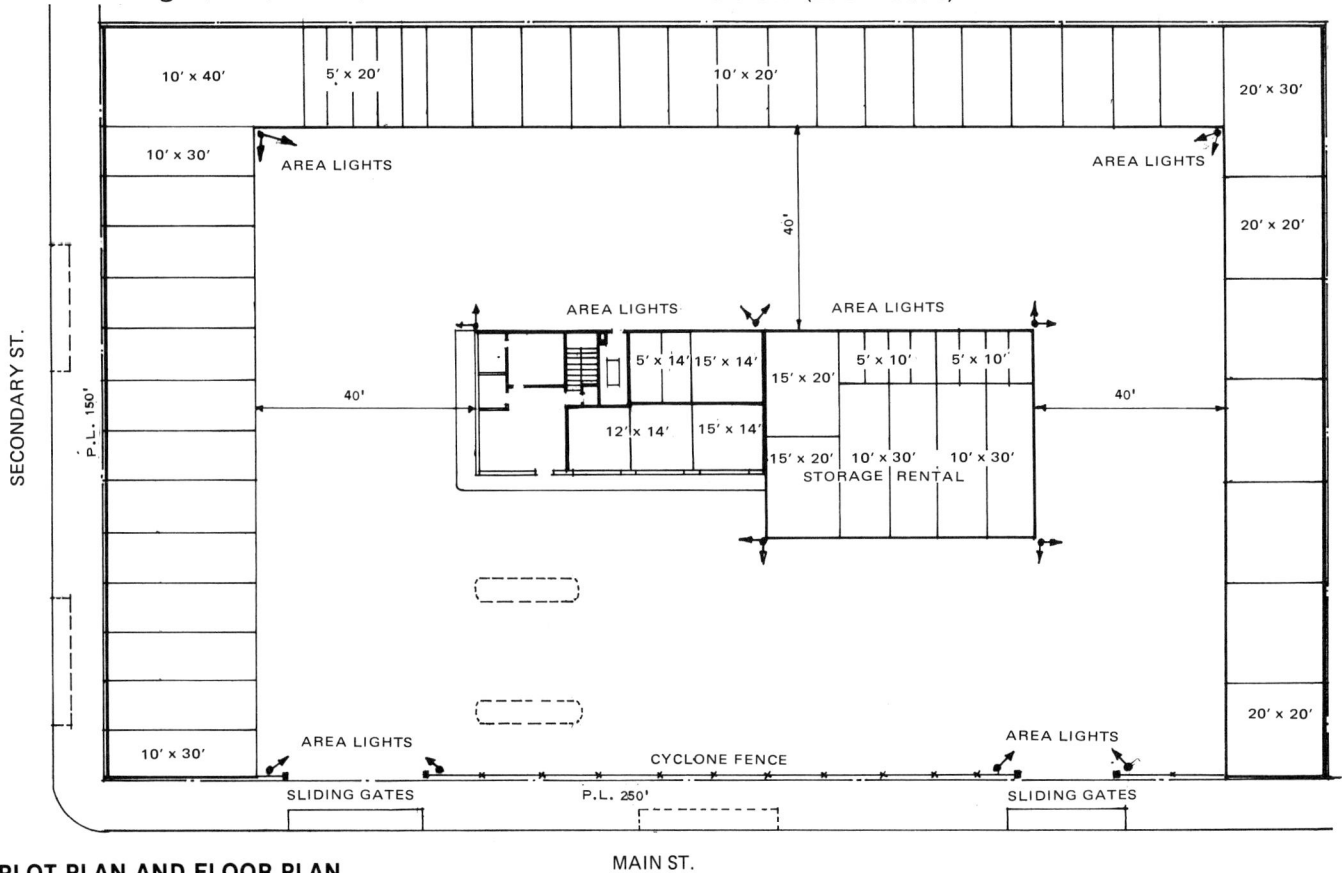

PLOT PLAN AND FLOOR PLAN

SCALE 1'' = 40'

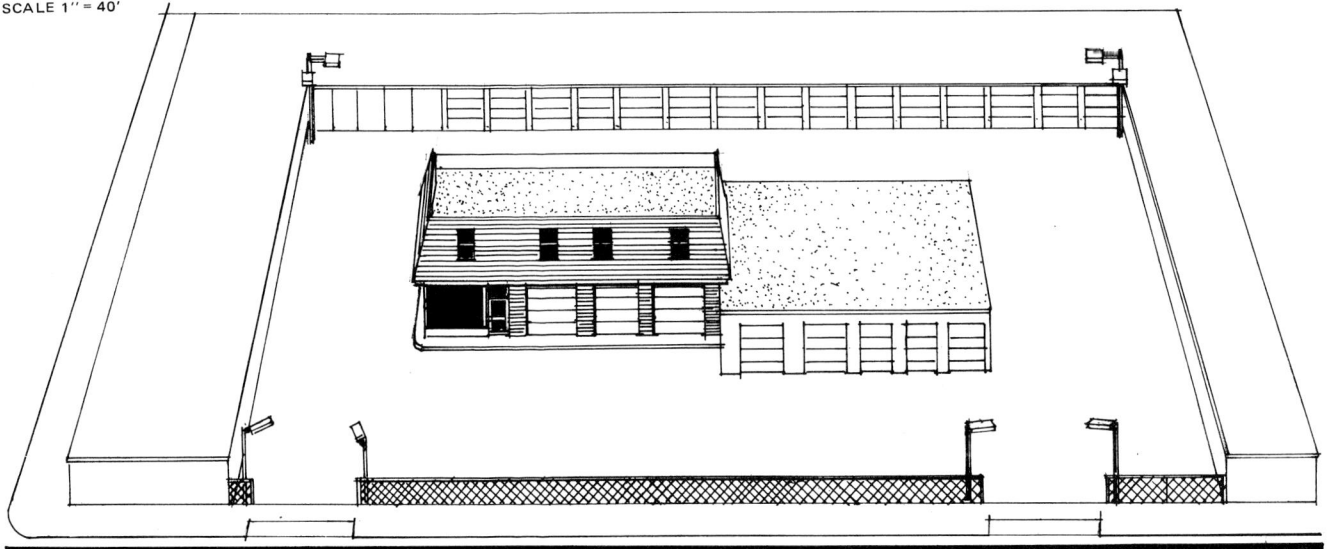

MINI - STORAGE

The addition of a second floor above the service station provides an excellent apartment for the resident manager. An attractive five room apartment consisting of 1800 square feet is illustrated. It may be necessary to increase the wall thickness and reinforce the roof beams of the service station building before adding a second floor. The lower floor can be subdivided into six storage units, a reception area and a manager's office.

The units are used by boat owners, apartment dwellers, professionals and contractors for the storage of supplies, furniture, store fixtures, inventory for businesses and as hobby workshops. The average cost for small closets is 25¢ to 35¢ per square foot and yields the best return. The large units rent for 60¢ per square foot and higher. High occupancy is dependent on the location. A site near apartments and/or a large university will yield the highest return and attain 80% to 90% occupancy. An excellent site will have 95% occupancy.

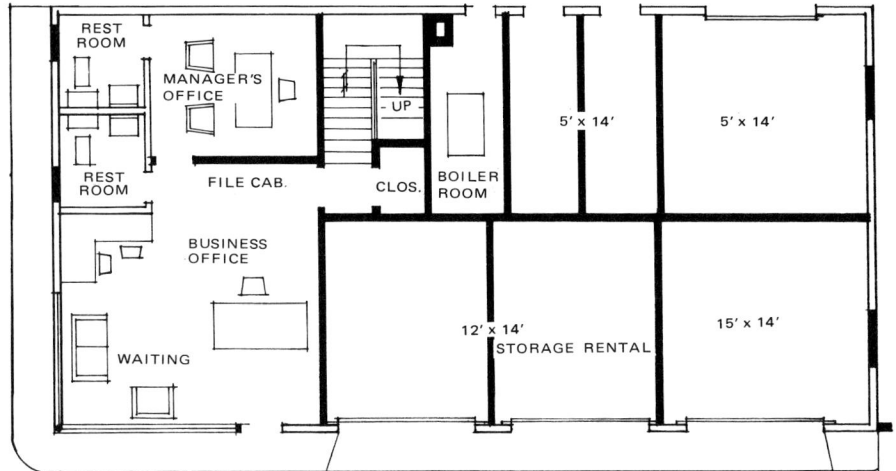

FLOOR PLAN

REST ROOM
MANAGER'S OFFICE
UP
REST ROOM
FILE CAB.
CLOS.
BOILER ROOM
5' x 14'
5' x 14'
BUSINESS OFFICE
WAITING
12' x 14'
STORAGE RENTAL
15' x 14'

SCALE 0' 5' 10' 15'

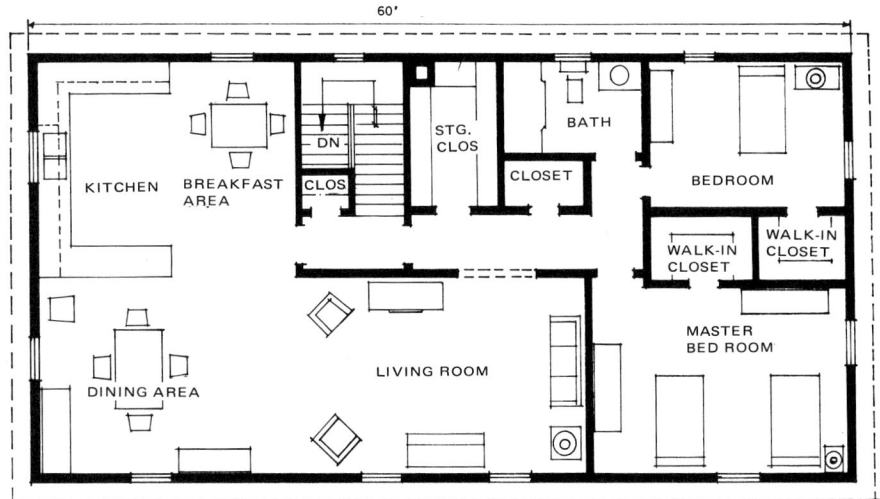

SECOND FLOOR PLAN

60'

KITCHEN
BREAKFAST AREA
DN
CLOS
STG. CLOS
BATH
CLOSET
BEDROOM
WALK-IN CLOSET
WALK-IN CLOSET
DINING AREA
LIVING ROOM
MASTER BED ROOM

LEFT SIDE ELEVATION

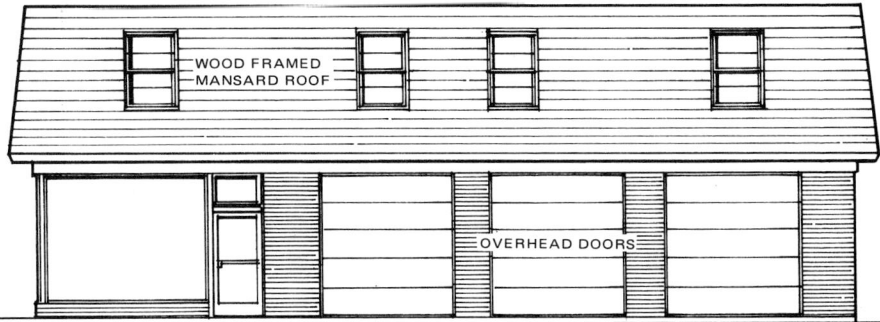

WOOD FRAMED MANSARD ROOF

FRONT ELEVATION

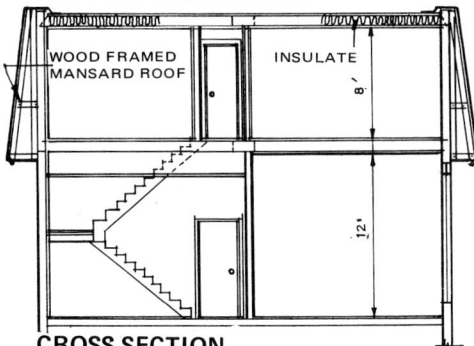

WOOD FRAMED MANSARD ROOF
OVERHEAD DOORS

CROSS SECTION

WOOD FRAMED MANSARD ROOF
INSULATE
8'
12'

LONGITUDINAL SECTION

KITCHEN
8'
BUSINESS OFFICE
10'
INSULATION
15' x 14'
12'
STORAGE RENTAL

SWIMMING POOL SALES CENTER

THE END BAY IS USED FOR STORAGE

TWO EXAMPLES OF LARGE BUILT-IN POOLS

AN EXCELLENT BUSINESS FOR THIS BUSY INTERSECTION

A FENCE ENCLOSES THE SWIMMING POOL AREA

SWIMMING POOLS

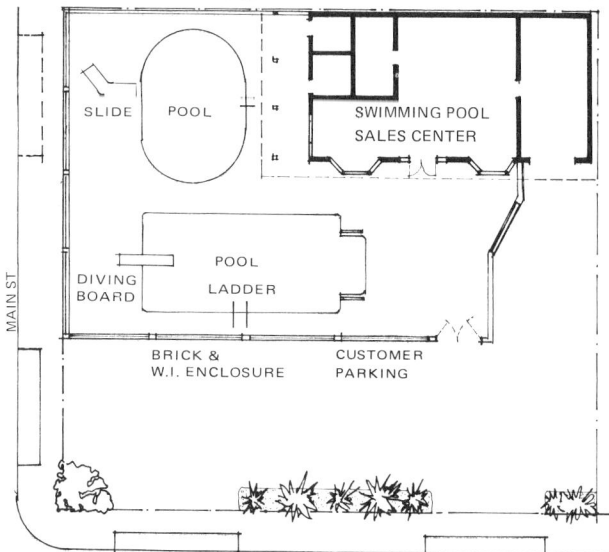

PLOT PLAN

In 1976, two entrepreneurs purchased a three bay ranch building, which had been abandoned for one year, for $65,000. The property is situated at an intersection with the highest traffic count in a capital city. The new owners converted the property into a novel pool center. The heavy motorist and pedestrian traffic has helped the business activity of this unusual use for an abandoned service station. The owners installed two in-ground pools for display purposes. The pools are enclosed with a wrought iron fence and brick wall.

Courtesy Acquarian Pools of Columbia, Columbia, S. C.

SWIMMING POOL SALES CENTER

A FIBERGLASS SPA

THE MANAGER'S OFFICE AND RECEPTIONIST

CHEMICAL STORAGE

THE RECEPTION AND DISPLAY AREAS

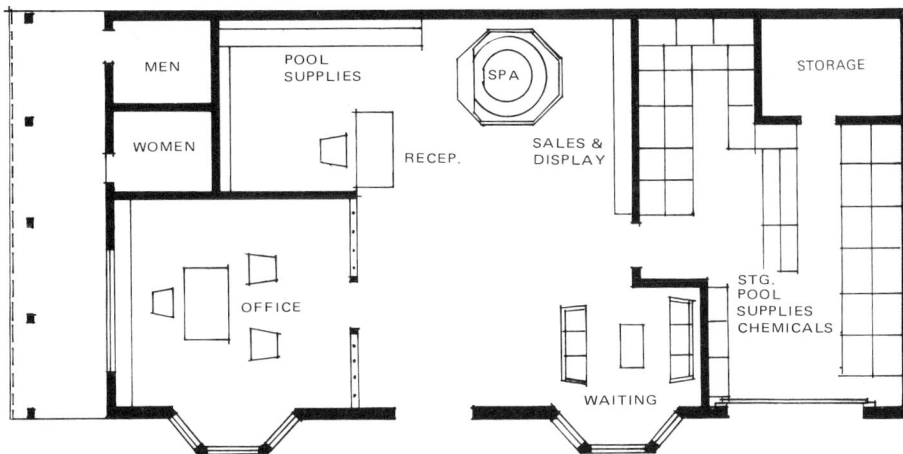

FLOOR PLAN

All of the work was performed by concrete and brick workers who install the pools, walks, platforms and landscaping for customers. A masonry wall was erected between the second and third lubritory bays, to provide a 600 square foot storage room for pool chemicals. The interior was panelled with cherry wood; attractive bay windows overlook the pool; and a distinctive new entrance was installed. A spa is positioned to the rear between a display of pool accessories. The sales of spas have increased substantially in recent years. A customer may sit in the waiting area and select a custom spa from the manufacturer's catalogs.

Adequate parking area is provided on the 108' x 100' plot for customers. In the opinion of the author, the 1981 market value of the converted building and land was $280,000. This is a business that will improve during periods of inflation and high petroleum prices. During a recession, consumers tend to spend vacations at home and improve their property with additions such as swimming pools.

Courtesy Acquarian Pools of Columbia, Columbia, S. C.

SMOKE SHOP

THE ABANDONED SERVICE STATION HAS BEEN CLOS-
ED FOR SEVERAL YEARS

THE COMPLETED REMODELING REPRESENTS A
MAJOR CHANGE IN APPEARANCE

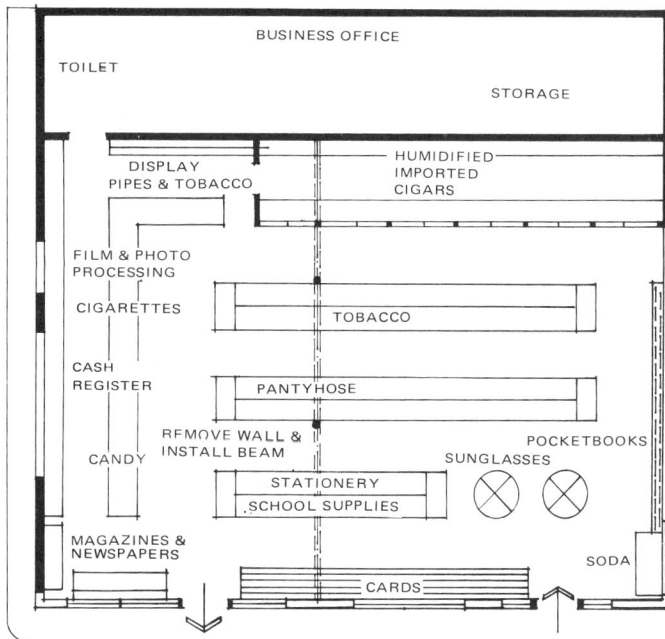

FLOOR PLAN

The remodeling of a forty to fifty year old frame abandoned service station, that had been closed for more than three years, is representative of the innovativeness and creativity of the small businessman. The cost to remodel normally was not justified. Everyone said, "tear it down, start from scratch, build a new store". Not so easy with the new setback requirements of the local community. The 100′ x 100′ parcel situated on a busy major artery on the outskirts of the business district, becomes a much smaller plot based on new zoning laws. The forty to fifty year old service station building was erected under the old zoning law, which permitted small side and rear yard setbacks. There was enough room to the rear of the building to add an extension and obtain the necessary square footage required for a successful business venture. The result is a super-store specializing in a variety of imported and domestic cigars, cigarettes, tobacco, pipes, candy, magazines, newspapers, paperback books, stationery, cards, ribbons, cold soda, health and beauty aids and many sundry items.

The exterior remodeling was accomplished by installing a new mansard roof on the front and sides of the building. Biege colored stucco was applied over the frame walls. Bronze store fronts were installed in the old overhead door openings. This conversion is an excellent example of a successful recycling of an old building to a new business use by the small businessman. It is to be noted that the store also features a humidified section for expensive imported cigars. This area maintains the proper humidity control, so important for cigars.

The new facility now provides employment in the community and is a valuable addition to the retail business district.

Courtesy Teamo Smoke Shop Ltd, Freeport, N. Y.

SMOKE SHOP

COSMETICS, A SODA COOLER AND GREETING CARDS

A HUMIDIFIED AREA FOR CIGARS

PACKAGED CIGARS

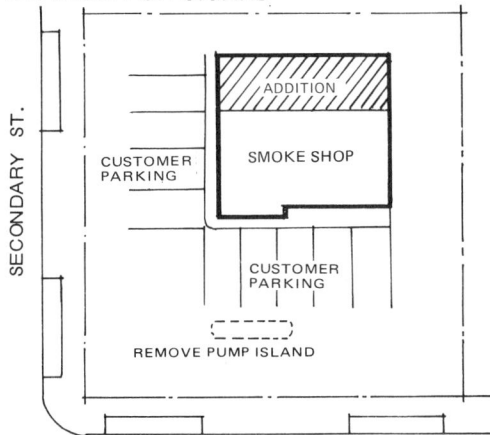

PLOT PLAN

SECONDARY ST.

ADDITION

CUSTOMER PARKING

SMOKE SHOP

CUSTOMER PARKING

REMOVE PUMP ISLAND

MAIN ST.

CIGAR DISPLAY

Courtesy Teamo Smoke Shop Ltd., Freeport, N.Y.

SMOKE SHOP

THE ABANDONED SERVICE STATION

A rapidly expanding and profitable business use is the smoke shop that sells a full line of tobacco and related products at reasonable prices. A quality smoke shop provides a humidified enclosure for storage of imported cigars. The cashier's counter contains cigarettes by the pack or carton, a large variety of candy, film, pipe tobacco, newspapers, magazines and many other walk-in items.

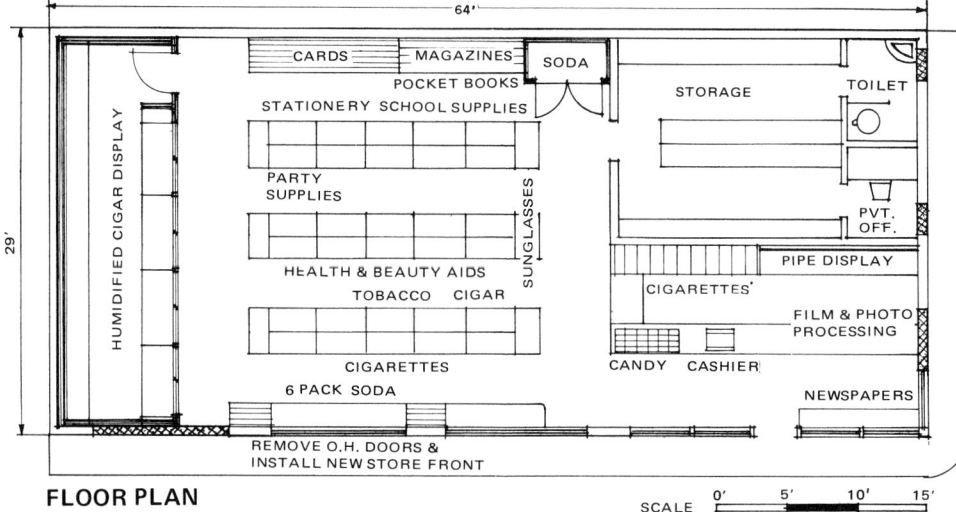

FLOOR PLAN

A pipe shop for quality pipes, bowls, cigar and cigarette lighters is situated adjacent to the cigar humidifier. Impulse products, such as soda by the case, chemical logs, and other seasonal items, should be located at the entry.

The canopy can be converted into a drive-up kiosk and an excellent profit center, selling cigarettes, cigars, tobacco, candy and newspapers, and a film drop for developing photos. The drive-up aspect appeals to the metropolitan motorists on the way to and from work.

FRONT ELEVATION

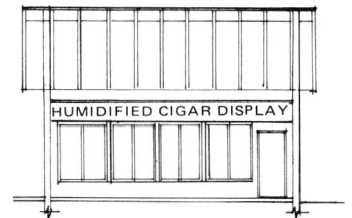

CROSS SECTION

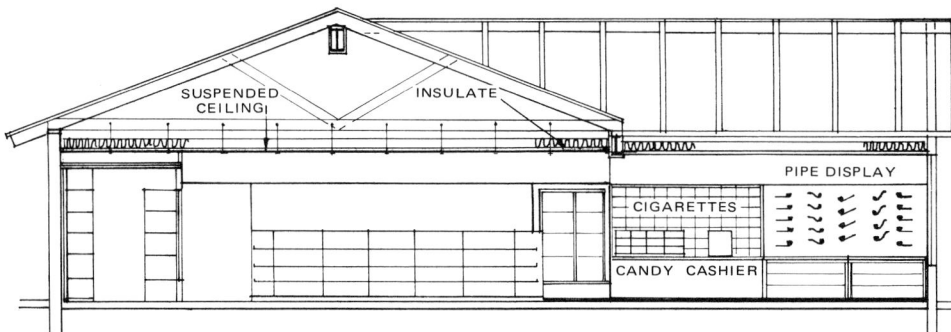

LONGITUDINAL SECTION

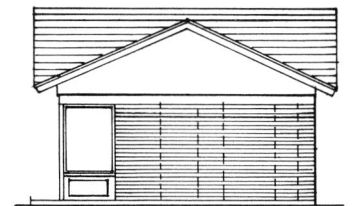

RIGHT SIDE ELEVATION

TOWEL STORE

Some service stations have closed due to the lack of residential or business density within the marketing area. These stations are generally situated on a main artery with a substantial traffic count. In the industrial mill areas, these buildings may be considered for conversion to towel stores. The lubritory area is ideally suited for this purpose and can be converted inexpensively. The installation of shelving on all interior walls, together with the fabrication of wood gondolas and display counters, can be accomplished by the journeyman carpenter. The overhead door openings must be closed in and small strip windows installed. This not only provides additional display and shelving areas, but conserves energy while permitting outside light into the store interior. The exterior facade treatment is attractive as well as practical. The materials are all pre-engineered, manufactured in a nearby plant and assembled and installed on a job site within one to two weeks. The materials require a minimum amount of maintenance. The metal fascia and shingles are all prefinished and can withstand the elements.

The store sales are generated by promoting the sale of quality towels at discount prices. The towels are stacked on the shelves and "seconds" are displayed in tubs and cartons on the floor. Special towels are placed on gondolas.

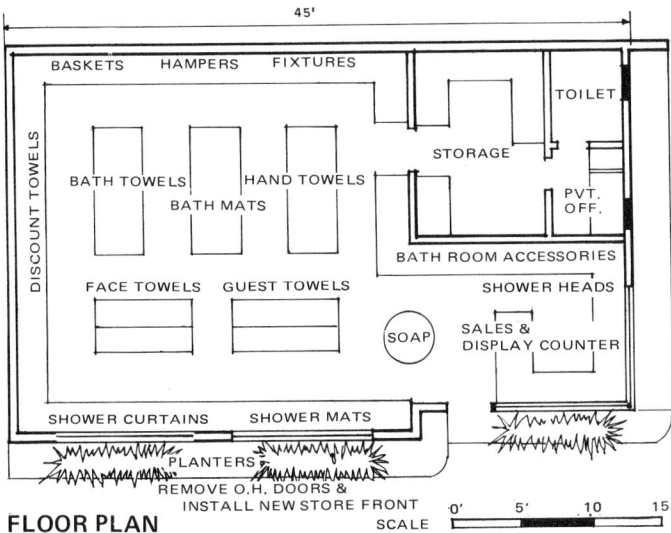

FLOOR PLAN

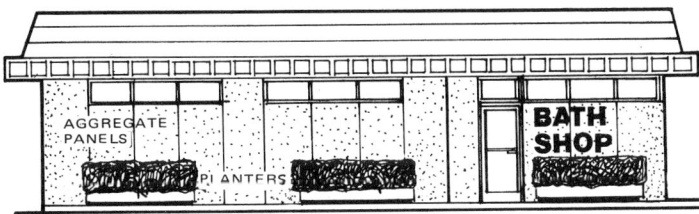

FRONT ELEVATION

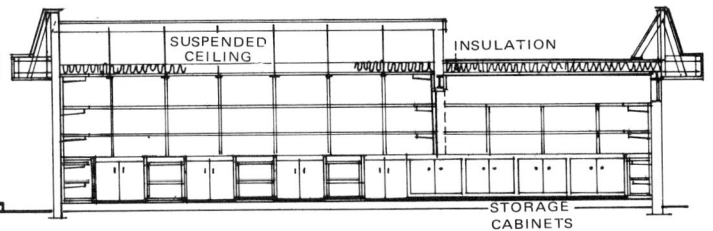

LONGITUDINAL SECTION

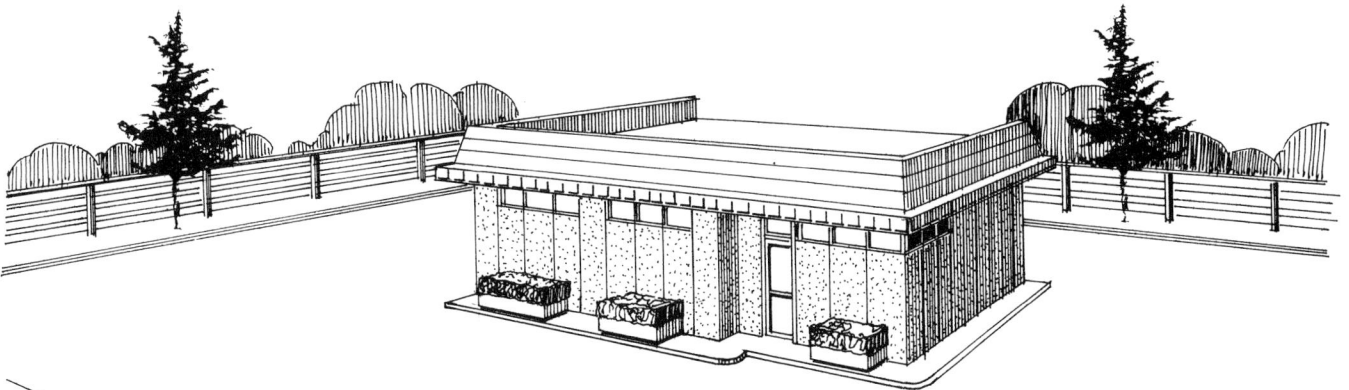

TRAVEL AGENCY

A closed two bay service station contains approximately 1200 square feet which is ideal for the establishment of a travel agency. The building shown in the photograph is a two level structure which has the potential of being converted with a contemporary facade. The removal of walls, as indicated, will provide an excellent floor plan arrangement for a travel agency. It will permit a counter area for computerized equipment. This expedites arrangements while the customer is waiting to finalize plans.

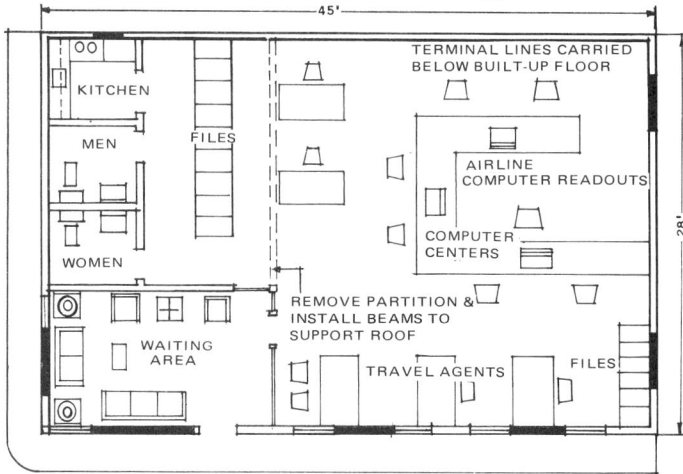

FLOOR PLAN

SCALE

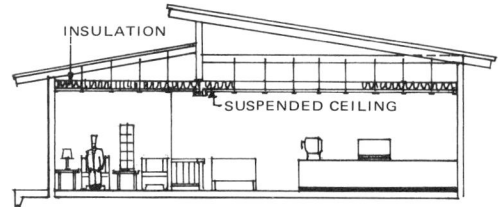

CROSS SECTION

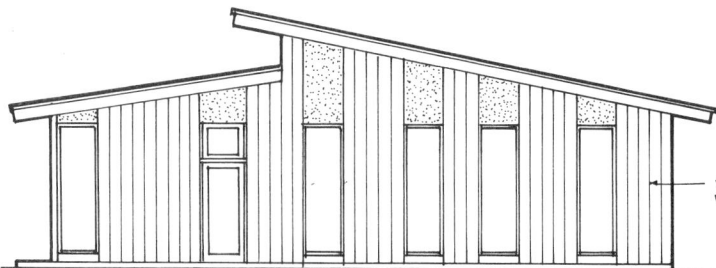

FRONT ELEVATION

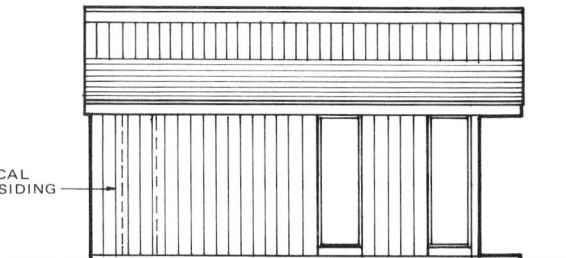

LEFT SIDE ELEVATION

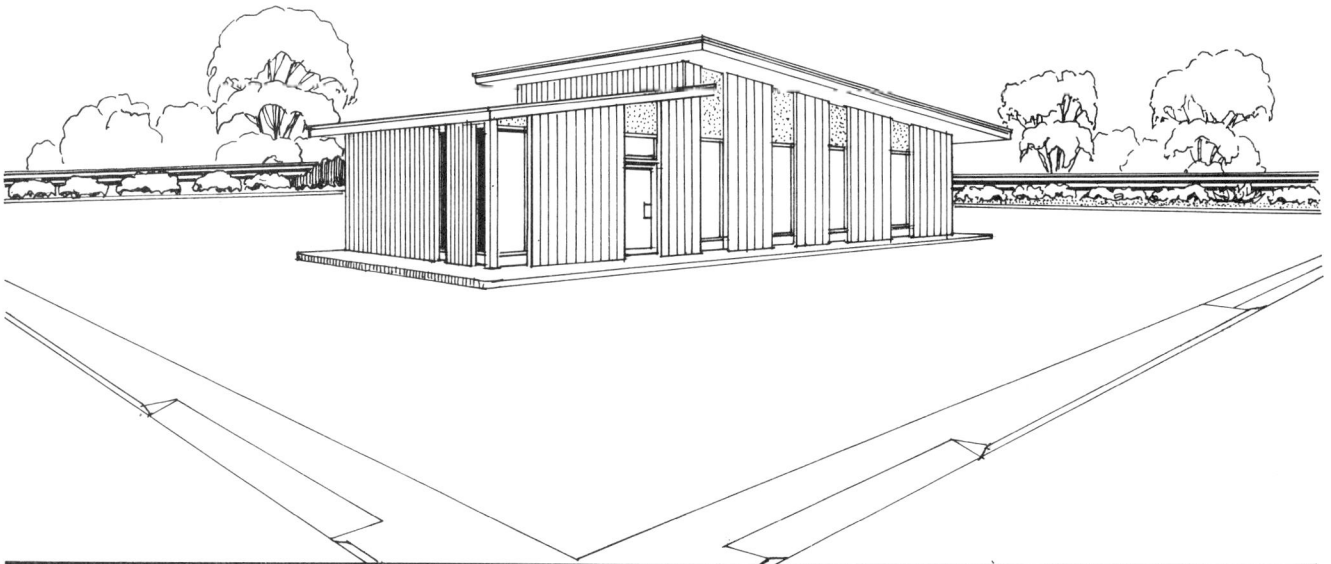

TROPHY STORE

In 1974, an owner moved his trophy shop to this closed two bay service station with a canopy. The service station was leased for $1000 per month, subject to the condition that the lessee would not alter the station structurally so that it could be changed into a service station if required.

The large sales room is perfect for the display of the trophies and plaques. The bays were converted into a storage section for the assembly of trophies and storage of the components. The interior of the sales room has a dark paneling which provides an excellent background for the trophies. Engraved bowling, baseball, school and league trophies are manufactured on a custom basis.

The interior sales room lighting together with the perimeter canopy illumination attracts the attention of the motorists. It is to be noted that this property is situated on a state highway with a traffic count in excess of 40,000 cars per day.

A NIGHT VIEW OF THE STORE

ILLUMINATED DISPLAYS

SALES COUNTER

A VIEW OF THE SALES COUNTER AND TROPHIES

DISPLAY OF WALL PLAQUES AND TROPHIES

STORAGE AREA AND WORK SHOP

Courtesy Little Falls Trophy & Engraving Shop, Little Falls, N.J.

TRUCK AND TRAILER RENTAL CENTER

Truck and trailer rental services are very popular due to the increased cost of moving a household across country. The do-it-yourself moving centers will continue to grow and prosper. During periods of gasoline shortages one major rental firm provided gas at their centers for customers as they travelled long distance.

Typical conversions of abandoned service stations to rental centers are illustrated in the photographs. Large properties in heavily travelled areas situated at intersections, or large interior parcels, are sought by auto rental centers. This company provides a total service to the do-it-yourself mover. Former lubritory bays are now being utilized for storage of moving cartons, dollies, pads, trailer hitches, padlocks, taillights and mirrors. All items are rented, but the small items may also be purchased.

This business should continue to grow as inflation and recessionary periods reduce the purchasing power of the consumer.

A CUSTOMER CHECKING IN

A NEWLY PAVED FRONT YARD

PARKING FOR A VARIETY OF VEHICLES

THE REAR OF THE PROPERTY IS FULLY UTILIZED

PARKING FOR TRAILERS AND TRUCK VANS

Courtesy U-Haul International, Phoenix, Arizona

TRUCK AND TRAILER RENTAL CENTER

ACCESSORIES, PADS, SIDE VIEW MIRRORS ARE DISPLAYED

A VERY LARGE PARKING AREA

A SMALL TRAILER AND PADS FOR MOVING

CARTONS FOR MOVING

THE OFFICE DISPLAYS RENTAL RATES AND MANY ACCESSORIES

A MOVING CENTER IN THE MIDWEST

A MOVING CENTER IN DENVER

Courtesy U-Haul International, Phoenix, Arizona

T-SHIRT & REPRODUCTION CENTER

There are many excellent business opportunities for the entrepreneur in a college town. Photocopying and T-shirts for college students can be successful businesses, either independent of each other or jointly operated by one owner. During periods of recession, it is more profitable to diversify with profit centers that complement each other. Customized T-shirts are extremely popular with college and high school students throughout the country.

Approximately 800 square feet of an 1800 square foot abandoned three bay service station building is required for the heat sealing equipment, transfers, letters, T-shirts inventory, and merchandise display and storage. The photocopy center should be designed with the self-service copier located at the entry where students will have easy access. This machine can become the most important source of income for the center. In addition, services such as sorting, collating, stapling and folding, will be in great demand by the schools and offices.

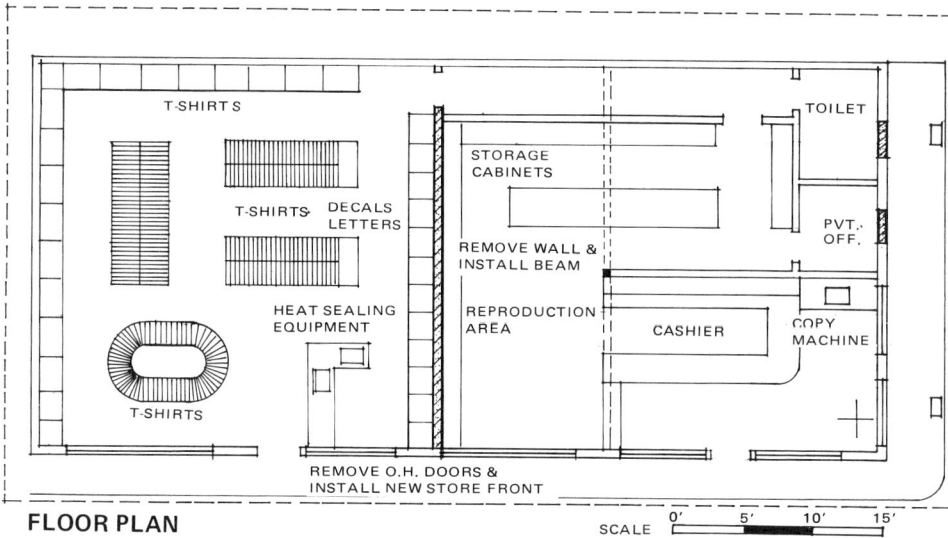

FLOOR PLAN

SCALE 0' 5' 10' 15'

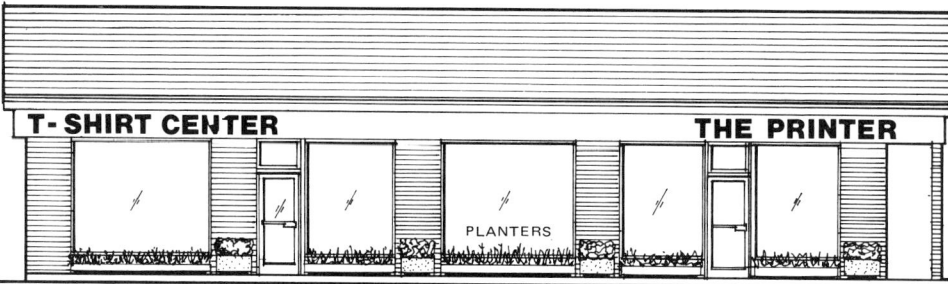

FRONT ELEVATION

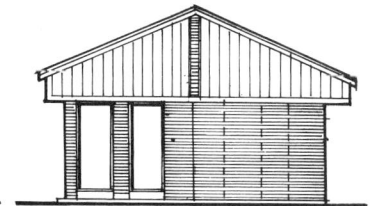

RIGHT SIDE ELEVATION

WEDDING CENTER

A bridal and tuxedo rental store can be a very successful business enterprise when the store is situated adjacent to a shopping center or in a business district where there is considerable pedestrian traffic. A three bay service station building can be remodeled into two 900 square foot stores. The layout must include space for fitting booths, mannequin displays, storage for wedding gowns, veils and accessories, shirts and dinner jackets. The exterior of the colonial building may be converted to a contemporary motif as illustrated on the drawings. The Wedding Center can be designed as two separate stores or as one large store for both bride and groom.

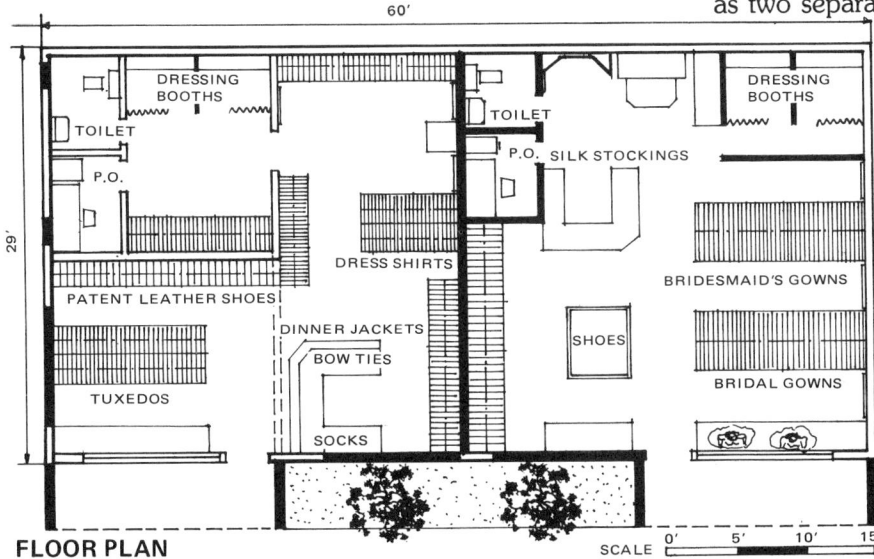

FLOOR PLAN

60'

29'

DRESSING BOOTHS

TOILET

P.O.

PATENT LEATHER SHOES

TUXEDOS

TOILET

P.O. SILK STOCKINGS

DRESS SHIRTS

DINNER JACKETS

BOW TIES

SOCKS

SHOES

DRESSING BOOTHS

BRIDESMAID'S GOWNS

BRIDAL GOWNS

SCALE 0' 5' 10' 15'

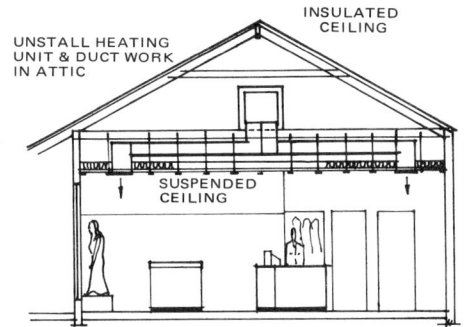

CROSS SECTION

UNSTALL HEATING UNIT & DUCT WORK IN ATTIC

INSULATED CEILING

SUSPENDED CEILING

FRONT ELEVATION

BRICK VENEER

REMOVE O.H. DOORS & INSTALL NEW STORE FRONT

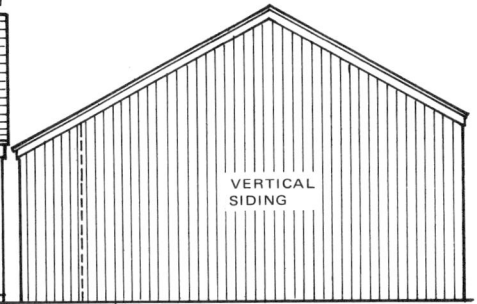

RIGHT SIDE ELEVATION

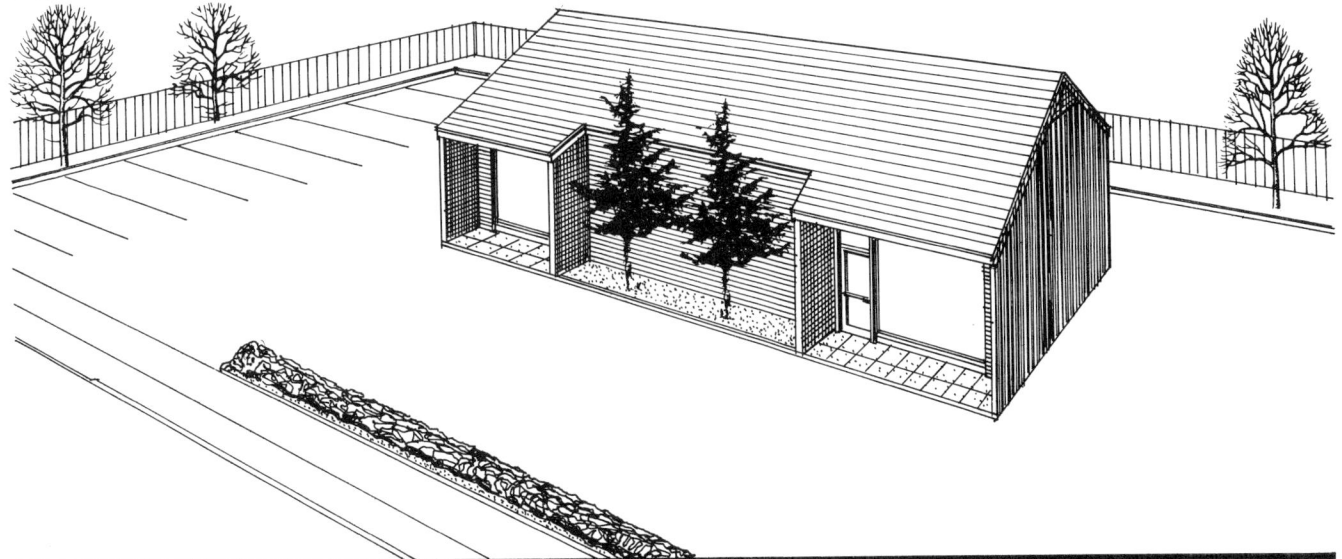

VERTICAL SIDING

INDEX